Charles Packe

A Guide to the Pyrenees

Especially intended for the use of mountaineers

Charles Packe

A Guide to the Pyrenees
Especially intended for the use of mountaineers

ISBN/EAN: 9783337289430

Printed in Europe, USA, Canada, Australia, Japan

Cover: Foto ©Lupo / pixelio.de

More available books at **www.hansebooks.com**

A GUIDE

TO

THE PYRENEES.

ESPECIALLY INTENDED FOR THE USE OF MOUNTAINEERS.

BY

CHARLES PACKE.

And the breath of the early morning, when the keen and nipping breeze
Came cold to the cheek from many a peak of the snowy Pyrenees.
(OLD BALLAD.)

WITH MAPS, DIAGRAMS, AND TABLES.

SECOND EDITION,
REWRITTEN AND MUCH ENLARGED
(May 1867).

LONDON:
LONGMANS, GREEN, AND CO.
1867.

PREFACE.

THE FOLLOWING PAGES are not written with any idea of giving an adequate description of the beautiful scenery of the Pyrenees. They are simply intended to serve as a *Pocket Guide* among the mountains, more particularly in the high chain of the Pyrenees. Mr. Weld, in his work on the 'Eastern and Western Pyrenees,' has given graphic descriptions of the scenery and structure of the whole range; but his book is clearly not intended as a guide to pedestrian excursions, and of some of the finest and by no means difficult ascents—such as that of the Pic du Midi de Bigorre—he makes no mention. In Murray's 'Hand-Book of France,' the principal routes are given with great accuracy, as far as they go; but I think they will bear amplification. Of the French guide-books on the Pyrenees, that by Johanne, which includes the whole range, and that by Dr. Lambron, which applies especially to the neighbourhood of Luchon, are both trustworthy and full of useful information; but they are too voluminous, and yet some points are omitted. Read at home, this little book will have no interest; but amid the mountains and gorges of the Pyrenees, it will, I believe, enable the tourist in a measure to be independent of professed guides—those iconoclasts of imagination and reverie—at all events to select for himself his own excursions, with a fair notion of the distances, and the time to be consumed in them. In the face of Nature itself, which speaks to its admirers in a language so far more eloquent than pen or pencil, description is worse than useless. I have, therefore, chiefly confined myself to mere indications of the route, with the heights and distances.

The times allowed from point to point, in the different excursions, are computed from my own experience, for the pace of moderate walking, including halts. If mounted, taking the rough and the smooth, I found the rate of progress to be about the same. Many of the excursions may, by a good walker, be performed in less time than that given; but I have thought it best to allow a sufficient margin, remembering the proverb—

> Chi va piano, va sano;
> Chi va sano, va lontano.

In the five years that have elapsed since the first edition of this guide-book, the author has had many opportunities of extending his own personal knowledge of the Pyrenees. He has also had the advantage of visiting many of the higher mountains with his friend, Count Henri Russell, and of consulting the very accurate work by him, entitled " Les grandes Ascensions des Pyrénées."

The present edition is therefore considerably extended, and is, it is hoped, an improvement upon the previous one.

The details given of the towns, promenades, and places of historical association, will perhaps be thought meagre; but it must be borne in mind that this is a guide-book especially intended for mountaineers and botanists. I have, therefore, preferred to curtail description, and to devote the pages so gained to tables which I think all travellers will find of use.

These have been necessarily somewhat amplified, as I have thought it advisable to give one or two of them both in French and English measures. Indeed, one of the main difficulties experienced in writing a guide is the conflicting system of measures in use in England and in France: those of the latter country being so much more simple, easy, and uniform; but it is very hard to get an Englishman to estimate heights and distances in metres.

The heights of the different mountains given in metres are generally taken from the maps of the État-Major; in most cases, I have also given the equivalents in feet. As conveying an excellent idea of the topography of the mountains, though, of course, none whatever of the grand feelings which they evoke, the relief model of the chain of the Haute-Garonne by M. Lézat, in the *Établissement* at Luchon, and that by M. Frossard, at Bigorre, are well worth a visit.

I have given a map of the central and principal portion of the Pyrenees on a larger and clearer scale, and for a general map of the whole chain would refer the reader to that lately published by M. Lézat (and to be bought at Luchon, *chez* Lafont, in the Allée d'Étigny, price three francs) as the best I have met with. For the excursions around Bigorre, I would refer the traveller to an excellent map just published by M. Pambrun, of Bigorre, to be had at Sajous' library. When my map of the Pyrenees was engraved, the maps of the État-Major were still unpublished. Of course, had I seen them, I should have produced a better map, and with less trouble to myself. I think, however, the traveller will not discover in it any important error. The additions on the Spanish side of the frontier are entirely my own. The map of the Maladetta is also taken from an original survey.

I must not conclude without a last word about the hotels, which are for the most part very good in the Pyrenees. There is one system of being pretty sure of being set down at the worst of these, when there are two in the same place, and against that system I would here say a word. Of course, the best guarantee of the goodness of a hotel is one's own previous experience, and the next best that of the recommendation of a friend. Too many travellers, in default of either of these, are apt to give themselves up blindly to their coachman, guide, or courier, to deposit them where he wills, i.e. at the house where he himself will be best entertained. It is far more safe to take the recommendation of any guide-book, or even to enter purely by chance the first hotel that presents itself.

BAGNÈRES DE BIGORRE:
1867.

CONTENTS.

THE PYRENEES FROM WEST TO EAST.

Section 1.

INTRODUCTORY.

	PAGE
Approaches to the Pyrenees	2
From the West	2
From the Centre	3
From the East	4
Bordeaux	4
Bayonne	5
Biarritz	5

Section 2.

Structure of the Pyrenees	9

Section 3.

From Bayonne to Pau		13
A. Bayonne to Pau by Orthez, 4 hours, rail		
Bayonne to Orthez	62 km.	
Pau	38 km.	
	100 km. = 62 miles.	
B. Paris to Bordeaux	538 km.	
Bordeaux to Dax	150 km.	
Dax to Pau	83 km.	
	816 = 507 miles.	

Section 4.

Pau to Eaux Bonnes	42 kil. = 26½ miles.	13
„ to Eaux Chaudes	43 = 27	13
Eaux Bonnes to Eaux Chaudes		16

Section 5.

Ascent of the Pic de Ger 16

Section 6.

Eaux Chaudes to Panticosa 17

Section 7.

To Urdos by the Col d'Iseye, returning by the Col des Moines . . 18

Section 8.

From Eaux Chaudes to Cambo and Biarritz by the mountains, including the ascent of the Pic Anie 20

Section 9.

Eaux Bonnes to Argelez 21
    By the Col de Tortes; on foot, horse, or carriage, a day's journey 21
        Eaux Bonnes to Arrens . . . . . 6 hours.
            to Argelez 3
                                9 hours.

Section 10.

* Valley of Azun and Pic de Baletous or Murmuret 23

Section 11.

Lac Miguelou—Lac d'Artouste 26

Section 12.

Arrens to Cauterets by Monné and Lac d'Estaing; on foot . 27

Section 13.

Arrens to Argelez 27

Section 14.

Argelez to Cauterets 27

* Those excursions not to be attempted without a good guide, except by an experienced mountaineer, are marked with an [*].

Section 15.

Cauterets 28

Section 16.

Lac de Gaube—Vignemâle . . . 29

Section 17.

* Vignemale 31

Section 18.

The Monné . . . 33

Section 19.

Cauterets to Gavarnie 33
    A. Viâ St. Sauveur 33
    B. Viâ the Val d'Ossoue . . . 33
Cauterets to Luz and St. Sauveur . . . 33
    A. By the carriage-road, Pierrefitte . 33
    B. By the mountains on foot or horse . 33

Section 20.

Pierrefitte to Luz . . 33

Section 21.

Luz 34
Luz to St. Sauveur 35

Section 22.

Pic de Bergons 36
    Pic Ardidieu . . . 36
    Pic Aubiste 36

Section 23.

Luz to Gavarnie—St. Sauveur to Gavarnie . 37
* Brèche de Roland 38

Section 24.

* Mont Perdu 40

Section 25.
* Lac de Mont Perdu, the Port de Pinéde . . 41

Section 26.
* Taillon and Fausse Brèche . . . 41

Section 27.
* Source of Cascade and Marboré 42

Section 28.
Mont Piméné. 43

Section 29.
Gavarnie to Boucharo by the Port de Gavarnie—Vallée d'Arras . 44

Section 30.
Gavarnie to Baths of Panticosa 46

Section 31.
Gavarnie to Cauterets by the Val d'Ossoue and Lac de Gaube . . 47

Section 32.
Gavarnie to Cauterets by the Valley of Lutour 48

Section 33.
Gavarnie to Héas 49
    A. By the Mountain. 49
    B. By Gèdre 50

Section 34.
Héas to the Cirque of Troumouse 50

Section 35.
Port de la Canaou 51

Section 36.
Cirque of Estaubé 51

Section 37.

	PAGE
* Cirque de Troumouse—Pic de la Munia	52

Section 38.

| Luz to Barèges | 52 |
| Lac Escoubous | 53 |

Section 39.

| Ascent of the Pic du Midi de Bigorre | 53 |

Section 40.

From Barèges to Bagnères de Bigorre	55
A. By the Tourmalet	55
B. By the Lac Bleu	55
C. By the Pène Taillade	56

Section 41.

| Bagnères de Bigorre | 56 |

Section 42.

| L'Heris—Puits de la Pindorle—Grotte de Judios | 57 |

Section 43.

| The Pic de Montaigu—Bigorre to Pierrefitte | 59 |

Section 44.

Bigorre to Lourdes	60
A. By the road	60
B. By the Croix Blanche	60

Section 45.

| Bagnères de Bigorre to Bagnères du Luchon | 60 |

Section 46.

| Val d'Aure | 63 |

Section 47.

| Valley d'Aure into Spain | 65 |
| Port de Plan | 65 |

	PAGE
Col d'Ourdisset	65
Port de Moudang	66
Port de Bielsa	66

Section 48.

* Aragnouet to Héas and the Pic des Aiguillons	67
To Gèdres by the Col de Cambiel	68

Section 49.

Col Badet	68

Section 50.

* Pic Long or Vierge	69

Section 51.

Vallée d'Aure to Barèges	69
A. By Col d'Aubert	70
B. By Port de Barèges	70

Section 52.

* Pic de Néouvielle	70

Section 53.

Bagnères de Luchon	71

Section 54.

Climate of Luchon	75

Section 55.

Lac d'Oo—17 kil.=11 miles; 4 hours to go, 3 hours to return; to be made on foot, on horseback, or in carriage	79

Section 56.

* Port d'Oo and Lac Glacé	82

Section 57.

* Portillon d'Oo	84

Section 58.

* Pic de Crabioules	86

CONTENTS. xv

Section 59.
* Lac d'Oo to Val de Lys, and so to Luchon	87

Section 60.
Port de Clarabide and Port d'Enfer	88

Section 61.
* Les Gours Blancs and Lac Caillaouas	89

Section 62.
Monségu, Écho de Néré, and Esquierry—11 hours, on foot, or on horseback	90

Section 63.
Superbagnères—7 hours easy to go and return, allowing one hour on the top; horses may be taken	92

Section 64.
Pic Céciré—11 hours to go and return; horses may be taken	93

Section 65.
Valley du Lys and Cascades d'Enfer	93

Section 66.
Lac Vert or Lac d'Île	95

Section 67.
* Tus de Maupas	96

Section 68.
* Pic du Boum	96

Section 69.
Pic Sacroux	97
Port Vieux	97

Section 70.
* Les Quinze Lacs	97

Section 71.

Cascade de Demoiselles and Cascade des Parisiennes . PAGE 98

Section 72.

Port de la Glère; on foot or horseback; 9 hours to go and return; 9 hours to Venasque 99

Section 73.

Port de Venasque and Port de Picade, on foot, horses, or chaise à porteurs, price 60 francs; 10 hours, not including halts, 23½ miles . . 100
Trou de Toro 101
Pic de Sauvegarde 102

Section 74.

Luchon to Venasque 104
  A. By Port de Venasque 104
  B. By Port de la Glère 104
  * C. By Port Vieux 104
  * D. By Portillon d'Oo, or Port d'Oo 105

Section 75.

Venasque to Viella and Las Bordes 105
  Route A. North side of Maladetta 105

Section 76.

  Route B. South side of Maladetta 106

Section 77.

Viella to Bosost and Luchon 108
Val d'Aran 108
  A. By the Portillon, 4 hours; 14·8 kil., on horse, or foot . . 108
  B. By La Bacanère, on foot 109
  C. By Pont de Roi and St. Béat; good road all the way, carriage or horses, distance 40 kil.=25 miles 109

Section 78.

Luchon to St. Bertrand de Comminges 110

Section 79.

Montauban, Juzet, and Tour of the Valley of Luchon . . . 112

CONTENTS. xvii

Section 80.
PAGE
Pic du Bacanère 112

Section 81.
Pic de Poujastou 114

Section 82.
Pic de Couradilles 114

Section 83.
Pic de l'Entecade 115

Section 84.
Pic d'Antenac 116
    A. By Cazaril, *ascent* 116
    B. By St Paul, *descent* 117

Section 85.
Pic de Monné and Vallée d'Oueil 117

Section 86.
* Pic de Nethou 120
    A. By the Rencluse 121
    B. By Malibierne 123
* Pic de Maladetta 123
* Pic Albe 123

Section 87.
* Pic Fourcanade 124

Section 88.
Luchon to Malibierne and Castaneza 125

Section 89.
* Pic des Posets 128
    A. From Cabane de Turmes 128
    B. From Cabane de Paoules 128

Section 90.
Caldas de Bohi and Pic de Montarto, Lac Becibère. 130

Section 91.
Les Mines du Cap de Guerri and Pic Crabère 132

Section 92.
Ariège and Eastern Pyrenees 133
Luchon to Castillon, A. by St. Béat, and B. by Melles 134
Castillon to St. Gaudens, 40 kil. carriage-road 134

Section 93.
Upper Val d'Aran 134

Section 94.
Viella to Conflens 135
    A. By the Port de Peyreblanca and Port de Salau . . . 135
    B. By the Port d'Aula 135

Section 95.
Montvallier 136

Section 96.
Esterri to Viedessos, by the Port de Tabascain 136
    Esterri to Tirvia, 4 hours
    Top of Port to Esterri, 3 hours
Esterri to Aulus 136

Section 97.
* Ascent of the Montcalm and the Pic d'Estats 137

Section 98.
Toulouse to Ax 138
Pic St. Barthélemy 138

Section 99.
Ax to the Valley of Andorre 139
    A. By the Port de Saldeu; to Andorre 139
    B. By the Col de Puymorins; to Puycerda 139
Andorre; historical sketch of the Republic 140

Section 100.
Urgel to Puycerda 143

Section 101.
Puycerda to Ax 143

Section 102.
Andorre to Foix by the Port de Siguier; Pic de Siguier . . 144

Section 103.
Ax to Montlouis and Cabanasse 1st day Lac Lanoux . . . 145
                                2nd day, Pic Carlitte—Cabanasse . 146

Section 104.
Ax to Cabanasse by Querigut, Llaurenti, Capsir 147

Section 105.
Vallée d'Eynes; Cambredase; and Pic de Puigmal . . 148

Section 106.
Perpignan to Puycerda 148

Section 107.
Perpignan to Vernet 149
Ascent of the Canigou 149

Section 108.
Vernet to Prats de Mollo, by the Plat-Guilhem . . . 151

Section 109.
Perpignan to Amélie les Bains 151

Section 110.
Concluding Remarks 152

Section 111.

	PAGE
Climate	154

Section 112.

Length of Day 157

Section 113.

Barometrical, Thermometrical, and Hypsometrical Tables, with Scientific Memoranda 159

Section 114.

Tables for reducing French Measures 176

Section 115.

List of Pyrenean Peaks and Passes and Lakes . . . 180

Section 116.

List of Ferns found in the Pyrenees 185

ILLUSTRATIONS.

LAC DE PORTILLON D'OO	*Frontispiece*
MAP OF THE PYRENEES, from the Valley d'Aspe to the Valley d'Aran	*In a pocket*
GEOLOGICAL MAP OF THE CHAIN OF THE PYRENEES . .	*To face page* 11
DIAGRAM illustrating Geological Epochs	,, 12
MAP OF THE CHAIN SOUTH OF LUCHON	,, 83
MAP OF LES MONTS MAUDITS	*End pocket*

GUIDE

TO

THE PYRENEES.

How it comes to pass that so many Englishmen and Englishwomen cross the Channel every summer for the sake of a holiday tour of a month or six weeks among the Alps of Switzerland, while so very few in proportion think it worth their while to pay a visit to the Pyrenees, that magnificent mountain barrier that separates France and Spain, has always been to me a matter of astonishment; and I can only account for it on the principle that my countrymen, like a flock of sheep, love to go 'non quâ eundum est, sed quâ est iter.' We are all very loud in protesting that the inundation of our own countrymen, who meet you at every turn, is one of the great drawbacks to the pleasure of a summer tour through Switzerland. Why, then, do we not give the Pyrenees a trial? —mountains inferior, indeed, to the Alps in height and expanse of barren glacier, but far more picturesque in form as well as colour, and which, preeminent for the rich verdurous beauty of their valleys, are not less conspicuous for the wild, savage, and untrodden scenery presented in many of the higher passes. The leg of a compass placed on London in the map of Europe will give the distance to the Pyrenees rather longer than that to the Lake of Geneva; but in these days of railroads one hundred miles more or less is matter of little moment, and it is quite possible to leave London in the morning and sleep the second evening at Pau, the capital of the Basses Pyrénées. This rate of progression is, however, rather too fast for enjoyment, and the rest required after such a forced march would probably cause more time to be lost than gained by such expedition. A long day or night in the train will take the traveller from Paris to Bordeaux; and after reposing there one or two days, he may proceed direct to Morcens or Dax by rail, and so on to Tarbes or Pau, or he may continue his journey south as far as Bayonne before quitting the train.

On the west, there are four principal railway termini, by which the Pyrenees may be approached, of which Tarbes is the most central; but perhaps those who visit the Pyrenees for the first time, and are bent on seeing as much of them as they can in a tour of a month or five weeks, will do well to enter from either Biarritz or Pau, and work eastward. The routes are as follows:—

Section 1.

APPROACHES TO THE PYRENEES.

From the West.—Route A.

	DISTANCE.		PRICES.		
	Kil.	Miles.	1st Class.	2nd Class.	3rd Class.
Paris to Bordeaux	583	362 } rail	65 fr. 25 cent.	48 fr. 35 cent.	35 fr. 90 cent.
Bordeaux to Bayonne	198	123 }	22 fr. 20 cent.	16 fr. 65 cent.	12 fr. 20 cent.
Bayonne to Biarritz	8	5 omnibus	1 fr.		
	789 kil. = 490 miles.		88 fr. 45 cent.	65 fr. 0 cent.	48 fr. 10 cent.

The railway station at Biarritz, being at La Negresse 2 good kilometres from Biarritz, travellers will probably find it more convenient to take the omnibus from Bayonne. The time from Paris to Bordeaux is 12 hours 30 min. by express, and 20 hours by ordinary trains. By express trains there is no second class, not even for servants. Those who can endure what the French call a *nuit blanche* will find the train leaving Paris at 8.15 p.m., reaching Bordeaux at 7.8 A.M., and Bayonne at 20 minutes after noon, the most convenient.

The rail is now open from Bayonne to Pau, 100 kilometres (62 miles), passing through Orthez, 62 kil. from Bayonne. Fare 11 fr. 20 cents first class, 8 fr. 40 cents second.

For those who have no wish to visit Biarritz, the most direct route to Pau is as follows:—

Route B.

	DISTANCE.		PRICES.	
	Kil.	Miles.	1st Class.	2nd Class.
Paris to Bordeaux	583	362	65 fr. 50 cent.	49 fr. 15 cent.
Bordeaux to Dax	150	93	16 fr. 60 cent.	12 fr. 25 cent.
Dax to Pau	83	52	9 fr. 30 cent.	6 fr. 95 cent.
	816 = 507		91 fr. 40 cent.	68 fr. 35 cent.

Should Tarbes be the point of approach chosen, you quit the Bordeaux line at the Morcens station, 109 kil. south of Bordeaux, whence the Chemin de fer du Midi takes you to Tarbes.

N.B.—The buffet at Morcens is excellent, whereas that at Bordeaux is very much the reverse.

Route C.

	RAIL.		TIME.	
	Kil.	Miles.	hrs.	m.
Paris to Bordeaux	583	362	12	30
Bordeaux to Morcens	109	67½		
Morcens to Tarbes	139	86½		
	831 kil. = 516 miles.			

From Tarbes the distance is but short, with excellent roads, and daily conveyances to either Pau, Luz, Cauterets, Bigorre, or Luchon.

	Kil.	Miles.
Tarbes to Pau, rail or diligence	39	24½
Tarbes to Luz and St. Sauveur, rail and diligence	51	31½
Tarbes to Cauterets, rail and diligence	48	30
Tarbes to Bigorre, rail	20	12½
Tarbes to Luchon, viâ Bigorre, the Col d'Aspin, and Arreau, rail and diligence (see *post*)	90	56
Tarbes to Luchon, by Montrejeau (50 kil.), and the valley of Luchon	89	55½

The rail from Tarbes to Montrejeau, which is only 40 kil., or 3 hours, from Luchon, is not completed, and will probably not be open till the year '68.

That from Tarbes to Lourdes is now open; and the line from Lourdes to Pau is to be open this season. The railway from Lourdes is intended to be carried on to Pierrefitte, to the great detriment of the fertile and beautiful valley of Argelez; as also will be the line projected from Montrejeau up the valley of Luchon. It will be still some years, I hope, before these two last are accomplished.

Route D.

	Kil.	Miles.
Paris to Bordeaux	583	362
Bordeaux to Toulouse	257	160
	840 =	522

The fare from Bordeaux to Toulouse is, 1st class, 28 fr. 80 cents; 2nd class, 21 fr. 50 cents. The time is 6 hours 20 minutes by express, and 10 hours 25 minutes by ordinary trains.

TOULOUSE.—Hôtel Souville, Place du Capitole, the best; Hôtel de l'Europe, Place La Fayette. Bankers: Messrs. Courtois.

	RAIL.	
	Kil.	Miles.
Toulouse to Carcassone	91	56½
Carcassone to Narbonne (junction)	58	36
Narbonne to Perpignan	63	39½
	212 =	132

From Toulouse there are trains three times a day to and from Montrejeau; distance 104 kil. 65 miles: fare, 11 fr. 65 cents first; 8 fr. 75 cents second class. At Montrejeau, where there is a good buffet, there are diligences in correspondence with the trains, ready to start for Luchon, distant 40 kilometres. There are generally also a good supply of carriages, and a return carriage may often be found, for which the price should be 20 francs.

At Portet St. Simon, the first station on the line, 12 kilometres from Toulouse, there is a branch line to Foix, the 'chef-lieu' of the department of Ariege, distant 83 kil.; fare, 9 fr. 30 cents first; 6 fr. 95 cents second class.

If entering the Pyrenees by Luchon the shortest and most direct route is by Perigueux, as follows:—

Leave Paris (Orleans Station) at 7·45 p.m.

1st Class. FARES.	2nd Class.	Kil.	Miles.		
55 fr. 90 cent.	,,	499	310	Paris	. . . . 7.45 p.m.
				Perigueux (buffet)	{ arrival 6.32 a.m. / depart. 6.44
72 fr. 90 cent.	,,	651	404½	Agen (buffet)	{ arrival / depart. 11.20
	,,	721	436¾	Montauban	. . . 12·45 p.m.
86 fr. 45 cent.	,,	772	480¾	Toulouse	. . 1·45
				Toulouse (leave)	. . 5.15 p.m.
11 fr. 65 cent.	,,	104	65	Montrejeau (buffet)	. 8.35
98 fr. 10 cent.	,,	876	544½	Total rail.	
	,,	40	25	To Luchon, by diligence, in about 4 hours.	
	,,	916 =	569½	Total distance from Paris.	

The return from Luchon is best made, leaving by the early diligence at about 2 A.M., to catch the train leaving Montrejeau at 6.30 A.M., reaching Toulouse at 9.45.

		Kil.	Miles.	Toulouse	. . . 10.55 a.m.
,,	,,	121	75½	Agen	{ arrival 1.28 p.m. / depart. 1.55
,,	,,	273	169¾	Perigueux (buffet)	{ arrival 5.40 / depart. 6. 5
,,	,,	772	480¾	Paris	. . . . 4·55 a.m.

GUIDE TO THE PYRENEES.

From the East.

The Eastern Pyrenees are best approached from Lyons as follows:—

	DISTANCE.		TIME.	PRICE.		
	Kil.	Miles.	Express Trains.	1st Class.	2nd Class.	3rd Class.
Paris . .						
Lyons . .	512	318	11 hrs. 20 min.	56 fr. 80 c.	42 fr. 60 c.	31 fr. 25 c.
Avignon .	230	143				
Tarascon [branch line to Montpellier & Cette]	21	13	5 hrs. 36 min.	28 fr. 10 c.	21 fr. 10 c.	15 fr. 45 c.
Nismes . .	28	17½				
Montpellier .	49	30½				
Cette . .	28	17½	3 hrs. 5 min.	11 fr. 75 c.	8 fr. 80 c.	6 fr. 45 c.
	868 kil.= 539½ miles.		20 hrs. 1 min.	96 fr. 65 c.	72 fr. 50 c.	53 fr. 15 c.

	DISTANCE.		TIME.	PRICE.		
	Kil.	Miles.	Express Trains.	1st Class.	2nd Class.	3rd Class.
Cette from Paris .	868	539½	20 hrs. 1 min.	96 fr. 65 c.	72 fr. 50 c.	53 fr. 15 c.
Narbonne [branch line to Toulouse 149 kil.]	71	44				
Perpignan .	63	39½	5 hrs. 5 min.	14 fr. 65 c.	11 fr. 0 c.	8 fr. 5 c.
	1,002 kil.= 623 miles.		25 hrs. 6 min.	111 fr. 30 c.	83 fr. 50 c.	61 fr. 20 c.

They may also be approached from Bordeaux as follows:—

	DISTANCE.		TIME.	PRICE.		
	Kil.	Miles.	Express Trains.	1st Class.	2nd Class.	3rd Class.
Paris . .						
Bordeaux .	583	362	12 hrs. 30 min.	65 fr. 25 c.	48 fr. 35 c.	35 fr. 90 c.
Toulouse .	257	160	6 hrs. 20 min.	28 fr. 80 c.	21 fr. 50 c.	15 fr. 85 c.
Narbonne .	149	93	3 hrs. 45 min.	16 fr. 65 c.	13 fr. 50 c.	9 fr. 40 c.
Perpignan .	63	39	1 hr. 54 min.	6 fr. 70 c.	5 fr. 50 c.	3 fr. 70 c.
	1,052 kil.= 645 miles.		24 hrs. 29 min.	117 fr. 35 c.	88 fr. 85 c.	64 fr. 85 c.

Perpignan to Vernet, 54 kil.= 33½ miles. Daily public conveyances. (Sec. 107.)

Perpignan to Amélie les Bains, 38 kil.= 23¾ miles. Daily public conveyances. (Sec. 109.)

Bordeaux is a fine city, but he would not suit me for a travelling companion who could be long detained by the prospect of its surnburnt quays and streets, with the fresh mountain air inviting him from so short a distance. (Hotels: De l'Univers, Nantes.) There is a handsome theatre; but we have not left Paris for the sake of going to the play. Though the see of an archbishop, the cathedral is not very remarkable; and the high stone houses, though solid and imposing, are devoid of the picturesque. As we stroll along the river, we see such a quantity of ships and steamers with the union-jack flying, that, except for the characteristic indolence of the French, which the increasing heat of the climate, as you go farther south, certainly does not lessen, we could almost fancy ourselves in the Liverpool docks. Along the quays I noticed a number of puncheons, which turned out to contain British corn spirit, sent out to Bordeaux for the purpose of being transmuted into genuine cognac, which is effected by redistilling it with the refuse of the vintage. In a course through the market that I made after breakfast in search of some peaches (which, by the kind disposition of Providence, seem to grow most abundantly in those countries where they are most required), I came upon a marchande who was selling snails (escargots), not, as you might suppose, to stock a vivarium, but for the cuisine. They are bred on purpose in marshes near Bordeaux, and very fine specimens of molluscs they are—green and glutinous-looking, and, I dare say, to the epicure, invaluable in enriching a *potage*, but to my unsophisticated palate they did not look inviting ; and the marchande assured me, with an earnestness that showed her repugnance, that, though she had dealt in them for twenty years, she had never tasted one.

Should the traveller stop at Dax (De aquis*), he should see the hot spring, which is of a temperature of 208° Fahr., higher than any other in France; also the salt works just established by a French company, on a valuable deposit discovered by M. Lyte, of Bigorre ; but, if he takes my advice, he will allow the train to carry him on to Bayonne, a town which surpasses all that I have seen in France in the picturesque beauty of its situation, which is, perhaps, not the less striking because you come upon it with the eye wearied by the monotonous and featureless plains of the landes. As you approach Bayonne, the stunted fir trees, which, though few and far between, ever since leaving Bordeaux have formed the main feature in the arid vegetation, are in a great measure replaced by cork trees ; and the red-brown trunks of those that have been recently stripped of their bark present a singular appearance among their rich green foliage, for the trees are not killed by the process. On alighting at the Bayonne railway station, the kissing of the ladies, children, and bearded men, was most amusing, from its exceeding vehemence, so different to the greetings of our phlegmatic temperaments ; the men, however, only kissed each other, as the Spanish ladies never salute in public.

The town of **Bayonne** (Hôtel de Commerce) is divided into 3 portions by the rivers Adour and Nive ; the best point of view is from the citadel, the fortifications of which, as of most French towns on the frontier, seem very strong, but strangers are not allowed to enter without an express permission from the governor. The panorama of the town from this spot, with the distant Pyrenees, might vie with that of Thun and the Bernese Alps from the Pavillon Bellevue, which it even exceeds in beauty, if, from the absence of the snow-capped mountains, it somewhat falls short in grandeur.

In these days no one thinks of leaving Bayonne without a visit to **Biarritz**, which is 5 miles distant. In bygone days this excursion was always made on the back of some animal, either a horse, mule, or donkey, on whose back was arranged an apparatus very like two panniers, and this mode of travelling was called riding *en cacolet*. The traveller seated himself on the one side, while on the opposite side of the animal, by way of balance, sat the cacolettierra, a sprightly, black-eyed Basque damsel of lively wit and quick repartee, whose engaging conversation no doubt contributed very materially to compensate the traveller for the length and weariness of the route. The progress of civilisation, and the necessity of more rapid and frequent communication with a

* So at Ax (aquæ) in the Eastern Pyrenees, the hot springs, which attain a temperature of 168°, have in the same manner given a name to the town.

place which the French Empress has selected as her favourite summer retreat, have, however, put an end to these romantic journeyings, and a good road now joins the two places. Various omnibuses and minibuses, decorated with the name of diligences, now carry on an active traffic, which is rendered more exciting by the competition of cabs and other carriages; but I have no doubt that many a traveller, when seated within these dusty and ill-adjusted vehicles, has often regretted the *cacolet* of former days. Nevertheless, a journey to Biarritz is well worth an hour's exposure to dust, and the jolting of a rough vehicle. This little town, seen in the glow of a September sun, is the very realisation of some lovely and picturesque scene in an opera; or, to compare it with something more real, the beach at Biarritz reminded me in many of its features of that at Atherfield, at the back of the Isle of Wight. There are the same variegated cliffs, and of about the same height; the same sparkling shingle of minute quartz pebbles, and, more than all, the same pure transparency of the sea, which comes tumbling in at both places with the same incessant roll, though, of course, in the magnitude of its billows, the Bay of Biscay has the pre-eminence.

Imagine a small plateau formed by a rent in the cliff, sloping rapidly towards the sea, upon which gardens and cottages stand interspersed with rocks and ravines, the whole scene being wild and abrupt, although on a miniature scale. Such was Biarritz before it became one of the most celebrated watering-places in the south of France. Its two lines of hills advance into the sea in the form of a double-horned cape. To the left, at the Point des Basques, begin a high range of cliffs, which stretch far towards the south, to the mountains of Fontarabie; while to the right the Coté des Fous is dotted over with the perforated rocks of the Attalaye, whose insulated pinnacles have been strangely fashioned into various shapes by the waves that are rapidly wearing them away. Between the Pointe des Basques and the rocks of Attalaye lies the Port Vieux, from whence many whalers were wont to set sail year by year in former times, although the harbour, which is constantly being contracted, can now do no more than give shelter to a stray fishing-smack. Encased, as it were, within this charming framework, lie the picturesque habitations of Biarritz. Some of the houses occupying the plateau and the bottom of the valley constitute the village and its principal street, while others are grouped here and there by chance as it were, and in accordance with the nature of the ground. The green blinds and dazzling white walls of these cottages give them an air of cleanliness and comfort which is well adapted to attract visitors, and each year the nomad population which flocks to Biarritz tends more and more to convert it into a fashionable watering place, to the prejudice of the invalids who formerly resorted to it.

The Villa Eugénie, the imperial residence, is a plain brick mansion, situate at the northern extremity of the town, with its garden reaching down to the beach. No one would take the house for the palace of an Emperor; it might be the abode of an English country gentleman, but for the scantiness of the grounds in which it stands, which do not exceed 2 or 3 acres, and which are altogether destitute of trees. In spite of this unpretending appearance, the proximity of the Court exerts a visible influence upon the manners and customs of this little town. The invalids and fishermen have been scared away by the influx of pleasure-seekers, and indolence and fashion seem to be now the presiding genii of the place. Indeed, modern life at Biarritz seems to be the nearest possible approach to the Homeric ideal of Elysium:—

τῇ περ ῥήϊστος πέλεται βίος ἀνθρώποισιν·
οὐ νιφετός, οὔτ' ἄρ χειμὼν πολὺς, οὔτε ποτ'
ὄμβρος·
ἀλλ' ἀεὶ ζεφύροιο λιγυπνείοντας ἀήτας
Ὠκέανος ἀνίησιν, ἀναψύχειν ἀνθρώπους.

There the human kind
Enjoy the easiest life; no snow is there,
No biting winter, and no drenching shower,
But zephyr always gently from the sea
Breathes on them, to refresh the happy race.
Cowper, *Odyssey*, book iv.

On the rocks of Aftalaye, facing the Villa Eugénie, a handsome building has just been completed, which, under the name of Casino, combines reading, billiard, music, and dancing rooms; and here every evening during the months of August and September, which constitute the season at Biarritz, a first-rate band from Paris is generally retained, to play all sorts of music adapted to the different tastes of the visitors, from Beethoven and the Stabat Mater to waltzes and mazurkas. Among the hotels, the Hôtel de France, kept by M. Gardères (brother of the landlord of the Hôtel de France at Pau), is the best. The Hôtels d'Angleterre and des Ambassadeurs are also good, and there are numerous *cafés*, *restaurants*, and *maisons garnies*.

Strangers coming to reside at Bigorre will, I think, find the Coté des Basques, the South Cliff, the best situation. Here there are several good lodging-houses, among which I can confidently recommend the Maison Maroye and the Maison Antoine. From either of these there is a charming view, looking over the Bay of Biscay to the coast of Spain.

In bathing, the primitive simplicity and customs of the olden time still prevail, and during the summer months the beach at Biarritz presents a scene of which Brighton and Margate can give but a very faint notion. From 3 to 5 o'clock on a fine afternoon, the northern beach, which is called the Coté des Fous, is thronged by the whole population of the place, who come here to take their promenade, some in the sea, and some on the land; and so accustomed do they all seem to this amphibious rendezvous, that it would be hard to specify which was their natural element.

A French writer thus describes the scene:—'Thanks to the patriarchal traditions of Biarritz, no artificial barrier separates the male and female bathers, who enjoy their daily recreation in the most social and familiar manner. Men and women have adopted a costume to which the most scrupulous prudery could not object, but which varies somewhat according to individual taste; and thus conversations, and even flirtations, are as agreeably conducted in the bath as on the promenade.' Owing to the configuration of the coast, the shores of Spain and France forming, as it were, a kind of funnel to the waters of the Atlantic, the waves of the Bay of Biscay often roll in with great violence, and there is a considerable swell even on the calmest day. For the sake of making a better stand against the onset of the breakers, the lady bathers join hand-in-hand, and not unfrequently have recourse to the aid of some attendant cavalier; and the fun of the lookers-on, who, seated in chairs, line the beach, consists in watching them as they vainly spring up to avoid the advancing mass of waters, which comes tumbling them over and over, and beneath which for a moment they occasionally quite disappear. There is not, however, any real danger or cause for alarm, as a life-boat well manned is always stationed at a short distance from the shore, ready to pull to the rescue at a moment's warning. Those of the bathers who swim, for the most part resort to the Port Vieux, which, being protected by the rocks on either side, affords a calmer sea for their evolutions.

North of the Port Vieux a harbour of refuge is being now constructed by order of the Emperor. Huge oblong blocks of Portland cement are formed in wooden frames, and when set they are run out on a tramway, and tilted into the sea, thus forming a breakwater from rock to rock, which is afterwards faced with masonry. Each of these blocks weighs 30,360 kilogrammes (over 30 tons), and costs 500 francs. Thousands of them will have to be sunk.

A mile north of Biarritz there is a water-worn cavern in the cliff, which bears the romantic name of 'La Chambre d'Amour,' from the sad fate of two lovers who were here overtaken by the rising tide, and drowned in each other's arms. Their bodies, carried out to sea, were never recovered; but often in the twilight of a summer evening, it is said, you may still see their spirits wandering hand-in-hand along that shore; and the sea-worn cavern in which they met is still known as the Chamber of Love.

The geologist feels, perhaps, even a greater interest than the poet in the 'Chambre d'Amour;' for the undulation of the soil above it marks the extreme boundary of the chain of the Pyrenees. At a few feet from this little bay, the cliffs sink to rise no more, their last rocks dipping beneath the sea of sand which extends as far as 'La Gironde,' exhibiting, in the midst of the richest provinces of France, a miniature representation of an African desert.

On the rocks beyond the lighthouse, on the Biarritz side of the 'Chambre d'Amour,' the maiden hair fern is abundant. To procure it go down to the beach. Higher up are some rather curious plants, the *Smilax aspera*, and a pretty little composite, the *Linosyris chrysocoma*, Goldilocks.

All the rocks are of the lower tertiary formation, abounding in nummulites, which may be collected to any extent on the north side of the Port Vieux. Gathering samphire here would not be the same 'dreadful trade' that it is described by Shakspeare. The rocks are covered by it down to the sea level, and here and there among the ordinary samphire (*Crithmum maritimum*) may be seen what is known in England as the golden samphire (*Inula crithmoides*), a plant of the composite order, with succulent and very aromatic leaves.

From Biarritz to **St. Sebastian**, in Spain, it is a very pretty drive of 61 kil. = 38 miles, passing through St. Jean de Luz, 20 kil. = 12½ miles. At Behobie, 11 kil. beyond St. Jean de Luz, the river Bidassoa, separating France and Spain, is crossed by a bridge. Here a boat should be taken, for a visit to the old Moorish town of Fontarabie, one of the most picturesque in the north of Spain. You can return to meet the carriage at Yrun, on the other side of the Bidassoa. The scenery, as you approach St. Sebastian, especially about Passages, is very striking. At St. Sebastian (Parador Real) the only thing to see is the view from the citadel, which affords a magnificent prospect over the sea.

The rail is now open, doing the disance in a little more than 2 hours; but the drive, especially about Passages, is so beautiful that it is a pity to miss it; and the rail gives no opportunity of visiting Fontarabie. Passports are now not necessary. At St. Jean de Luz, 17 kil. by rail from Biarritz, the Hôtel de la Poste is excellent. From here the mountain of La Rhune may be ascended in 3 hours, and something less to return. Though only 900 metres, it affords a fine view over the mountains, ocean, and plain;

and the ground is interesting from the fierce struggle which here took place between the English and French in 1813. Traces of the redoubts may still be seen on the mountain.

Section 2.
Structure of the Pyrenees.

Regarded in its largest extent, the Pyrenean range may be said to extend from Cape Creux on the Mediterranean, to Cape Finisterre on the Galician coast, a distance of about 650 miles, comprising the Asturian as well as the Isthmian part of the chain; but under the name Pyrenees, the latter only is generally understood, which, for 270 miles in a direct line, forms a mountain barrier between France and Spain. The mean altitude of this mountain wall is 1,782 metres, or nearly 6,000 feet, and the maximum height is attained nearly midway in the Isthmian range, where the Pic de Nethou rises 11,168 feet above the sea. Between this and the Pic de Midi d'Ossau, 70 miles west, are the highest peaks of the chain, many of them above 10,000 feet, and 4 or 5 little inferior to the Pic de Nethou. It is through this portion of the range that the following pages are intended principally as a guide; and I do not think more can be comprehended in a summer tour. It is always advisable to see a small portion thoroughly, rather than hurriedly to scamper over a large tract of country, which latter plan increases the expense and fatigue, without yielding the same amount of pleasure, or allowing objects and scenery sufficient time to be impressed on the mind.

About the centre of the Isthmian chain, west of the Maladetta, there is a sudden break or fault, the western portion of the chain coming to an end in the Pic de Fourcanade, at the head of the Val d'Aran, while the eastern portion commences about 16 miles in a direct line north, with the Pic Tentenade and Tour de Crabère east of Pont du Roi. The interval between the two chains, the Val d'Aran, is Spanish territory, though certainly, if the principle of natural boundaries, at present in fashion, is to be carried out, one may expect ere long to see it attached to France.

In surveying the Pyrenean chain, it is to be borne in mind that there are three different aspects under which it presents itself. First, there is the granitic axis, or *medial line of elevation*; secondly, there is the *culminating line* along the highest points of the ridge, which forms the watershed; and thirdly, there is the *frontier line*, as fixed by the Treaty of the Pyrenees, A.D. 1659. These do not necessarily and invariably coincide: thus, while the culminating point of the granitic axis and of the whole chain is the Pic de Nethou, 11,168 feet, the limestone of the Cylindre de Marboré, 10,893 feet, east of the Brèche de Roland, is the highest point of the frontier line. Connected with the watershed of the Pyrenees there is this peculiarity—that while four-fifths of the waters that rise on the French side have their outpouring in the Atlantic Ocean, as tributaries of the Adour and Garonne, all the streams on the Spanish side are received by the Ebro, and flow into the Mediterranean. The culminating crest of the chain generally coincides with the granitic axis, which forms the heart of the mountains; but here and there the central axis is dominated by mountains of superior height, both on the northern and southern sides of the main chain; some of the mountains on the Spanish frontier, as the limestone of the Mont Perdu (3,351 metres = 10,994 feet), and the granite peaks of Posets (3,367 metres = 11,046 feet), and Nethou (3,404 metres = 11,168 feet), being the very highest of the Pyrenees; while on the French side the lofty granite ranges, from which spring the Pics Long (3,193 metres = 10,476 feet), Neouvielle (3,092 metres = 10,145 feet), and Cambiel (3,173 metres = 10,410 feet), form a third transverse chain, enclosing between them such a mass of snow and ice as makes this far the most Alpine portion of the French Pyrenees, and the most worth visiting.

At the Pic d'Anie (2,504 metres = 8,215 feet), 87 miles in a direct line from the Atlantic, in the western portion of the Isthmian Pyrenees, the granite first appears on the highest ridge; and a little farther to the east the mountains first attain the elevation of perpetual snow, in the forked peak of the Pic du Midi d'Ossau, 2,985 metres = 9,793 feet. From this point to the Pic de Fourcanade (2,882 metres = 9,455 feet), at the extremity of the Val d'Aran, eastward of the Maladetta, the chain preserves a mean height of upwards of 10,000 feet. In the eastern portion of the chain, commencing northward of the Val d'Aran from the Tour de Crabère (2,630 metres = 8,629 feet) to the Pic d'Estat (3,141 metres = 10,305 feet), a distance of about 35 miles, the mean elevation exceeds 9,000 feet; from the Pic d'Estat to the Canigou (2,787 metres = 9,144 feet), some 60 miles farther east, the mean elevation is less than 8,000 feet; and at Canigou the chain suddenly sinks, so that in its remaining portion of 32 miles, the Monts Albères, it does not attain 3,000 feet above the sea.

Few mountain chains present such a regularity of disposition as the Pyrenees. From the higher mountains spurs are thrown out on either side, 20 or 30 miles towards the plain, so that the crest of the Pyrenees may be said to resemble the spinal column, the lateral ridges which separate the valleys representing the rib-bones. There are but five carriage-roads across the chain, all lying to the extreme east or west. The gaps, or 'ports,' as they are called, in the main wall which give passage between France and Spain, are generally higher than the ordinary Alpine passes, and present exceedingly wild and grand scenery. The mountain passes over the lateral ridges, called *Cols*, sometimes *Hourquettes*, are also of extraordinary beauty and some difficulty. Those who cannot make the ascent of these lose the finest scenery, and have to retrace their steps over the same ground; but to a tolerable pedestrian all the 'cols' and 'ports' are accessible, and the majority of them may be traversed on horseback. A remarkable and very interesting feature in the Pyrenees are the natural rocky basins—*cirques*, or *oules*, as they are there called. They are situated in the transverse valleys lying between the buttresses of the principal range, and are generally surrounded on three sides by lofty walls of rock, opening into the valley by a narrow gully. The scenery of these cirques is peculiar, combining much sublimity with great pastoral beauty. The Pyrenean valleys on the French side are much lower than the Alpine, few being more than 2,000 feet above the level of the sea, whereas those in the Alps are for the most part nearly double that height; so that the mountains in the Pyrenees, when seen from the valleys, frequently assume a more imposing appearance than those in Switzerland of higher elevation. In consequence of the Pyrenees being nearly four degrees south of the Swiss Alps, the line of perpetual snow is much higher on the mountains. I should be inclined to place it not lower than 8,900 feet, or about 700 feet higher than the snow-line on the Northern Alps, though snow lies all through the summer in sheltered places at a much lower elevation.* Thus the

* The snow line is that line of elevation at which the annual fall of snow exceeds the

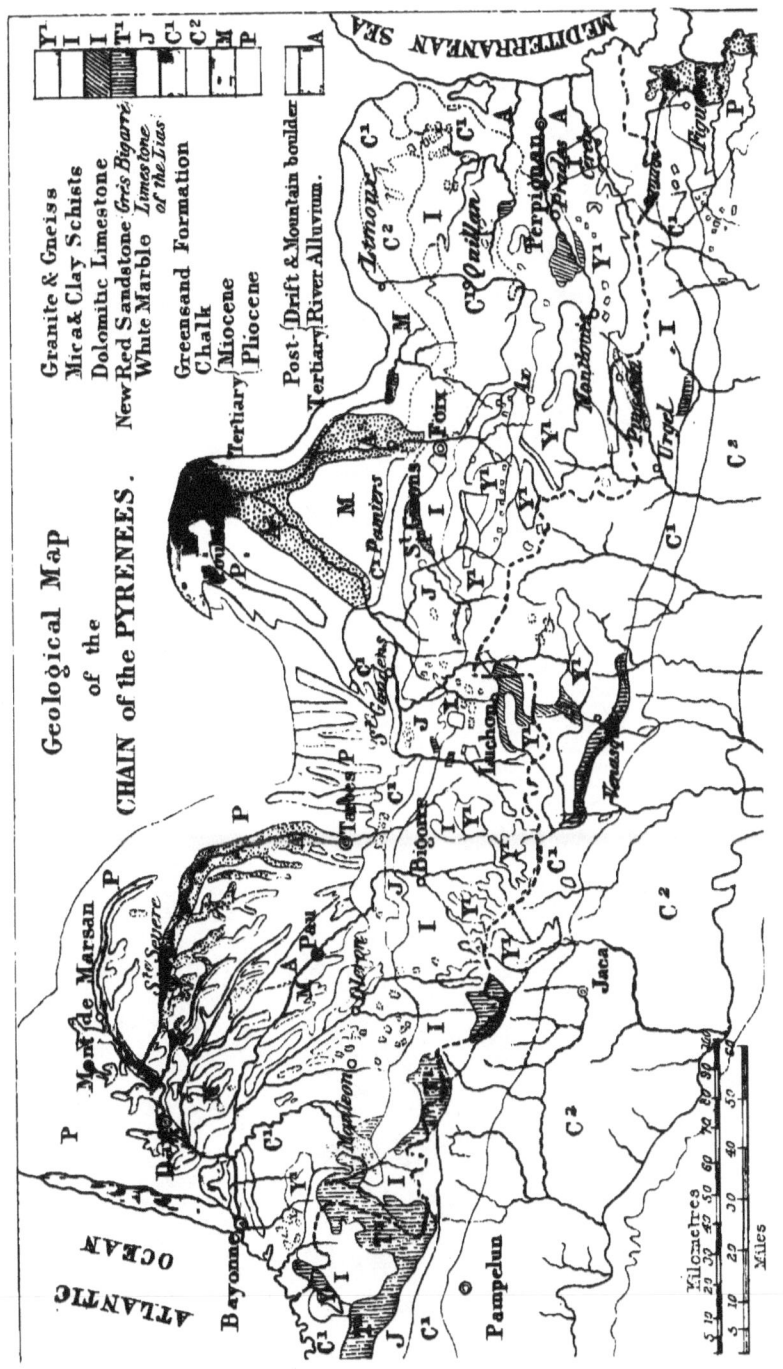

grand glacial features which are characteristic of Alpine passes are frequently missed in the Pyrenees, even on elevations which in the Alps are covered with ice and snow; but there are passes, such as that from the Lac de Gaube to Gavarnie, the Brèche de Roland, the Port d'Oo, and the Portillon d'Oo, where the sterner scenery of snow-field and glacier are not wanting; and in these high 'ports' the weather is generally so wild, and the path so bad, that the natives have a proverb, 'In the port, when the wind rages, the father waits not for his son, nor the son for his father.' From the scantiness of glacier there is also this advantage: the streams, not turbid like those of Switzerland, but clear and bright, which gush from every annual waste. Dr. Lambron, in his work on the Pyrenees of Luchon, places the lower limit of the 'permanent snows' at 2,730 metres = 8,956 feet. At this height on the French side of the mountains, one sees all around patches of unmelted snow, even in the hottest months of summer, wherever the inequalities of the ground have allowed it to accumulate in any mass; for above this elevation it is only during the months of July and August that the winter snows begin at all to disappear. On the Spanish side of the Pyrenees, by reason of its steeper declivity, which is on an average nearly one-third greater than that on the French, and also its being exposed to the full midday sun, in addition to the radiated heat being greater from the higher level of the plain, scarcely any snow is to be found during the month of August. Towards the end of August there is least snow to be seen in the Pyrenees, and during that month, especially if the season has been at all dry and hot, many peaks considerably higher than this snow-line are altogether free, as Mont Canigou, the Pic du Midi de Bigorre, and some others. This depends, however, on certain special conditions, as the steepness of their sides, and their advanced position on the outposts of the central chain—the one being unfavourable to any very thick accumulation of the snow-beds, and the other giving a less rigorous climate to these isolated peaks by bringing them less under the influence of the central masses metres of snow and ice. Above the height of 3,000 metres = 9,843 feet, says M. Lezat, rain seldom if ever falls in the Pyrenees, but the moisture descends in snow or hail.

hollow, and water every valley, impart an exquisite verdure to the lower lands, and nourish an almost endless variety of wild flowers. The dark fir forests which clothe the mountain sides, and contribute so much to the beauty of Pyrenean scenery, are chiefly varieties of the silver fir, interspersed on the lower hills with beech, chestnut, and other trees.

The geological structure of the Pyrenees proves them to have been among the most recent of mountain elevations; and though they did not probably rise to their present height *à un seul jet*, as the French geologists formerly maintained, it seems pretty clear that the tremendous subterranean throes that finally upheaved them, and erupted, perhaps at a single coup, such mountains as the Mont Perdu and the Marboré, must have taken place in the comparatively brief interval between the deposition of the chalk and certain tertiary strata; for while some of the highest peaks of the ridge are formed of marine calcareous beds of the age of our chalk and green-sand series, the tertiary strata at the foot of the chain are for the most part horizontal, and reach only to the height of a few hundred feet above the sea.

The last and principal upheaval of the Pyrenees would seem, then, immediately to have preceded that of the Jura, and to have been anterior to the formation of the principal Alps, though long subsequent to the Cumberland, Welsh, and Scottish mountains, and also to the more recent systems in Brittany and central France. It was not till this last upheaval that the Pyrenees presented such an effectual barrier, separating the two countries of France and Spain; but the system of these mountains, if we may judge from the disposition of the rocks of which they are composed, would seem to be the result of four successive upheavals, alternating with the same number of immersions, or partial subsidences beneath the sea. The formation of the Pyrenees will perhaps be understood by referring to the subjoined diagram.

Fig. 1. Represents sundry isolated points of primary and transition rock, which formed the commencement of the Pyrenees.

Fig. 2. Then became a partial subsidence and immersion beneath the Triassic sea, which covered with its waters most of the western portions of the chain, especially those on the southern side: in this sea the *Grès Bigarré*, the red sandstone peculiar to the Pyrenees, corresponding to our new red, was deposited.

Fig. 3. Represents the chain as it emerged from the Triassic sea, many of the southern peaks in the western part of the range being composed of the red Pyrenean sandstone.

Fig. 4. Once more came submergence, and the Jurassic sea, the deposit of the lias, not only covered the lower spurs of the chain, but with its waters found a channel between the eastern and western portions.

Fig. 5. Then followed the third upheaval, most energetic in the western portion of the chain, adding considerably to the general elevation, and leaving evidence of the central intersection of the chain by the Jurassic sea, in the white lias formation to be seen NW. of Luchon in the Pics de Gar and Cagire, and also south of the Port de Venasque in the Peña Blanca.

Fig. 6. The next submersion was beneath the waters of the sea that held in deposit the chalk formation, and this probably was of considerable duration, to allow the enormous thickness to be accumulated which the Cirque de Gavarnie and mountains contiguous present.

Fig. 7. Lastly came the final and most considerable upheaval, to which the Pyrenees owe their present elevation, and general direction as a continuous chain, which is 18° north of west. Since that period the Atlantic and Mediterranean have never again united their waters to sweep across the southern provinces of France. The northern spurs of the Pyrenees have indeed yet once again been bathed by the waters, but those waters have been fresh, the accumulation of lake basins and mountain streams; in these waters were deposited, abutting horizontally on the last uptilted Eocene strata, those tertiary sands and clays of the Miocene, whose beds formed the graveyard of the huge primeval animals immediately preceding man.

In great mountain chains the lower elevations are commonly composed of secondary and transition rocks, through which the granite pierces, and forms the highest peaks; but it is not so in the Pyrenean system: the great mass of the range is composed of secondary rocks (chiefly argillaceous schists) and limestone (in many places altered), with the primitive rock cropping out at intervals in bands or zones parallel to the chain, but not always, as we have seen, appearing on the highest part of the central ridge. Between Mont Perdu and the Maladetta, on the south side of the range, there is a very long granitic zone; but it is in the Eastern Pyrenees around Mont Canigou that the granitic formation most preponderates.

To this geological structure, and the frequent points of junction of the granitic and sedimentary rocks, the hot sulphurous springs of such constant occurrence throughout the whole chain are to be attributed. It is not settled whether these hot sources should be referred to chemical or volcanic agency, but probably to the latter. Many of their phenomena remain to be explained. Sources not only totally different in mineral composition, but also in temperature, are sometimes found issuing at dislocations in the rock within a few yards of each other. Three points, however, seem to be established: 1. That in the eastern part of the range, where the granitic formation preponderates, the springs are generally warmer than those issuing in the western part of the chain. 2. Throughout the Pyrenees, the springs containing most sulphur are found near the axis of the chain; whereas those containing little or no sulphur rise in localities at the base of the mountain spurs on the northern side, as at Bigorre. 3. Not only the flowing, but also the temperature of the springs, is affected by any serious convulsion in the earth's

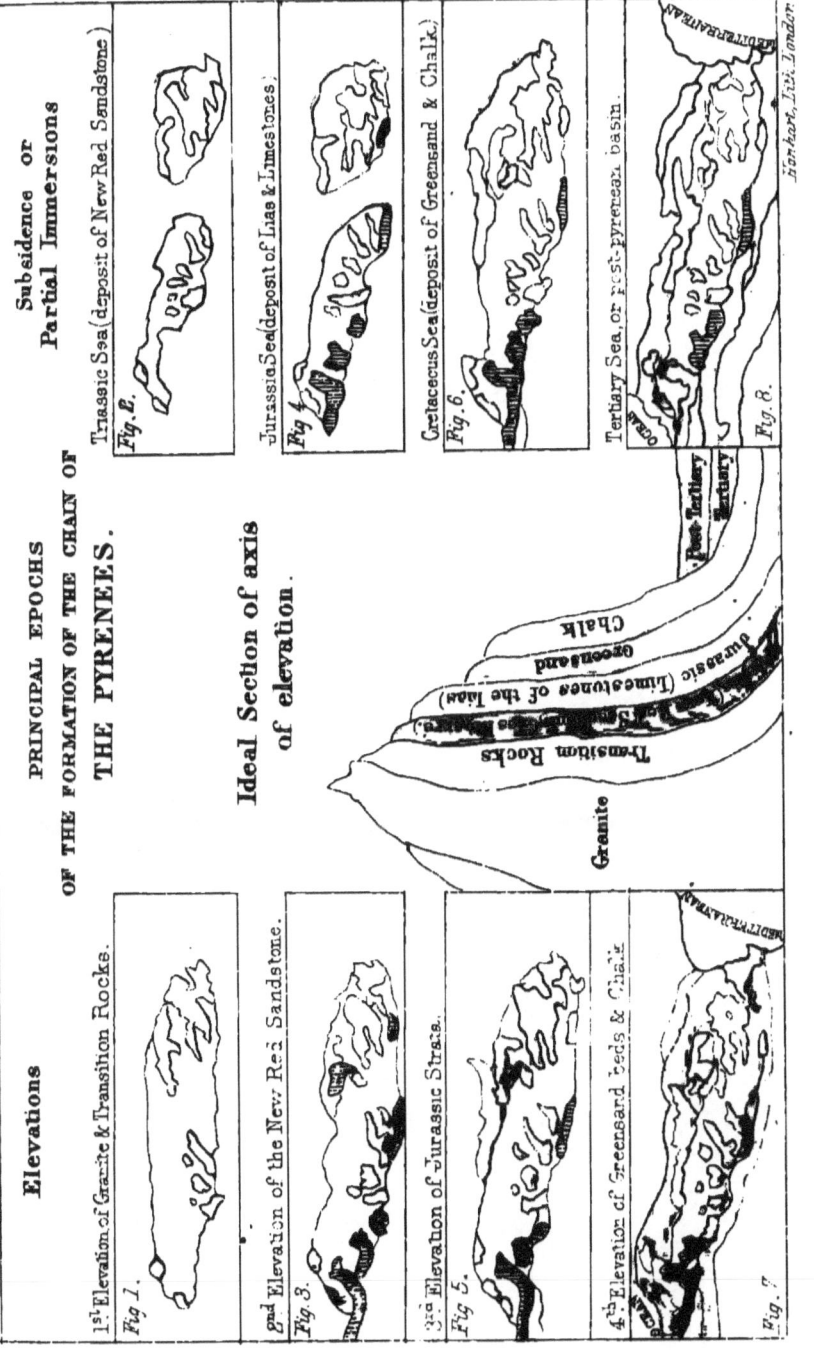

crust; the Source de la Reine, at Luchon, rose from 110° to 175° Fahr. during the great earthquake of Lisbon. Small earthquakes are not uncommon in the Pyrenees; in the last five centuries 117 have been noticed, but of these only seven have been violent enough to do any mischief. Earthquakes occur most frequently at the extremities of the chain, and especially on the side towards the Atlantic; this agrees with the remark made by Ramond, that the limestone formation in this chain is more liable to commotion than the granite.

Section 3.

FROM BAYONNE TO PAU.

By Orthez, 100 kilometres = 62 miles. Rail 4 hours.

Orthez 62 kil.
Pau . 38 kil.
100 kil. = 62 miles.

There is but little of interest on the journey from Bayonne to Pau, except the bridge at Orthez, which is picturesque and of historical interest. At Pau, though the Hôtels de la Poste, Europe, and Angleterre are all good, I can most recommend the Hôtel de France, not only for its good cuisine and civil landlord (M. Gardères), but as far the best in situation, as from your bed-room windows at this inn you command a view of the principal chain of the Pyrenees, the Pic du Midi de Bigorre on the east, and the Pic du Midi d'Ossau on the west, rising most conspicuous from the serrated ridge. The spurs of the Pyrenees extend as far as the left bank of the Gave de Pau, and the low vine-clad hills which here form their northern extremity, produce a strong and highly flavoured white wine, known as Jurançon, or vin d'Henri IV., which, though somewhat too powerful for a summer beverage, is the only wine to be met with all through the Pyrenees, at least on the French side, at all drinkable.

The season at Pau is during the winter, when house-rent is very dear, owing to the quantity of English who resort there for the sake of the mild climate; but in August the town is almost too hot for enjoyment, and the tourist, after having paid a visit to the chateau of Henri IV., is glad to pass over to the mountains, and secures a carriage to convey him to Eaux Chaudes or Eaux Bonnes, two charmingly situated watering-places at the head of the Valley d'Ossau, and in the heart of the mountains.

Section 4.

PAU TO EAUX BONNES.

42 kilometres = 26¼ miles.

PAU TO EAUX CHAUDES.

43 kilometres = 27 miles.

There is an excellent carriage-road to each.

After leaving Pau by the stone bridge which crosses the gave, at the 26th kilometre you reach the village of Louvie-Juzon (Hôtel des Pyrénées), where the road crosses the gave. From Louvie-Juzon there is a carriage-road to Oloron on the west, 21 kilometres, passing through Arudy, and another on the east to Lestelle, 23 kilometres, 2 kilometres beyond Nay. From Oloron to Pau 38 kilometres: from Lestelle to Pau 24 kilometres. Here begins the Valley of Ossau or Valley of the Bear, though the animal which gave its name to the valley is now become somewhat rare, and it is not often that the bear-hunts are rewarded by a successful chase. The costume of the inhabitants in this valley is very picturesque, a scarlet jacket and brown woollen stockings being the holiday costume of the male peasants; and on those occasions the women are equally conspicuous with a bright scarlet hood. The costume and manners of the people may be well studied at Laruns, a large village, *chef lieu de canton*, which the road passes 38 kilometres from Pau. Immediately on leaving Laruns, the *old* road to Eaux Chaudes branches off to the right, but passing this about half a kilometre farther you come to the bifurcation of the new road; that to the right leading to Eaux Chaudes, distant 4½ kilometres, and that to the left to Eaux Bonnes, at a distance of 3½ kilometres. The extreme head of either defile is blocked up by a mountain of very imposing appearance; that of Eaux Chaudes terminating in the Pic du Midi, while the Pic de Ger towers grandly over the long street of hotels and shops that constitute the gay and fashionable watering-place of Eaux Bonnes. The Englishman, however, who only travels in search of the picturesque, will do better to make Eaux Chaudes his principal quarters in this valley, as more consonant to our ideas of a mountain village than a spot like Eaux Bonnes, of which the following, by a French author, M. Taine, is a fair description:—

'In the Pyrenees one counts, at any rate, on seeing country, and when you do come on a village, you look for thatched roofs, weatherworn walls, and crazy door-posts, with the court-yard full of old carts, fagots, pots and pans, and domestic animals, jumbled together in all the charming and picturesque confusion of rustic life. But here you meet one of the second-rate streets of Paris, and promenades that remind you of the Bois de Boulogne. Never, certainly, was country less pastoral; the eye runs along a line of houses drawn up like soldiers on guard, pierced at regular intervals with regular-shaped windows, which, with their smart notices and well-kept pavements on either side the street, are but too suggestive of furnished lodgings in a fashionable watering-place.

'These uniform buildings, with their hard straight lines and formal architecture, form a striking contrast to the wild mountains that rise behind. Strange it is to think how a little hot water has brought civilisation and the art of the cuisine to these out-of-the-way caverns. Year by year this singular village is striving to extend itself, but with great difficulty, so completely is it shut in, and almost smothered in the deep ravine. They blast the rock, and on the platform thus excavated houses are built above the torrent, on either side of the mountain, with their chimneys carried up among the roots of the overhanging beech trees. Behind the principal street runs a narrow alley, winding along the inequalities of the rock, and in this quarter the guides and workmen have their lodgings. It is a wonder how the wretched mud-built little huts and wooden cabarets contrive to cling to the almost perpendicular side of the mountain; but so they do, now sunk in a cave, and now perched on a rock, till at last they dip down to the stream in a corner behind the baths, where all the linen is hung up to dry, after having been washed in the place that serves also for the pigs to bathe in.

'Of all places in the world, Eaux Bonnes is the most disheartening on a rainy day, and, unfortunately, rainy days are here frequent. The clouds first descend between the walls of rock that form the Valley of Ossau, and roll slowly upwards along their sides. The summits are soon hidden, and the detached masses of mist become united, and accumulate till they are brought to a check by this pent-up gorge. Here they have no outlet, and descend in a fine cold and continuous rain. The village becomes a prison, and the lowering clouds which overspread the houses deprive them of what little daylight the mountain tops had left them. An Englishman would suppose that he was in London.'

Such is the account of the place given by a native writer, who may be supposed to have no desire to paint the place worse than it is; but this affectation of Parisian life is a decided drawback to Eaux Bonnes.

Eaux Chaudes, on the other hand, is a place of much less pretension, and at either of its two hotels (de France and Baudot) the traveller will be very comfortable.* Moreover, Eaux Chaudes will be found a much more convenient station than Eaux Bonnes for excursions to the Pic du Midi and the Plateau de Bioux Artiques, from which the view of the mountain is inconceivably grand. The whole distance to the Plateau from Eaux Chaudes is only 8 miles, and on no account must this be left unseen, though probably few will care to make the ascent of the Pic itself, which is as fatiguing and difficult as its appearance leads one to expect. The walk up the valley as far as Gabas, 5 miles, is very picturesque, and the sketcher might here find subjects to occupy him for at least a month; indeed, it was with great difficulty that one of our party could tear himself away from this spot. From Eaux Chaudes an excursion is frequently made to the baths of Panticosa in Spain, and from these baths the return may be made into France either over the Col de Marcadou to Cauterets (9 hours), or to Gavarnie by the Brèche de Roland; but all these excursions are very long, and the enjoyment afforded by the scenery scarcely repays for the fatigue, especially when the bad and dirty accommodation of the Spanish inns (very inadequate to restore the traveller's strength after his hard day's work) is taken into consideration. There is no difficulty, however, for the traveller who is prepared to incur all this; only let him remember that the expedition from Eaux Chaudes to Panticosa is one of those for the *enjoyment* of which an early start is indispensable. The track turns off to the left at Gabas, that to the right leading to the Plateau de Bioux Artiques. Between these two rises the precipitous Pic du Midi d'Ossau to the height of 2,985 metres = 9,793 feet, and the pedestrian who cares to scale this mountain commences his climb from Gabas up its steep wooded sides; but the whole ascent is very laborious, and the last 2,000 feet, after leaving the grassy table-land that forms the shoulder, is rather trying to heads subject to giddiness. Within the last few years, the Préfet of the Basses Pyrénées has caused iron bars to be fixed in the most difficult cheminées of this peak. Camy of Eaux Chaudes, and Orteig and Lanoux (father and son) of Eaux Bonnes, are good guides.

The curative properties of the baths at Eaux Chaudes and Eaux Bonnes, I suspect, are much the same, though at the present day those of the latter place seem decidedly to have the call, probably for no better reason than that with which a certain doctor is said to have convinced his patients— 'Employez vite ce remède, pendant qu'il guérit encore: les médicaments ont des modes comme les chapeaux.' The doctors at the waters really do

---

* At the Hôtel de France, the prices are, bedroom, 2 fr.; table d'hôte déjeuner at 10, 2 fr.; table d'hôte dîner at 5.30, 3 fr.

At the Hôtel Baudot, strongly recommended and chiefly resorted to by English, un thé complet, 1 fr.; un déjeuner avec œufs, 1 fr. 50 c.; dîner, 3 fr. 50 c.; room, 2 fr.; a family may be boarded with private sitting-room, and everything included, at 7 fr. a day. [July 30, 1860.]

At Eaux Bonnes, decidedly the best hotel is the Hôtel de France (M. Taverne), which is in all respects perfect. Though the prices are somewhat higher than those of Eaux Chaudes, they are not extravagant; bed and board being about 10 fr. a day, including service.

believe in the waters. It is their normal state; but it is curious to observe how each of them, with a kind of paternal predilection, has only faith in the water of his own place, though, by way of compensation, he assigns this every virtue, and pronounces it specific against every kind of disease.

From Eaux Bonnes, height 748 metres = 2,454 feet, to Eaux Chaudes, height 675 metres = 2,215 feet, the distance by the road is 8 kil. = 5 miles; but there is a pass over the mountain by the Col de Gourzy, which in clear weather presents no difficulty. The path leads up through the woods south of the town, emerging on a grassy plateau, from the highest point of which, 1,839 metres = 6,034 feet, there is a fine view over the Valley of Laruns.

In descending to Eaux Chaudes, the best route is to cross to the other side of the ravine, and then descending through a perfect forest of box trees, the path strikes into that leading to the grotto, 30 min. above Eaux Chaudes.

The passage of this col may be made on foot or horseback, occupying from 4 to 5 hours.

Section 5.

ASCENT OF THE PIC DE GER.

2,613 metres = 8,573 feet.

10 to 12 hours; on foot.

The ascent of the Pic du Midi d'Ossau is excessively laborious, and the panorama in no respect equals that from the Pic de Ger, which may be made without difficulty in fine weather from Eaux Bonnes. The best route is as follows:—Follow the left bank of the Soude rivulet, leaving to the right the Gorge de Balourd, and after passing by the side of an enormous piece of fallen rock some 40 feet long, you reach the origin of the valley, the Coume d'Aas, at the fountain of Gesques, whose water trickles into the hollow trunk of a fir tree. 300 steps beyond this, at the bottom of the gorge, you turn to the right and plunge into a fir forest, through which there is a steep and rugged path covered with stones, which for 2 hours skirts the base of the 5 threatening aiguilles of the Ger, known as the Quintettas. 30 minutes after this you reach the Cabane de Ger, where shepherds are nearly always to be found. The hardest part of the route now begins. From here, for three-quarters of an hour, you have to climb a rapid slope, sometimes over turf, sometimes over slippery white rocks, and by this, in 30 minutes, you circumvent a buttress of the Ger (the Pic du Caperan or du Moine), from whence issues a remarkably clear and limpid fountain. You now soon reach the foot of the arête of rocks that shuts in the gorge of Eaux Bonnes on the east, and unites the Pic de Ger on the right to the Pic de Pembecibé on the left. This arête must be scaled to arrive at the base of the peak itself, and a rough ascent it is; and when you think you have nearly reached the top, you see two new summits rise before you: that to the right is a truncated cone, easy to climb, and generally known as the Pic de Ger proper. To reach the second summit you take a sort of pont to the left, which connects the two summits. The return may be made by the Plateau d'Anouillasse, the table-land to the west of the Pic de Ger, and thence over the pastures of the Col de Gourzy either to Eaux Bonnes or Eaux Chaudes.

Section 6.

EAUX CHAUDES TO PANTICOSA.

12 hours; on foot or horse. Guide not absolutely necessary.

From Eaux Chaudes to Gabas, 5 miles = 8 kils. 1½ hour.

On leaving Gabas, striking to the left, the track mounts the stream of the Broussette, first along its left bank, and about half-way crossing to the other side of the stream. 2 hours 10 minutes from Gabas brings you to the Casa de Broussette, formerly used as a hospice, but now a perfect ruin, consisting of four bare walls, standing on the left bank of the stream, at a height of 1,382 metres = 4,534 feet; and sometimes during the summer months occupied by shepherds as a cheese cabane. From here the top of the Pic du Midi is in sight. 30 minutes above the Casa de Broussette, the shorter and more direct path branching off from the main track (which is continued to the right, and leads to the Col d'Ancou) mounts steeply by the side of a small stream, and on attaining the watershed, passes for about 30 minutes over a grassy tableland. 1 hour 30 minutes from the Casa de Broussette brings you to the top of the col, the Pierre de Claude, where a low wall of loose stones indicates the frontier of France and Spain.

[From the top of this col, the Port de Peyrelue (1,847 m.), some mountains of very striking form and a deep red colour are seen in the SW. on the other side of the valley. They have never been explored, and are probably well worth the attention of any botanist. The most conspicuous pyramid of this range is the Pic Anajet (2,817 m.). On either side of this mountain there is a pass leading from Sallent to Canfranc. That to the north is the easier, the Col de Canaourouye, (something over 2,000 m.), between the peak of the same name to the north and the Pic Anajet, south. It leads into the valley of the Aragon, 2 hours above Canfranc. Both these passes from Sallent occupy about 5 hours, and are said to be practicable for mules. I have not been over either, and cannot say which is the best, but, for any one who visits Panticosa from Eaux Chaudes, the return by one of these passes and the Col des Moines would be an agreeable change.]

In descending from the Port, leave to the left a tarn bedded among the rocks of the Pic de Peyrelue, and avoiding the somewhat shorter but more laborious passage of the Pierre de Claude, you pass over arid and uninteresting rocks, having the station of the Spanish douanes below you on the right, and in about 3½ hours from the col reach Sallent, a village situate on the left bank of the Gallego at a height of 1,252 metres = 4,108 feet.

The imposing and precipitous mountain that overhangs the village is the Peña Foratata. [From Sallent you may pass over the mountains to the baths of Panticosa by the Col du Bondellos in 6 hours; a guide may be taken to the col, whence you are in view of the baths. Mounting NE. for 1 hour 30 minutes by the path leading to the Pierre St Martin; at a desolate sort of cirque, at the foot of the Pic de Bondellos strike E., in 20 minutes a cabane; leave two cascades to the right, and continue to mount. One hour from the cabane brings you to the first col (2,500 m. ?), rather difficult, and exposed to avalanches. From here steer S.E. for a little basin, in which two little tarns for 9 months of the year repose beneath the snow; and beyond them is a gentle ascent to the Col de Bondellos (2,600 m. ?),

C

immediately south of the Pic de Bondellos (2,900 m.), the summit of which may be easily reached in 1 hour from here. From the col the baths and lake are seen 1,000 metres beneath your feet, in the S.E., like a little oasis in the stony desert. Descend S.E. over the snow slopes, then over granite, and lastly over grass, making for the north side of the lake.]

Twenty minutes from Sallent you traverse the little village of Lanus, and here the valley of the Roumigas debouches into the larger valley of the Tena, which stretches southward towards the plains of Aragon. The path continues to wind round the base of the mountains on the left, keeping high above the stream, and on its left bank, till it enters the gorge in which, amidst some magnificent trees, chestnuts and walnuts, is built the wretched little town of Panticosa, 1 hour 30 minutes from Sallent.

From the town of Panticosa to the baths takes the best part of 2 hours.

On quitting the town, the road lies through a narrow gorge, known as l'Escalar (escalier), in the bed of which the torrent Calderas comes tumbling down from the lake of Panticosa. As you ascend, the gorge keeps narrowing, until at last there is hardly room for a horse to pass.

At last, after passing a mineral spring, which leaps down from the rocks on the left, about 50 minutes from the town, the road suddenly turns, and you come upon a sort of cirque, formed by granite rocks, almost entirely bare of vegetation. On the right of this basin, which is but about half a mile in diameter, are erected some nine or ten straggling edifices, which constitute the thermal establishment, at a height of 1,616 metres = 5,392 feet. To the left lies a little blue lake, into which cascades are streaming on all sides from the lofty walls of rock. During the three summer months, the baths of Panticosa are crowded to repletion with invalids, chiefly Spanish; but the traveller must not be under the delusion that he will here see real Spanish life, or meet with Spanish fare. Waiters, chambermaids, and cooks are all French; and the accommodation is much the same as that of the 'thermes' on the French side. About September 20th, there is a general clearance of the regular staff of attendants; but all through the winter there is always some one remaining, so that food and lodging can always be had. No English traveller will, I think, care to pass more than one night at Panticosa. The exercise of the habitués of the baths of Panticosa is very limited, being restricted to a promenade of 200 metres before the houses, and a row on the lake, which is but 200 yards across. Those who are not content with this may ascend the Punta de la Machimaña, the northernmost of the three peaks on the west side of the lake. From this point, at a height of 2,752 metres = 9,029 feet, there is a magnificent prospect of the southern side of the chain, as far as the Mont Perdu. The Punta de Bondellas, the centre of these three peaks, is about 100 metres higher than that of Machimaña.

Section 7.

To Urdos by the Col d'Iseye, returning by the Col des Moines, on foot, 2 days' excursion.

A lofty lateral ridge, running due north from the main chain, separates the upper valley of Ossau from the parallel Vallée d'Aspe; and over this chain those who pass some days at Eaux Chaudes may make several pleasing excursions, either to Oloron and the Vallée d'Aspe, or to Canfranc and Jaca in Spain.

From Eaux Chaudes, following the Gabas road for 2,500 metres, you cross the bridge known by the name of the Pont d'Enfer, though it has nothing in it to justify its savage appellation. Immediately beyond this a kind of rude tramway is seen winding up to the forests on the right. Along this track the path lies. Mounting among the trees for 1 hour and 30 minutes, and then emerging on some open pastures at the foot of the Pic de Sesque (2,487 m.), the path, after passing a very pretty waterfall on the right, becomes very bad, in some places entirely disappearing; but the stream will form a guide, keeping it on the right. About 3 hours brings you to the top of the **Col d'Iseye**, or de Sesque, 1,669 metres = 5,476 feet above the sea; and from here an easy, though stony, zigzag path leads down in $2\frac{1}{2}$ hours to the village of Accous, 27 kilometres from Oloron. In descending, the views of the valley d'Aspe, spotted with hamlets, are very beautiful. From Accous, following the main road, 12 kilometres of gentle ascent brings you to the Fort of Urdos, always garrisoned by French troops, and apparently a place of considerable strength, and 1 kilometre farther brings you to the town of Urdos, the last in the French territory. The town is dirty, and uninviting enough in its appearance; but very fair quarters for the night and excellent food may be obtained at the house of one of the principal inhabitants, who entertains for a consideration, as there is no inn. The next morning, starting from Urdos, 11 kilometres = 7 miles along the new road in course of construction by the French Government, after passing by the Auberge de Paillette, where good bread, cheese, and white wine may be procured, brings you to **Somport**,

or Summus Portus, the confines of France and Spain. From Somport 11 kilometres leads down to the pretty village of Canfranc, situate on the right bank of the river Aragon, at a height of 989 metres = 3,240 feet. From this place, following the descent of the Aragon river, the town of Jaca on its left bank, at a height of 790 metres = 2,589 feet, may be reached in about 3 hours.

The tourist, however, who is tempted to make a digression to either of these two places, will do well to remember that he is no longer in France, and that in Spanish posadas it is always necessary to make your bargain beforehand, to prevent being imposed upon. At Canfranc especially is this caution necessary. From Somport the return to Gabas and Eaux Chaudes is effected without any difficulty by the **Col des Moines**; though the first time I passed I could not induce my guide to go, because there was a little mist on the mountains. My companion and myself were obliged to cross, trusting to our own instincts. Steering by the compass you can hardly go wrong. From Somport, striking up the valley, with the crest of the Pic d'Arnousse on the N., and keeping the left branch of the stream, in about 2 hours you gain the col, and come upon a lake. Skirting this on the right, and bearing to the SE., you again mount for a short space over a slope partly covered with snow, and are on the highest point of the pass, 2,204 metres = 7,231 feet. When once this ridge is gained, the imposing mass of the Pic du Midi, with the intervening pastures of the upper plateau of Bioux-Artiques, lies before you. From this point you descend in about 2 hours to the plateau of Bioux-Artiques, and thence passing through the forest of Gabas, which produces some of the finest firs in the Pyrenees, you reach Gabas in another $1\frac{1}{2}$ hour. At Gabas carriages may always be found to carry you, if tired, down the remaining 8 kilometres of road to Eaux Chaudes.

A shorter cut over the mountains from Urdos to Gabas may be taken; but it is not easy to find. The path turns off about 2 kilometres from Urdos, striking over the mountains N. of the Pic d'Arnousse in a SE. direction right for the Col des Moines, avoiding altogether the Col de Somport.

There are many pleasant afternoon strolls to be made from Eaux Chaudes, up the mountains on either side of the valley. The village of Goust perched on the mountain side above the Pont d'Enfer, on the west side of the valley, and the grotto on the opposite side at about the same height, are either of them an easy walk or ride for those who do not desire hard exercise; also the waterfall before mentioned to the left of the pathway leading over the Col d'Iseye may be reached in 1 hour 30 minutes from Eaux Chaudes, and is situate in the midst of the most delightful forest scenery, affording a cool shady retreat for a hot day.

There are many mountain paths between the vallies of Ossau and Aspe. One from Laruns to Bedous, passes the village of Aydius 5 kilometres ENE. of Laruns; and thence to Bedous over the Col de las Arques (1,700 m.), occupies about 5 hours walking. A still easier one, starting from the village of Bielle, 9 kilometres lower down the valley than Laruns, conducts in 4 hours over the Col de Maricblanque (992 m.) to Escot, in the valley d'Aspe, 11 kilometres below Bedous, and 13 kilometres from Oloron.

Section 8.

From Eaux Chaudes to Cambo and Biarritz by the Mountains, 7 days, including the ascent of the Pic Anie.

From Eaux Chaudes a very pleasant excursion in the spring may be made to Cambo. On foot it requires about 6 days, as follows:—

1st day over the Col d'Iseye, 6 hours, to Accous (sec. 7.)

[From Accous the ascent of the **Pic Anie** is worth making in fine weather. The Pic d'Anie is the last eminent peak on the W. distinguishable from Pau, and the last mountain at this end of the chain, attaining a height of 2,500 metres. Five kilometres above Accous the verdant valley of Lescun is seen on the right, opening SW. Here quit the main road, and cross the gave of Aspe by the bridge of Lescun. Thirty minutes from the bridge, on the left bank of the stream, is seen a very fine cascade and cabane, after which 15 minutes of a detestably bad path leads to the village of Lescun (902 m.), perched above the stream on its left bank. At Lescun there is a very decent little inn, chez Cazou. From here there is a fine view: the Pic Larraillé (2,323 m.) in the S.W. heading the gorge and stream of the Ansabe; and in the W., over sundry other peaks, rises the snowy pyramid of the Pic d'Anie (2,504 m.) The N. side of this peak is inaccessible; to gain the summit from Lescun requires 4 hours. Mount to the W. by the green and cultivated valley of the Hourque de Lauga, and then inclining to the NW., the rocks become more barren and escarped; the stream of the Anaye, descending from the Port d'Insole, is passed on the left; and a little farther on the right, the path leading to the baths of Laberou. From here, making westward, traverse the wood of Braca d'Azuns; whence you emerge upon the cabane of the same name. S. of the Pas d'Azuns is the little col, 30 minutes above it. From here follow the stream SE., which issues from the side of the mountain; and after passing on the right the diminutive lake d'Anie, you soon reach the cone of moving schists and gravel forming the summit, and which makes the last 30 minutes a steep scramble. The Pic d'Anie, not only in height, but in legendary lore, is the Olympus of this part of the Pyrenees. Even so late as the present century the inhabitants of Lescun, fearing to irritate the genius who has his habitation and garden on the summit, violently opposed an attempt to ascend the mountain. To-day, here as elsewhere, the 'auri sacra fames' has triumphed over every other genius, and the traveller will have no difficulty in finding a guide who for 5 francs will dare to profane with him this fairy-haunted summit. The descent may be made to the NE. by the village of Lées.

From St. Engrace the ascent of the Pic d'Anie is a little longer, but easier. Striking up the gorge E. of St. Engrace, beyond the church, in 2 hours you reach the first col, the Port de Sescous; descend for 5 minutes by a stony little ravine, and on the left opens the green valley of Baretous with uniform fern-covered hills on either side. Avoid descending into this valley, and strike obliquely through the beech forest SE. for 50 minutes. On emerging from the forest, continue across the grass, bounded on the right by a ridge of blueish rock: in 20 minutes you reach the Pas de Guliers, and to the right of this col you have a rocky defile, through which you must mount, and make across the pastures 1 hour 30 minutes to the Cabane d'Arlas, near the Pierre St. Martin, which marks the frontier. Thence turning to the left, across easy pasture-land, 30 minutes to the cabanes of Pescamon; whence there is a 2 hours' climb over stones and débris up the W. side of the cone of the Pic d'Anie.]

2nd day.—Accous to St. Engrace, nearly due W., passing over three low ranges of hills: the first, forest; the second, grass; the third, forest.

The road is not very easy to find without a guide, and the peasants here speak only Basque, so that it is not easy to ask your way. The best inn at St. Engrace is 1 hour beyond the village, on the road to Tardets, near the douane station. The church is a remarkable Roman edifice, dating from the 11th century. The heavy square tower has nothing very attractive; but the columns in the interior are curiously sculptured, and there are some pictures representing the legend of St. Engrace.

3rd day.—St. Engrace to Tardets 3 hours; by path along right bank of the stream; thence to Ahusky 4 hours, by a stony mule track. At Ahusky there are two small rivers, and a source of water remarkably pure, said to possess healing properties.

4th day.—From Ahusky to St. Jean Pied de Port, 6 hours. Four hours of mule path as far as Lecumberry; thence 10 kilometres along a carriage road to St. Jean Pied de Port (Hotel de la Pomme d'Or, de France).

5th day.—St. Jean Pied de Port to Cambo, 8½ hours' walk by Bidarray, and the Pas de Roland, a small tunnel overhanging the river Nive, by which the road passes: distant from Cambo 4·5 kilometres. At Cambo the best inn is the Hotel St. Martin in Lower Cambo, about 1 kilometre distant from Upper Cambo.

6th day.—Cambo to Bayonne, 18 kilometres, on the left bank of the Nive. Diligences daily throughout the year, fare 2 francs.

Five kilometres from Cambo, at Ustaritz, the road falls into the diligence road leading to Pampeluna.

Section 9.

EAUX BONNES TO ARGELEZ.

By the Col de Tortes; on foot or horse a day's journey, 10 hours. By carriage, 42·4 kilometres.

Eaux Bonnes, being situate at the

very top of the Val d'Ossau, is in a cul de sac as far as a carriage-road is considered, but for the traveller who can dispense with more luggage than can be taken by a horse, a guide, or still better on his own back, there is a magnificent walk over the mountains as far as Argelez. A new carriage-road has just been constructed between Eaux Bonnes and Arrens, passing over the mountain ridge to the west of the Col de Tortes. Between the Col de Tortes and the Col de Saucède, by a wonderful piece of engineering, this road is carried along the almost vertical face of the Pic de Gabisos; after which it again bends to the northward, and, passing a little to the north of the old foot-track over the Col de Saucède, descends by a long series of very gradually inclined zigzags to the town of Arrens. From Argelez to Eaux Bonnes by this route the distance is 42·4 kilometres = 31 miles by the carriage road. The carriage road starting from Eaux Bonnes mounts by long zigzags to the Col d'Aubisque (height, 1,710 m.) at some distance to the north of the Col de Tortes; thence sweeping round the gorge de Litor, under the Pic Gabisos, it remounts to the Col de Courret (height, 1,470 m.), north of the Col de Saucède, whence it winds down to the village of Arrens (at a height of 900 metres, 12 kilometres from Argelez), and thence along the old road through the villages of Marsous, Aucun, Gaillagos, Ancizan Dessus, and Arras, to Argelez (height, 458 metres).

If on foot, the course to be taken is over the Col de Tortes, 1,799 metres = 5,901 feet above the sea, passing by the needle-shaped rock on its summit; and after descending into the valley, and crossing another grassy col sufficiently steep to try the condition of man or beast, the road gradually descends to the village of Arrens; and here from the platform behind the church there is a fine view of the snowy mountains. The walk from the Eaux Bonnes to Arrens will take about 6 hours, and 3 more should be allowed for the remainder of the journey from Arrens to Argelez. It is a longish day's work, but by no means a difficult one, and, provided the weather is tolerably fine, a guide is not necessary. The needle-shaped rock which marks the summit of the **Col de Tortes** presents to the traveller an unmistakeable landmark; the road passes close under this, a little to the right, with the lofty Pic de Ger overhanging on the other side, and then, descending into the valley, again rises to surmount a second grassy col, the Col de Saucède, 4,902 feet, which, if less rugged, is scarcely less fatiguing than the Col de Tortes itself. Till reaching the top of the second col, the route lies close under the principal mass of mountains on the right, passing close to an icy cavern that is usually to be seen in the bottom, at the extremity of the valley that separates the two cols. This cave, the result of the accumulated snows of the Pic de Gabisos, 2,577 metres = 8,468 feet, is a favourite retreat for the cattle from the burning rays of the midday sun; but the situation, although inviting, is not without some danger; and the first time I passed, two vultures started up from the carcase of an unlucky cow that had been killed by the falling in of a mass of ice. This ice-cave, however, and indeed a great part of the wild charms of this pass, have been abolished by the newly-constructed road. If the weather be unsettled (on a day decidedly bad, of course no one would think of setting out on an excursion like this), or the traveller diffident of his own powers of pioneering, a guide may be taken to the top of the second col, and there dismissed, as from this spot a well-marked track leads down to Arrens in 1 hour and 20 minutes of easy walking.*

Section 10.

VALLEY OF AZUN AND PIC DE BALETOUS OR MURMURET.

The little eminence south of Arrens, on which stands the convent and chapel of Notre Dame de Poey la Hun, presents a grand view of the snowy mountains at the head of the Port d'Azun, and conspicuous above them all are the glaciers of the Pic de Baletous, which rises west of the Port d'Azun to the height of 3,145 metres = 10,318 feet.

The Pic de Baletous is the highest mountain in this part of the chain; and one of the grandest and most difficult mountains in the whole Pyrenees; but it lies so completely away from the route of the ordinary traveller, that the Eaux Bonnes guides seem quite at a loss as to its exact whereabouts, as a friend who started from Eaux Bonnes under their guidance found to his cost; for after passing two wretched nights in the mountain cabanes of the shepherds (a lodging which few Englishmen would prefer to the open air), he failed to attain even the foot of the Pic Baletous, the object of his search. After an unsuccessful attempt on this mountain in the year 1862 (for an account of which the reader is referred to the first edition), in 1864, taking with me the guide Jean Pierre Gaspard, I made a second trial, and after seven days spent in exploring the mountain, I succeeded in reaching the top; and ten days later my friend Count Henri Russell ascended by the same route. He as well as I was enchanted with the view, and I strongly advise any mountaineer staying at Argelez to make this ascent; it requires camping out for one night, but not more; and there are excellent quarters on the mountain.

The foot of the Pic de Baletous or Murmuret may be reached in a single day from Eaux Bonnes or Eaux Chaudes; but the ascent is better made from Arrens, or perhaps more easily still from Sallent, on the Spanish side. The route from Arrens up the valley d'Azun is as follows:—Starting southward by the path that is carried above the left bank of the stream, 1 kilometre beyond the chapel of Poey la Hun, which is passed on the right, the stream on the left forms a cascade, the Cascade Bourridis; and just beyond this, in the meadows on the opposite (i. e. east) side of the stream, there is a remarkable hanging block of granite, which touches only in a single point the rock of grey marble on which it is suspended. Close to this, on the same side of the stream, a vein of gold, my guide told me, was anciently worked, though whether there are yet existing traces of the precious metal I cannot say. In this, the lower part of the valley, the rocks on either side are of grey limestone, but deeply striated, and covered with numerous roches perchées of granite, which have been borne down by ancient glaciers from the upper part of the valley. Nowhere in the Pyrenees, or indeed elsewhere, are the traces of glacier action more distinct. About 1 hour's walking from this point brings you to the Gorge du Labas, opening to the west, by which there is a pass to Eaux Bonnes, passing the col at the foot of the Pic Arrè. Here you come into full view of the Pic Baletous, which was before hidden by the lateral chain on the east. 'Voilà votre

---

* N.B.—Eaux Bonnes and Cauterets are both fashionable places, so that, if the traveller make any stay at either of these places, he will probably not like to be without his luggage; otherwise all except his knapsack might be sent on at once from Pau, either to Cauterets, Luz, or Bagnères de Bigorre. If heavy luggage is taken to Eaux Bonnes there will be some difficulty in sending it round.

ennemi qui vous attend,' was the exclamation of my guide; and we halted to speculate as to the best way of surmounting the ice-clad ledges of rock that guard its summit. Half an hour beyond this you see a shepherd's cabane on the opposite side of the stream, where cheese and milk, and generally also fresh trout, may be procured. The fisherman is only too glad to be spared taking them down to Arrens, his usual market; and by all means take a few of these, if you can get them, as an addition to your evening meal upon the mountain. The rocks here on the right (west side) change into granite, and the valley is covered with huge fragments of débris from the spurs of the Pic d'Arrieu Grand.

Soon after this the pathway mounts and passes over some ice-planed terraces of granite rock, on descending from which the path crosses to the right side of the stream, and 1 hour and 40 minutes from the fisherman's cabane brings you to the Lac or Gourgue de Suyen. The opaque creamy hue of the waters of this lake at once proclaims their origin in the melted glaciers, and also that these glaciers are bedded in granite rock; for it is only a rock containing felspar, that is ground by the ice into the impalpable fine powder that produces this peculiar colouring. This lake is neither large nor deep, and with the sun upon it the water was not at all too cold for a bathe—the only drawback being the soft mud-like sediment that formed the bottom, as is always the case with these glacier-fed lakes when the ice-channel is formed over protogene rock. Just above the lake the stream separates into 2 branches. That from the south, of a pearly white, comes down from the Port d'Azun, which above this is called the Val Castierry, and has its origin in the north-eastern and largest glacier of the Baletous. The stream from the west, unlike its sister, is of pellucid transparency, being fed by upland tarns and fountains, whose waters, whatever may have been their origin, have filtrated and been purified in their subterranean course. The pic immediately east of the Gourgue de Suyen is the Pic Cristail, and a little farther than this there is a col, sometimes used by the contrabandists, leading over to the Marcadou, and so down to Cauterets. From this lake to the Port d'Azun is about 3 hours, passing on your right the two small lakes and cabane of Rimonées, and so to the Pierre St. Martin, a pile of stones marking the summit of the port (2,295 m.). At the Pierre St. Martin, which you pass on your left, the Spanish and French shepherds used to meet annually to settle disputes about cattle-liftings; and all who pass this way, in honour of St. Martin, contribute one stone to the pile. This port is very much frequented by contrabandists, and I suspect there is no pass in the Pyrenees through which so much smuggling is carried on, chiefly of silk and tobacco. From the Pierre de St. Martin, about 4 hours brings you down to Sallent, from whence you may reach Panticosa and the Baths, making the return into France by the Marcadou to Cauterets, or by the Col d'Ancou to Eaux Chaudes (sec. 6). [From the Pierre St. Martin the path to Sallent descends 30 minutes to a little lake, and thence crossing the valley mounts on the opposite side the Port de Piedra Fitta, from whence there is a rapid and stony, but not dangerous descent to the village.] Beyond the Lac Suyen there is a cabane, best known as the Cabane du Chevalier, the owner having been decorated by the Prince de Joinville in some naval action. This cabane is about 6 hours from Argelez, height 1,580 metres = 5,184 feet. From this cabane the path to the Port d'Azun mounts

S. up the barren gorge of Castierry; that to the Pic de Baletous strikes up the gorge of Arribit to the west following the stream in its course amid the rocks. It is a rough path for the first 30 minutes, when it becomes rather better; and an hour from the lake Suyen, after passing some very pellucid and calm pools formed by the gave, a very grand chaos of granite rocks is seen among the stunted pines on the left bank of the stream. The most colossal of these blocks is so perched on its edge as to form vast caverns between the ground and the impending rock. The most commodious of these is occupied by Aragonese shepherds for about 3 weeks in the year; and here you will do well to make your quarters for the night. This rock is at an elevation of 1,800 metres, and is called the Tour d'Arribit. The second day, start at earliest dawn, cross the stream, and follow up the torrent coming down from the SSW. An hour brings you to the morsel of table land, known as the Place d'Arribit, at a height of 2,020 metres = 6,626 feet, with its cabane and source, but no wood. From here the pyramid of the Baletous is in full view. Mounting a little, and passing S. through a small brèche, you find yourself in the savage gorge of the Bacradére, which forms a sort of cirque. Passing 3 small lakes, the direction to be taken is nearly S. over the rocks and snow to the arête connecting the Pic de Baletous with the Pic Pallas. There are 3 passages in this arête: the westernmost, the Port de Lavedan; then the Hourquette de Balan; and the eastern one the Port de Connoir (2,580 metres), which is the one to be taken. This brèche is marked by a single obelisk of rock which stands up on the W. side of it. If the Bergschrund here presents difficulty, take the Hourquette de Balan. Once through the brèche, continue ESE. nearly on the same level, for 20 minutes; when you reach the foot of an enormous rock, which forms a convenient station for breakfast. From this spot to the top requires 1 hour 40 minutes, and the difficulties here begin. Mount first NNE. over a couloir of loose débris, and at the top of this turn more to the right (east), and climb upwards as best you can, clinging to the obelisks of dislocated rock. My dog Ossoüe, is celebrated for her activity and powers of spring; but here they were put to the test, and it was long before she could summon up strength and courage to pass these rocks. A little higher, it is better to take to the south side of the arête; and, finally, a couloir of rather slippery schist, terminating in an awful precipice, leads to the summit.

The summit (3,178 metres = 10,427 feet) is a tolerably ample table-land consisting of schistose pebbles of various form and composition, and ending to the east in a precipitous arête, called the **Montagne Fermée**. The view is of course magnificent. Pau, Tarbes, and Agos in the valley of Argelez are well seen; and in the NW., a little to the right of the Pic de Pallas, the Atlantic Ocean is visible on a clear day. The Pic du Midi d'Ossau, due W., and the Pic d'Arrieu Grand NNW., which is the remarkable forked mountain so conspicuous from Arrens, are here dwarfed into insignificance; but the noble mass of the Vignemâle on the ESE. still dominates by 120 metres, completely masking the Mont Perdu and a large portion of the horizon on this side.

The principal glacier is on the NE. of the Baletous, furrowed with crevasses, and separated from a smaller one on the NW. by the crête de Fachon.

It was by this crête that on my first attempt I endeavoured to reach the summit. A stone cairn set up on it about 200 metres below the top marks the spot where I turned back. The Pic de Baletous (or Bat Laiteux, milky source, from the colour of the glacier waters) was ascended many years ago by the officers of the Etat Majeur, probably by M. Saget, as is attested by an ample pyramid constructed by them; but since then it has only been visited by Count Russell and myself, and some guides. Among these I can recommend two as well acquainted with the mountain, Jean Pierre Gaspard of Arrens, who went with me, and Jacques Orteig of Eaux Bonnes. The last has, I believe, ascended the mountain by the great glacier and the eastern arête, which, though difficult, must be an exceedingly interesting course. The descent from the summit of the mountain to Sallent occupied M. Russell 5 hours. From the rock recommended for breakfast, where the difficulties are over, but not the toil, follow down the gorge SSW., and after passing 5 small lakes, you are once more in the region of pines. Here, leaving to the left the path that leads up to the Port de St. Martin, cross to the left bank of the stream, where the path at first remounts for 200 metres, but after that descends, to become practicable for mules. From Eaux Chaudes the Pic Baletous may also be well ascended, sleeping the first night at the cabane of the Lac d'Artouste (2,092 m.), 6 hours from Eaux Chaudes. Thus far it is just possible to arrive on horseback. On the second day, from the Lac d'Artouste 2 hours of easy ascent, after passing the small lake of Arrèmoulit, places you on the summit of the Col d'Arrèmoulit (2,455 m.), between the Pic Arriel (2,823 m.) on the SW., and the cime de Pallas (2,976 m.) on the NE. From here descend into the hollow, crossing a large dyke of green porphyry, and then ascend NE. by a small stream; in 2 hours 30 minutes you reach the Port de Balan, whence 2 hours and 20 minutes to the summit of the Baletous. I must be allowed to correct my friend Count Russell on one point. He says that the summit of the Pic Pallas is accessible on the SW. from the Col d'Arrèmoulit. I have tried it on this side, and could not make much progress; I believe it might be attacked with a better chance from the east.

Section 11.

Lac Miguelou.—Lac d'Artouste.

The traveller stopping at Argelez may make an interesting course of 3 or 4 days visiting both these lakes, as follows: First day to Arrens (12 kil.), here take Gaspard, and thence up the Vallée d'Azun, as far as the Pont d'Asté, 3 hours from Arrens, where the torrent on the west comes down from the lake Miguelou. Follow up this torrent for 2 hours, and you reach the Lac Miguelou, a large desolate sheet of water at an elevation of 2,267 metres, surrounded by granite cliffs of easy access. Skirt the east shore of the lake, and make for the **Col d'Artouste** (2,502 m.) above the SW. corner of the lake. By this col you pass without difficulty in 3 hours from the Lac Miguelou to the Lac d'Artouste, placed at a height of 2,092 metres on the other side, one of the largest sheets of water in the Pyrenees. The cabane where you may pass the night is quite at the northern extremity of the lake. From here, 4 hours descent, following the gave along its right bank, the path emerging in the Gabas valley at the saw mill 4 kilometres above Eaux Chaudes. The return may be made by the Col de Labas, between Eaux Chaudes and Arrens, 8 hours; or by the usual route of Eaux Bonnes and the Col de Tortes.

Section 12.

ARRENS TO CAUTERETS BY THE MONNÉ.

8 hours on foot.

North of the Pic Midi d'Arrens (2,268 metres = 7,441 feet), and ESE. of the town, there is a very easy and low grass col, by which the adjacent valley of Labat may be gained, and thence you may pass over to Cauterets. The path from Arrens strikes off eastward just below the chapel of Poey la Hun, and emerges in the valley of Labat 6 kilometres below the Lac d'Estaing. After passing the Lac d'Estaing (1,264 m.), which is left on the right, the path mounts rapidly by the small stream coming down from the SE. to the Col de Lis, S. of the Pic of the Monné, whence the descent to Cauterets is made by the Lac Ilhéou and the Valley of Cambasqué, or more to the north by the valley and stream of the Lis. From this col the summit of the Monné is easily reached. The Lac d'Estaing is 16 kilometres = 10 miles from Argelez; there are some good trout in it; else it is hardly worth a visit.

Section 13.

ARRENS TO ARGELEZ.

13 kil. = 8 miles. Carriage-road along the left bank of the Gave d'Azun.

2 kil. from Arrens to the village of Marsous.
1 kil. village of Aucun.
2 kil. Gaillagos, village.
2 kil. Arcizan Dessus, village.
2 kil. Arras, village.
4 kil. Argelez, a town of 1,670 inhabitants, situate on the left bank of the Gave d'Azun, near its confluence with the Gave de Pau. The Hôtel de France, and the civility and attention of the host and hostess, M. and Madame Pierrefitte, cannot be too highly commended.

From Argelez to Lourdes the distance is 11 kilometres = nearly 7 miles.

From Lourdes to Pau (Hôtel de France), passing through Lestelle and Betharran, the distance is 40 kilometres = 25 miles. Rail now open.

From Lourdes to Tarbes (Hôtel de la Paix), 19 kilometres = 12 miles.

The sight of the open and fertile plain of Argelez is very refreshing to the eye after some days spent among the mountains, so that I should always counsel resting there for one night; though, for the traveller who does not care to descend the valley as far as Argelez, there is an exceedingly pretty walk or ride to Pierrefitte, by crossing the Gave d'Azun just below the confluence of the Gave de Labat, below the village of Arcizan Dessus. The road then mounts the opposite, or eastern, side of the gorge, and, passing through the villages of Sireix, St. Savin with its picturesque monastery, and Adast, brings you to Pierrefitte. The distance from Arrens to Pierrefitte by this route is 18 kilometres = 11½ miles.

Section 14.

ARGELEZ TO CAUTERETS.

15½ kil. = 10 miles. Good carriage-road.

On leaving Argelez, the road skirts, at a little distance, the left bank of the Gave de Pau as far as Pierrefitte—6 kilometres. The abbey and church of St. Savin stands on an elevation a little south of the road; but there is a byway leading to the village of Balagnas, which passes the church of St. Savin, the château of Miramont, the village of Adast, and the chapel of Pietad, and so reaches Pierrefitte by a little longer but much more picturesque route. On the opposite or right bank of the Gave de Pau is perched the château of Beaucens, the property of M. Fould, the Minister of Finance: with the village of Beaucens nestling at its foot. At Pierrefitte, the Hôtel de la Poste, though tolerable, will not compare with the comforts of that at Argelez.

At Pierrefitte the road splits into two branches; the left, still ascending the Gave de Pau, leads to Luz and Barèges; that to the right is carried along the Gave de Marcadou to Cauterets. From Pierrefitte to Cauterets the distance is 10 kilometres, and the road, hewn out of the rock, is carried along the left bank of the stream as far as the Cascade de Limaçon, about half-way, where it is carried over a stone bridge to the other side of the gave. A little above Pierrefitte, the maiden-hair fern is to be found growing in the fissures of the rocks to the right of the lower, which is the old road.

The height of Cauterets above the sea is 992 metres = 3,254 feet, while that of Pierrefitte is 507 metres = 1,663 feet, so that the road is a continual ascent, which, though gradual, is sufficiently steep to leave rare intervals of unbroken water in the boiling torrent which hurries along beneath; though an occasional deep blue pool, bubbling up like soda-water beneath some gushing fall, offers to those who can swim an inviting, though not very secure, bath. From a very narrow personal escape that chanced to myself some years ago, when I was only rescued by the exertions of my companions, I would caution the wayfarer along that dusty road not rashly to trust himself to the treacherous naiads, whose emerald eyes gleam so brightly from the green pools of the Marcadou.

Section 15.

CAUTERETS.—Hôtel de France is the largest, and most frequented, but doubtful whether the best. Hôtel des Princes (chez Derrey fils) recommended. Hôtel des Ambassadeurs, d'Angleterre, &c. At all these hotels, during the season, for a small room 4 or 6 francs per day is charged; and a separate bill is made out for the restaurant. In my experience Cauterets is the dearest place in the Pyrenees, and perhaps the most melancholy. My friend Russell well describes it as 'Ville de luxe et d'elegance mais non de plaisir, étranglée entre des hautes montagnes.' The native population is about 1,300; and of guides, carriages, and *chaises à porteurs*, there is no lack; and a tariff hung up at the bathing establishment and in the principal hotels regulates the price of each course. There are numerous diligences running daily to Tarbes, Pau, Luz, and Barèges.

The hot baths of Cauterets are said to have been known to the Romans, and to have been employed by Julius Cæsar; they were certainly visited by Marguérite of Navarre, sister of Francis I., with the ladies of her court. There are twenty-three sources set apart, or 'dedicated,' as the French would say, to the use of the public, and these may be divided into two groups—those of Mont Perrauté, to the east of the town, and those more south, which emanate from the junction of the granite and limestone near the confluence of the Gaves de Lutour and the Marcadou. Of this group the principal establishment is that of La Raillère, which is far the most fashionable and imposing of all the thermal edifices at Cauterets, and second only to that of Luchon throughout the Pyrenees. The establishment of La Raillère is built on a terrace at the foot of Mont Peguère, and overhanging the left bank of the Gave de Marcadou, at a height of 1,110 metres = 3,642 feet above the sea. The distance from the town is 1,800 metres by the grand route, but only one kilometre in a direct line. A little below the baths is a stable for the imperial horses, which are sent from the stud at Tarbes, that they may avail themselves of the waters. Higher up the stream are the smaller thermal establishments of the Petit S. Sauveur, Le Pré,* Mahourat (mauvais trou), Les Œufs, so called from the strong odour of rotten eggs that it emits, and of which the water attains a temperature of 60 cent. = 140° Fahr.; and last of all, the Bains du Bois, at a distance of 3 kilometres from Cauterets. The rocks above this pass from limestone into granite, and are much glacier-worn.

CAUTERETS in itself is not a very fascinating place. The climate is rainy, and the mists frequent. Though abounding in hotels, they are none of them very comfortable; and in August there is some difficulty in getting apartments, owing to the number of invalids who frequent the baths and tend to moderate the festive appearance of the town. The waters of Cauterets are very hot, and in great repute with consumptive patients. The ladies here seem to verify what was said by their gallant countryman, that the last sighs of a Frenchwoman are, not so much for the loss of her

* In 1857 I passed this way, before this source had been utilised, as the French naïvely call it—by which they mean enclosed in pipes and reservoirs, and so become a source of profit to the proprietor. A thermometer with which I inconsiderately ventured to test the temperature of the water was quickly broken, being graduated only to 120° Fahr. The stream issued from the living rock exactly at the point of junction of the granite and limestone; exemplifying what I believe is the established rule throughout the Pyrenees, that the hot and sulphurous waters have their source at the point of contact of the igneous and stratified rocks.

life, as for the loss of her beauty; and the fair '*poitrinaires*' who use the waters have such regard for their toilettes, that the last fashions from Paris and the most extravagant crinolines are seen on the promenade before the baths of this little out-of-the-way mountain town. At the table d'hote they are no less conspicuous; but here, at any rate, display is not their only object, and one lady apologised to me for her voracious appetite: 'Je mange beaucoup, n'est-ce pas, Monsieur? mais, mon Dieu! je suis poitrinaire.' In addition to the fine dresses of the ladies, the streets (or rather street, for there is only one) at Cauterets are enlivened by many a picturesque figure of Spanish contrabandistas and muleteers, resembling those of Rosa Bonheur, who here cross into France by the Col de Marcadou and the Pont d'Espagne. Tobacco and skins of wine are the usual burdens of those entering France; in the place of which they carry back sacks of empty glass bottles—an article requiring a higher civilisation than that of Aragon in its manufacture.

Section 16.

LAC DE GAUBE—VIGNEMÂLE.

If there is little to attract in the town of Cauterets, the scenery around it is exceedingly beautiful, and there is one excursion, that to the Lac de Gaube, which, for a very easy expedition, perhaps surpasses any other in the Pyrenees. It may be made either on foot or horseback, and excellent horses may be hired at Cauterets for 4 or 5 francs a day. The road passes La Raillère and the other bath establishments, leaving on the left the valley of the Lutour, and mounting through the woods along the right bank of the Marcadou stream, which comes foaming down among trees and rocks in a series of leaps, some of which are falls of considerable height. Passing the Cascade Cerizet, over which, during the morning sun, there is a constant rainbow, the path winds under shady trees and amid moss-grown rocks abounding in wild strawberries, and a great many varieties of fern, till it reaches the Pont d'Espagne, where a wooden bridge is carried high above the stream, leading into Spain over the Col de Marcadou, a pass abounding in wild and savage scenery. The height of the town of Cauterets above the sea is 3,254 feet, and that of the Pont d'Espagne 1,530 metres = 5,150 feet; the distance is about 6 miles, requiring 2 hours for the ascent; but the return downhill may be made in much less time. From the Pont d'Espagne a rough and less marked track, over trunks of trees and shattered stones, strikes off to the left, leading up the mountain, in a walk of about an hour, to the Lac de Gaube, a lonely basin of green water, 1,788 metres = 5,866 feet above the sea level, which, though not more than 2½ miles in circumference, is imagined by some the largest lake in the Pyrenees. Its superficial extent is 16·4 hectares = about 50 acres. There is a monument on the shore of the lake to an Englishman and his wife, of the name of Paterson, who were here drowned on their wedding tour. The Lac de Gaube presents to the artist one of the most perfect pictures in the Pyrenees, with the triple-peaked and snow-clad mass of the Vignemâle lying behind it and repeated by reflection in its glassy waters. The Petit Vignemâle, the eastern or 4th peak, may be ascended from here without much difficulty, by crossing the lake in a boat, and following the stream and then the glacier to the Col de Vignemâle, from whence to the summit of the Petit Vignemâle the ascent is not difficult; though it must be very hazardous, if not impossible, to pass over the chasm that separates this from the Pic Longue, the western and highest peak. On the borders of the lake is a cabane, at which a very tolerable *déjeuner* of trout and *poulet* is provided; and those who prefer fresh mountain air, and grand scenery with solitude, to a noisy town and the babble of invalids, will here find a clean bed and very comfortable night quarters from June till the middle of October. The trout in the lake are 'saumonée,' and among the best in the Pyrenees, and on a favourable day rise freely to the artificial fly. The *livre des voyageurs* here is unusually full of sentimental effusions. One specimen will suffice:—

Dans le bleu du ciel le temps éclaire,
On sent partout la fraîcheur de l'air ;
Le Vignemâle se reflet dans les eaux si calmes,
Comme ta visage chérie se répète dans mon âme.
Dieu fasse que les nuages jamais font obscure
Cette vue à mes yeux, ta présence à mon cœur

Numberless times have I now visited the Lac de Gaube, and each time have come away more impressed with the beauty of the scene, which, of course, depends mainly on the view of the Vignemâle being clear. The Vignemâle, '*la couronne de Cybèle*,' as Victor Hugo calls it in *Les Misérables*, has an elevation of 3,298 metres = 10,820 feet, and is the fourth highest of the Pyrenees, being the highest within the French frontier.

To reach the Pic Longue, the highest point, from the Lac de Gaube, it is necessary, after mounting towards the Col de Vignemâle as far as the foot of the glacier, to turn to the right, and, crossing the western ridge of mountains by the Col de Ratila, or Oulette, which separates France from Spain, to pass into the valley of Serbigliana. From this a very steep climb up the rocks lands the traveller on the upper plateau of the great glacier forming the saddle between the Montferrand and the Vignemâle. The summit of the Vignemâle was first reached by Cantouz, of Argelez, in 1834, with an English lady; and afterwards, August 10th, 1838, by him and other guides, with the Prince de Moscowa. The route pursued by the Prince was by the valley d'Ossoüe and the Plan d'Aube into the valley of Serbigliana, where quarters must be taken up the first night. Horses may be taken thus far. Starting the next morning at 5.30 a.m., M. de Moscowa, after scaling the rocks of the Montferrand, reached the summit of the Vignemâle at 2.30 p.m. By this route there is only one difficult *pas*, a sort of *cheminée*, some 18 or 20 feet high, so narrow that the body can scarcely enter. Above this passage you come in sight of the principal snow masses. On the left a vast '*cirque*' is opened to view, and towards the left of this amphitheatre lies the road on to the grand glacier. Then begins a fatiguing and monotonous walk of upwards of 2 hours, cutting steps as you go along, in a series of short zigzags up the steep *talus* of ice and snow. The summit of the glacier forms the southern brink of a great basin of snow, from which rise the four rocky summits of the Vignemâle, of unequal height. When you reach the edge of this basin, the westernmost and highest, the Pic Longue, 3,298 metres = 10,820 feet, has the form of a triangular pyramid; it is easily reached in an hour's walk over the snow, The Prince of Moscowa (who published an account of his ascent in the 'Revue des deux Mondes,' vol. xv.) mentions, as a curious instance of the distance to which sound may be transmitted, that in answer to a halloo raised by them on the summit of the Vignemâle, they heard the shout of the boatman on the Lac de Gaube. *Credat Judæus!* M. de Moscowa says that the mass of the mountain is the 'calcaire primitif' of the Marboré; but is in error, as the heart of the mountain is undoubtedly granite, and the summit of a dark clay slate containing iron pyrites. It is true, however, that changed limestone is a prevalent rock of the moraine of its northern glacier.

Section 17.
VIGNEMÂLE.

The ascent of the Vignemâle from Cauterets is made by the valley of Serbigliana; but from Gavarnie this mountain may be well ascended in a single day, by the great glacier. An account of an ascent I made with Laurent appears in the 'Alpine Journal,' vol 1, p. 131, but the following notes of this route may be useful.

The first part of the excursion is the same as that to the Col de Vignemâle, described post, sec. 31. The path lies westward up the Val d'Ossoüe: starting from the back of Belou's hotel, in 10 minutes cross the stream to its left bank, and in 15 minutes more recross to the right bank, by a wooden bridge. Just below this is a beautiful clear pool—a most inviting bath, of which the traveller making any stay at Gavarnie will do well to avail himself. From this spot about an hour of gentle ascent, the path winding among some stunted birch trees, brings you to an open plateau of grassy pasture, with some Spanish cabanes, known as the 'Source de Bat.' Here the Vignemâle first comes into full view; and the start should be made sufficiently early to witness the effects of sunrise from this spot. The rocks, which at starting are of clay schist, soon change into marble limestone, conspicuous in the Pic Blanc on the north side, and the Pic de Ségre on the south side of the gorge. The precipices on the north side are studded, high as the eye can reach, with the white pyramidal bosses of the *Saxifraga pyramidalis* and *longifolia;* monarchs of their order, some of whose nodding racemes attain the length of 3 feet. Beyond the Source de Bat, the rocks again pass into schist, and leaving on their left the singular 'Cascade de Tapou,' issuing from an orifice in the eastern buttress of Mont Ferrand, and on the right a waterfall known as the 'Saute de l'Espagnol,' 3 hours from Gavarnie, brings you to the Cascade d'Olette. Here, still keeping on the right bank of the stream, you must mount, by a sort of staircase, some steep reddish rocks; and then descending above the fall, you come upon the first snow. 30 minutes from this brings you to the eastern foot of the Vignemâle, which has to be scaled; but before doing so it will be well to halt here for breakfast. Thus far the road is the same as that to Cauterets, by the Col de Vignemâle; and it was in descending from these rocks that two travellers, some years ago, missed their way, fell, and lost their lives. To make for the Col de Vignemâle, you must descend again on the other side of this buttress; but to reach the great Vignemâle continue to mount along the ridge which forms the south-western boundary of the great glacier, terminating in the Mont Ferrand. Continuing along this arête in a direction 10° S. of W., with the séracs and ice-caves of the great glacier on the right, and an awful precipice falling away to the smaller glacier of Mont Ferrand on the left; on attaining to a height where the slope of the glacier is less rapid, you must take to the ice. Here, should there be but little snow, an axe must be used, which, of course, will impede the march; but should the snow be favourable, 50 minutes of easy walking across the glacier in a NW. direction brings you to the foot of the rocky pyramid of the grand Vignemâle. The difficulty of surmounting these last rocks, which are steep, again depends on the state of the snow; but the scramble up them will probably take about 20 minutes; and you are then on the top. The Vignemâle consists of five several pyramids of rock rising to the NW. of the great glacier and running from ENE. to WSW.; the highest summit, that of the great Vignemâle, being the fourth or westernmost but one; and attaining 3,298 metres = 10,820 feet. It will probably be too cold to remain long on the summit. On 22nd Sept., with a bright noon-day sun, I found the thermometer in the shade mark— 7° centigrade. The time for this expedition is about 13 hours; 7 to ascend, and 6 to return.

This route may be slightly varied, and rendered still more interesting by combining the exploration of the Vignemâle glacier with the ascent of the mountain. Instead of mounting by the rocky buttress of the Mont Ferrand, continue NW. skirting the lower moraines of the great glacier; till you reach the rocky precipice, which constitutes the S. side of the Petit Vignemâle. Here a rough scramble up the moraine will place you on the lower extremity of the glacier, amid a labyrinth of séracs and crevasses, which are nowhere in the Pyrenees seen to so much advantage as here. A rope is necessary, and also an axe, in case the snow should be hard, as the first part of the glacier is considerably inclined. In clear weather there is no difficulty in finding a way to the upper part of the glacier, though of course it would not be prudent to attempt it in a fog. At the upper part of the glacier the slope is much more gentle, and there are no crevasses: the rope and axe may be here left; and the climb up the last rocks is the same as that from the Mont Ferrand. This route is very little, if at all, longer than the last; but those who wish to have more leisure to study a Pyrenean glacier can sleep at the Spanish cabane of

Les Olettes. Though the territory is French, the shepherds of the Upper Val d'Ossoüe are all Spanish; some former unpatriotic Mayor of Gavarnie having sold the right of pasturage for Spanish gold.

Section 18.
THE MONNÉ.

2,724 metres = 8,937 feet, 4 hours to ascend, 3 to descend, on foot (or on horse as far as the Plateau des Cinquets), whence to the summit 1 hour 30 minutes on foot.

This is one of the finest points of view among the secondary peaks, and well repays the labour of the ascent.

Leaving Cauterets by the bridge which crosses the Marcadou opposite the Mayoralty, you take the path that leads westward up the Gorge d'Arresto: follow up the course of the Gave de Cambasqué, and then, turning to the right, mount, amid thickets of broom and rhododendron, to the Plateau des Cinquets, whence you gain the Monné by its north-west flank. At the Cinquets there are shepherds' cabanes and a fountain of excellent water.

South of the Monné is a col by which the pedestrian may descend upon the Lac d'Estaing; and thence pass over to the valley of Argelèz; but it is a rather uninteresting route.

Section 19.
CAUTERETS TO GAVARNIE.

A. Viâ St. Sauveur, 43 kilometres = 27 miles. Good carriage-road.

(Secs. 20, 23.)

B. Viâ the Val d'Ossoue, 11 hours on foot.

(See *post*, Sec. 31.)

CAUTERETS TO LUZ AND ST. SAUVEUR.

A. By the carriage-road and Pierrefitte, 24 kilometres = 15 miles.

B. By the mountains on foot or horse. Guide not necessary. 4 to 5 hours.

Leaving Cauterets by the thermal establishment of Pause Vieux, the path mounts eastward to the Col de Lisey, passing a little house known as the Grange de la Reine Hortense, from Queen Hortense having there stopped to take shelter from a storm in crossing over from Luz to Cauterets. From the Grange de la Reine, 45 minutes brings you to the Col de Lisey, a grassy plateau at the height of 1,943 metres, situate a little to the south of the Pic de Viscos, whose height is 2,143 metres = 7,031 feet. The paths down the eastern side all lead to the village of Grust, four kilometres distant from Luz. From this spot a good road slants down the flank of the mountain to the villages of Sazos and Sassis, where it joins the main road, about two kilometres above Luz.

Section 20.
PIERREFITTE TO LUZ.

12 kilometres = 8 miles. To St. Sauveur 13 kilometres; good post road.

For the Route to Pierrefitte, see Sec. 14.

On leaving Pierrefitte the road crosses the gave by a picturesque single-arched bridge, the Pont de Villelongue, and leaving on the left the village of that name, and also the ruined abbey of St. Orens, winds its way along the stream at the base of precipitous walls of rock. The whole of this gorge of Pierrefitte, is very striking, the road being escarped in the rock, and in some places almost

D

suspended above the torrent, which it crosses 3 times between Pierrefitte and Luz. As the road approaches Luz, the gorge opens out into a fertile and well-irrigated plain, with the cultivation extending up the hill-sides wherever they are not too steeply inclined, but not reaching so near the snow-line as I have noticed in many parts of Switzerland. Where the depth of the soil will support the crop, maize is the staple produce; but higher on the mountain this is replaced by millet and buckwheat (*sarrasin*). All along this road, as indeed on every one throughout the Pyrenees, the traveller is pestered by beggars, who spring up at every turn. You never meet a child who does not ask you to give him something 'pour l'amour de Dieu,' and from the ages of 4 to 15 years, mendicancy seems to be the universal and only occupation. No one seems to be ashamed of it. You pass by a cottage-door, where a group of chubby-faced children are playing; you stop to look at them; they can scarcely walk, but each, as if by instinct, stretches out his tiny hand. In some retired valley you come upon a young shepherd tending his cows; he immediately accosts you, and asks you to give him 'quelque petite chose.' A strapping young girl passes, carrying with ease on her back a huge truss of hay or bundle of fagots which you would stagger under; she halts when she sees you, not to take breath, but to ask for 'quelque petite chose.' A cantonier is busy mending the road, or rather pretending to do so; when he sees you, down goes his pick. 'See what a good road I'm making you,' he says; 'give me quelque petite chose.' As you enter a village a dozen young scamps are on the lookout for you, and as soon as they see you they seize each other by the hand, and having performed what they call 'the dance of the country,' they whine out a request for 'quelque petite chose.' But the climax of begging was reached when a decent-looking gentleman, mounted on horseback, condescended to ask 'quelque petite chose' from me as I was tramping along on foot. Set a beggar on horseback, says the proverb, and he will ride to the devil; and I certainly consigned him mentally to that destination. Disinterestedness is certainly not a mountain virtue. In a poor country the first of all wants is the want of money, and the bump of acquisitiveness is proportionately developed. Perhaps, however, we ought not to say that the natives look upon travellers as a prey, but rather as the most profitable of all their harvest crops to be collected during the months of August and September.

Section 21.

Luz, a little town of considerable antiquity, stands at the point of meeting of three valleys, those of Barèges, St. Sauveur, and Pierrefitte; and of these, in old times, when they constituted a little republic, it was the capital, as Bigorre was the *chef lieu* of another small federation. The only object now remaining in the place of any antiquity is the old church of the Templars, a contrast to the unsubstantial whitewashed houses, which reflect with intolerable glare the sun's rays, here concentrated. At midday there is scarcely a spot of shade; the only alleviation of the heat is caused by the number of mountain torrents which intersect and irrigate the town and the adjoining pastures; and their ice-cold clear blue waters cleanse the streets, the houses, the inhabitants, the pigs, and the horses, and indeed perform unknown good deeds. The sole employment of the *gamins* of the district, when they are not pestering tourists, appears to be basting the pigs in these streams with long-handled wooden spoons provided for the purpose.

The best hotel at Luz is the Hôtel des Pyrénées, chez Madame Cazaux: the table d'hôte is excellent. The Hôtel de France at St. Sauveur is also good.

The chief curiosity in the town of Luz is the ancient church of the Templars, built in the 12th century, at the entrance of which there is a narrow doorway set apart for the accursed race, the Cagots.

North of Luz, on the other side of the river Bastan, on a monticule at the foot of the Pic Leviste, are the ruins of the ancient château of Ste. Marie, which is said to have been a fort occupied by the English, and taken from them in 1404 A.D. by Jean de Bourbon. Behind the town to the south is a little mamelon of rock, the site of the old ruined hermitage of St. Pierre, which in the year 1860 was replaced by a very pretty modern church, built by order of the Emperor. From this spot there is a fine view over the valley of Luz.

LUZ TO ST. SAUVEUR.

1,400 metres (1 mile), of a good carriage-road carried over the Gave de Gavarnie by a solid marble bridge, leads from Luz to St. Sauveur; Hôtel de France good. The waters here, though milder than those of Barèges, Cauterets, and Luchon, are in high repute, and have been more than once patronised by the present Emperor. There is a constant rivalry between the waters of St. Sauveur and those of Eaux Bonnes, but of late years, the fresh water naiads of both these places have fallen into disfavour with the Empress, who has given all her patronage to the sea nymphs and tritons of Biarritz. The town is a straight row of clean and well-built houses, ending, to the south, in that irregular sort of wilderness, half flower-garden half wood, known as a *jardin anglais*. In this garden used to stand a small circular chapel, bearing the inscription 'Vos haurietis aquas e fontibus Salvatoris,' and hence the name of the town. This chapel was built by one Gentien d'Amboise, Bishop of Tarbes, who had taken a retreat here from the Protestant insurgents. In the year 1859 both the Emperor and Empress were staying here, and derived so much benefit from the waters, that the Emperor in gratitude erected a new and more commodious little church of white marble, nearly on the site of the old one. The architecture is very plain, and the interior is unusually devoid of ornament for a Roman Catholic church, but it is in very good taste, and the spire is a striking feature in the picturesque view which the town presents from all points. The cost of this church was 60,000 francs, defrayed by the Emperor out of his private purse.

In some respects St. Sauveur is a more agreeable residence than Luz during the summer months, but in the autumn all its hotels are closed, and in winter the main street blocked up with snow. The great lion of St. Sauveur is the magnificent stone bridge—the Pont Napoleon, built by the Emperor in 1860, at a cost of 300,000 francs. A work of art so beautiful, and at the same time so completely in harmony with the surrounding nature, is seldom seen in any country. The arch is semicircular, and its span 21 metres. The height of the bridge is, from the stream to the top of the arch, 61 metres; from the arch to the balustrade, 5 metres; making total height 66 metres = 216·5 feet.*

---

* Those who care to make the experiment may amuse themselves by dropping a stone from the top, and accurately *observing* the time of its descent to the water, from which they may calculate the height. As this rule may be conveniently applied to measure heights elsewhere, especially precipices and crevasses, I here give it.

Section 22.
PIC DE BERGONS.

Height 2,070 metres = 6,791 feet. 3 hours to ascend, 2 to descend, on foot or horseback. Guide not necessary. Chaise à Porteurs 20 francs.

The Pic de Bergons, the mountain behind Luz, may be ascended either from Luz or St. Sauveur. The best course is to follow up the stream of the Lise, and then turn to the right; but there are several paths, the mountain being impracticable only on the S. side.

The Pic de Bergons is a promontory forming the NW. buttress of the Neouvielle, to which it is attached by the rocky isthmus of Maucapera, of which the highest point attains 2,592 metres = 8,238 feet. The Bergons presents a magnificent view of the grand limestone mountains of Gavarnie; the Coumelie, and over that the cone of Mont Perdu, being due S. A little to the left, close at hand, is the double-peaked portal of the Fourche du Brada. Those who do not undertake the Pic du Midi should on no account omit this easy ascent.

Pic Ardidieu, 2,988 metres = 9,803 feet, and

Pic Aubiste, 2,791 metres = 9,157 feet.

These two peaks, on the western side of the valley of St. Sauveur, may also be ascended, and present magnificent views. The Pic d'Aubiste may be reached on horseback, but the summit of the Ardidieu is only attainable on foot. The track mounts W. of St.

The actual descent of a body dropped in vacuo, in a given time, is ½ the velocity it has acquired at the moment of striking the ground.

In London, lat 51° 30′ at the sea level, an unresisted body would fall 16·09541 feet = 4·90578 metres in one second of time.

At Gavarnie, in lat. 43°, the same body would fall 16·08291 feet = 4·90198 metres, the final velocity being 32·16582 feet = 9·80396 metres; and this velocity diminishes ·003 metres for every 1,000 metres of elevation above the sea level.

As, however, the descent is not in vacuo, but in still air, we shall be pretty nearly correct if we take 16 feet = 4·876736 metres as the fall in the first second: this will give 1 foot = 0·304796 metres as the fall in the first quarter second; and the rule thus becomes very simple, viz.:—Count the number of quarter seconds in the time of the fall as *observed*; square them; and the result is the number of feet—e.g., time of fall observed, 3¾ seconds = 15 quarter seconds × 15 = 2·25 ft = 68·579 metres.

It may be noted also that the height fallen in each successive equal interval of time increases in a ratio of which the indices are in arithmetical progression; *i.e.*—

	Inches	Foot		Second
The fall is	1·92	=0·16	for the first	0·1
id × 3 =	5·76	=0·48	,, second	0·1
id × 5 =	9·60	=0·90	,, third	0·1
id × 7 =	12·74	=1·12	,, fourth	0·1

And so on.

If the time is calculated by *sound*—i.e., from the commencement of the fall to the termination, as announced by the sound of the object striking; from the number of seconds we must deduct that portion of time which sound takes to traverse the distance; or that portion of space which is given in excess, on account of the lapse of time as *heard* being greater than the real time *occupied in the fall*.

Assuming sound to travel in air of ordinary density, temperature, and dryness, at the rate of 1100 feet per second, from the height as calculated *by sound* of the falling body striking the ground or water, deduct as follows:

For feet				Deduct
16	.	.	.	0·468
50	.	.	.	2·58
100	.	.	.	6·77
150	.	.	.	11·56
200	.	.	.	18·32
250	.	.	.	24·8
300	.	.	.	32·07
350	.	.	.	40·78
400	.	.	.	49·1
450	.	.	.	58·3
500	.	.	.	67·9
550	.	.	.	77·7
600	.	.	.	87·8
650	.	.	.	98·2
700	.	.	.	109·12
750	.	.	.	120·31
800	.	.	.	131·65
850	.	.	.	143·3
900	.	.	.	155·2
950	.	.	.	167·43
{ 1000	.	.	.	179·5
{ =7·91 seconds.				

Sauveur, over the grassy plateaux of Trazères and Aragnouède.

Section 23.

LUZ TO GAVARNIE.
20 kilometres = 12½ miles.

ST. SAUVEUR TO GAVARNIE.
19 kilometres = 11¾ miles.

Good carriage-road. At Gèdre, 12 kilometres, the road bifurcates, that to the right leading to Gavarnie, that to the left to Héas. The path to Héas is as yet only available for horses, but that to Gavarnie is a magnificent wheel-road, constructed in 1860 by M. Singulet. The ordinary price for a carriage from Luz or St. Sauveur is 20 francs, not including the *pour-boire*. From Argelés or Cauterets the fair price is 30 francs. There is an omnibus every day in the season from Luz to Gavarnie, starting from the Hotel de l'Univers; and a crowd of horses and donkeys are to be hired at Gavarnie to carry those travellers to the cirque who have the indolence and bad taste to appreciate these adjuncts to the scenery.

Before the year 1860, it was only by a rough foot-path descending from the Jardin Anglais that you could gain the carriage-road to Gèdre from the south end of St. Sauveur; but in that year the beautiful stone bridge, which spans the stream with a single arch, was constructed by the Emperor's order; which makes the road from St. Sauveur shorter by one kilometre. No Pyrenean traveller omits this expedition, which the scenery along the road would well repay, without the attraction of the celebrated Cirque de Gavarnie. After passing the new bridge, you soon come to the Pas de L'Echelle, and a little farther to the triple bridge of the Pont de Sia; of which the uppermost one was completed in 1858. Near this, on the W. side of the valley is the Cascade de Sia. After passing this cascade, the road continues along the left bank of the stream as far as the bridge called Desdouroucat, on passing which the Casque de Marboré comes in view; on the right is the Pic d'Ardidieu.

The gorge soon opens into the basin of Pragnères, with its dirty village, 8 kil. from Luz, near which two affluent streamlets from the lateral gorges on either side pour into the gave. That from the east is the Gave de Pragnères, which comes down from the Pic Long. That from the west has its rise in the Lac Cestrède near the Lac d'Estom Soubiran, and descends between the Pic d'Aubiste on the N. and the Pic de Soumaoute on the S. To the gorge of Pragnères succeeds the still more savage gorge of Trimbareille; and 50 minutes (5 kil.) from Pragnères brings you to the village of Gèdre, 980 metres = 3,214 feet above the sea, at the point of junction of the valleys of Héas and Gavarnie. At Gèdre a path branches off to the left to the chapel of Héas, and the Cirque de Troumouse. See *post*, Sec. 33.

On leaving Gèdre, the path winds round the base of Mount Coumelie, and the Cascade de Saussa is seen on the opposite side of the ravine, a fine waterfall, not unlike Scale Force in Cumberland, though on a larger scale. Shortly after this the road cuts through the huge masses of fallen mountain, rightly named Chaos. The snowy peaks of the Marboré Cylindre and Taillon with the Brèche de Roland, as seen from this spot, make a pretty sketch, with these huge blocks for the foreground. On reaching Gavarnie the road crosses the torrent some 50 metres above the inn, by a very pretty white marble bridge just completed.

The village of Gavarnie, at an elevation of 1,346 metres = 4,416 feet is of the humblest kind; but the inn *chez Belou* affords capital board, and very tolerable lodging.

From the many and varied excursions to be made in every direction, and embracing every degree of difficulty and fatigue, I have long been confirmed in considering Gavarnie as the most favourable head-quarters in the Pyrenees, for all those who really prefer fresh air and grand scenery to the heat and frivolous life of 'les Eaux.' At present the new stable constructed by M. Verges Belou, the owner of the the inn, is the most imposing edifice in the village; but it is to be hoped within two years at the farthest, he will have constructed a good but unpretending inn, similar to that at Argelés; when this is done, I am sure that in the eyes of English travellers, no place in the Pyrenees will compete with Gavarnie, as head-quarters during the summer months; though the accumulation of snow, and sharp frosts, combined with the scarcity of wood on the French side of the mountains, interfere sadly with its comfort as a winter abode. The douane station that existed here in 1865 was wisely abolished by the administration. Some few contraband goods, or 'ballons,' as they are called, principally cotton stuffs and clocks, are from time to time smuggled over into Spain; but the import duties into France are so light, and the commodities of Spain so indifferent, that the receipts at the Gavarnie customhouse scarcely amounted to 100 francs per year, including the proceeds from the sales of seized goods. Live animals seem to be the objects of the greatest jealousy between the two nations, and for every sheep, ox, horse, or mule (not dogs), that passes the frontier, one of the principal inhabitants acts as a *caution* that the animal will return within a certain time, and should the animal die, his skin is obliged to be produced. I may observe that, contrary to the usual belief, France, and not Spain, is the country in which the mules are bred: and nearly all those used in the north of Spain are bought at the annual fairs of the French towns, and marched across the frontier; of course paying a duty.

Those who come to enjoy the grand scenery of the cirque, and not merely to do the place, should by all means sleep here for at least one night; and 2 or 3 days here, if fine, will not be regretted. There are many fine points of view attainable in 30 minutes from the inn at Gavarnie. For those who are not capable of the longer excursions the entrance of the Val d'Ossoüe affords very grand rock scenery for the sketcher, with the noble mass of the Vignemâle closing in the picture; and the village itself is a very pretty object from this spot. The botanist will find equal occupation. On the schistose rocks above the left bank of the stream, 300 yards below the village of Gavarnie, the beautiful purple flowers of the *Ramondia pyrenaica* grow in great abundance. Here and in the neighbourhood of Gèdre many rare Pyrenean plants are to be found; and the botanical tourist is recommended to make the acquaintance of M. Bordères, who is to be heard of at the Hôtel des Voyageurs at Gèdre.

Gavarnie is beyond the line of douanes, so that, if horses are brought an 'acquit à caution' ought to be taken at Gèdre, otherwise they may be seized. But the guide will arrange this.

Sunset is the time for seeing the cirque to advantage; words cannot describe the weird splendour of those rocky battlements as they peer above the mist, illumined by the setting rays. Three hundred yards from the inn there is a good point of view; but to reach the extremity of the cirque and the far-famed waterfall, 3 miles more must be travelled, over a rough boulder-strewn plain. This part of the journey it is as well to defer till morning, if the weather is settled, as the fall is seen to more advantage when the sunlight is on it. On the 21st of August, 1857, I observed that it was not till 12.30 that the sun had attained a position to throw its beams on the fall; the effect then was really magnificent, ὑψόθεν ἐκ πέτρης κατελείβετο δάκρυεν ὕδωρ.* And, with the sunlight on it, I think no one can be disappointed.

Horses may be taken within half a mile of the fall, and there left at the cabane which faces it and presents the best point of view. From the small amount of water, which appears yet more insignificant from its stupendous height, and the vast space of the cirque into which it falls, for a moment perhaps the spectator is disappointed, and the reality falls short of his conception of the highest known waterfall in the world, its descent being 405 metres = 1,326 feet; but taking the scene with all its adjuncts, it is certainly one of the most wonderful in the Pyrenees. In the late spring, when the snows are melting, there is the greatest body of water; and it then sometimes clears the ledge, coming down in a single leap. In summer the stream is seldom sufficiently copious to do this, but in the afternoon of a hot sunny day has considerable volume. The basin of the fall is reached by scrambling up a rough *talus* of loose stones, and here the valley absolutely finishes. A perpendicular wall of marble rock rises before you, with the snow lying in deep layers on successive ledges, and its summit crowned with glaciers. On the other side is Spain—a different kingdom and a different race; but they are not entirely cut off from each other, for on the right of the cirque a track, to appearance scarcely practicable, leads up to the cleft in the rock known as the **Brèche de Roland**, 2,804 metres = 9,200 feet above the sea. The ascent of this scarcely compensates for the fatigue, as many elevations of inferior height afford a finer view, which is here much interfered with by the walls of rock that rise on either side. It is just practicable for ladies who are good walkers and tolerably firm of foot. From the little inn at the foot of the cirque it occupies 3 hours to ascend and 2 hours to descend. The track is roughly marked most of the way; but it requires a practised mountaineer to undertake this ascent without a guide. An axe must be taken, to cut steps in the glacier on the top, as if there is no snow on its surface, it is too steep to traverse without this precaution. The principal prospect is over the mountainous ranges of Aragon, rising one behind another; and on a clear day you may even see the blue Sierra de Moncayo which separates Aragon and Castile. A portion of the Vignemâle also comes into view; but the shoulder of the Taillon shuts out the greater part of this fine mountain. Those who care to prolong their expedition, and to obtain what is really a fine view, may ascend the Tour de Marboré in 1 hour 30 minutes from the Brèche de Roland, or the Casque in 30 minutes, but the last is the most difficult.

From the Brèche de Roland you may descend in 7 hours to the Spanish village of Fanlo. For the ascent of the Brèche and all other expeditions in the neighbourhood of Gavarnie, I can confidently recommend the guides Hippolyte and Henri Passet, the brother and son of the late Laurent Passet. They are two of the most willing and competent guides in the Pyrenees, and I do not think any one who has employed them will think that they are overpaid with 10 francs a day for their services. Pierre, the garde forestier, is also much to be recommended, should the Passets be engaged. He is to be heard of at the inn at Gavarnie; he also demands 10 francs a day for his services, and is worth the money.

---

* Simonides.

Section 24.
MONT PERDU.
3,351 metres = 10,994 feet.

The Mont Perdu, the loftiest of the limestone range of Gavarnie, and the third highest mountain of the whole Pyrenees, is a favourite ascent, inasmuch as it is most favourably placed for a magnificent panorama, and may be ascended without any excessive difficulty. It may be ascended most easily from Gavarnie. A. by the Brèche de Roland. B. by the Brèche d'Astazou; but the last is more difficult.

ROUTE A.

		hrs.
1st day.	Gavarnie	
	Brèche de Roland	5
	Cabane de Gaulis	3
	(Here pass the night.)	
2nd day.	Summit of Mont Perdu	3
	Halt on summit	1
	Cabane de Gaulis	2
	Brèche de Roland	2.30
	Gavarnie	3
		11.30

For the ascent to the Brèche de Roland, see Sec. 23. To the left of the Port on the Spanish side, an inscription cut on the rock records the ascent of the Duchesse de Berri, and the ladies of her suite, on August 19, 1828. To the right of the Port also in Spain, there is a cavern in the rock, which will afford the traveller shelter in case he is surprised by bad weather in this elevated region; but few probably would care to make here his bivouac for the night, as was done by my friend Count Henri Russell, who, after an ascent of the Mont Perdu, was surprised here by night all alone. On passing through the Brèche, a steep descent down the glacier on the Spanish side, which may be got over in a glissade, and then a weary scramble over glacier-worn rocks to the Cabane de Gaulis at the foot of the Mont Perdu. Here the scene is desolate and savage beyond description —not a tree in sight, or wood of any kind to be had for miles, and you must pass the night as you best can, without a fire. The cabane is one of the usual dirty little stone boxes, just capable of holding four packed side by side. This is one of the spots where a sleeping bag is highly desirable. On this side of the Mont Perdu there are no existing glaciers; but the mountain consists of 3 échelles which have to be scaled in succession. A sharp ascent over rough ground, skirting the base of the mountain known as the Tour de Gaulis, brings you to the first of these. Up this you must scramble hands and feet, as well as the two succeeding ones, the last of the three being the most difficult, and liable to be wet and slippery from the water running down from the melted snow above. By leaving the cabane early you may reach the top in time for sunrise, and Mr. Gill mentions having seen a curious effect at that hour—the shadow of the Mont Perdu being projected in the air as a still loftier mountain to the west. Though the Mont Perdu presents a magnificent panorama of all the central Pyrenees, it is itself placed so far back as to be well seen from but few points. From the Col des Aiguillons, on the road to the Valley d'Aure (Sec. 48), there is a magnificent view of its N.E. side and glaciers. The return is most easily made by the Brèche de Roland, though sometimes, if you have a steady head and a good guide, by the Brèche d'Astazou. Henri and Hippolyte Passet, of Gavarnie, are excellent guides for the ascent of the Mont Perdu and other excursions in this neighbourhood. The ascent of the Mont Perdu may also be made in a single day, going and returning by the Brèche d'Astazou to the left of the cirque; but by this route there are considerable difficulties to be surmounted. The time required to reach the summit from Gavarnie, is about 7 hours, and about 6 to return.

Starting at 4 A.M. on a summer's day gives ample time for the ascent, including 2 hours on the top. After attaining the Brèche d'Astazou, steer SE. across the glacier for the col that unites the Mont Perdu to the Cylindre. One of the most difficult passages is in getting off this glacier on to the rocks of the col. The Lac de Mont Perdu is left below to the left On attaining the col, follow the arête; but where the rocks are bad descend a little to the snow on the S. side; passing by a little lake, in height and situation similar to the Lac Couronnè on the Maladetta. After passing this the summit is attained by scaling a dome of ice and snow, much resembling that of the Nethou. From the summit

Pic de Nethou bears	111°
Pic Posets	109°
Pic Malibierne	115° 20′
Pic Cotiella	143°
Pic Baletous (just visible)	313° 40′
Vignemâle	321° 30′

The summit of the Mont Perdu may also be reached by the Valley d'Estaubé, and the Port de Pinéde, passing by the W. side of the Lac de Mont Perdu, and thence mounting by the glacier to the Col de Cylindre.

Section 25.

LAC DE MONT PERDU, AND THE PORT DE PINÉDE.

This expedition is one of the finest in the Pyrenees, and with a Passet for a guide presenting no excessive difficulty. The time is as follows:— Starting from Gavarnie at 6 A.M. along the road to the cirque; 2 hours' good walking brings you to the Brèche d'Allanz, the gap in the transverse chain of the Piméné. From the Brèche d'Allanz 1 hour 20 minutes across the 'gravier' at the foot of Mont Paillas brought us to the foot of the Glacier de Canaou, which is marked by an outjutting pinnacle of rock that rises from its extremity.

Here we halted 20 minutes for breakfast, and then reached the top in 50 minutes by scaling the glacier—cutting steps where the snow was not sufficient to secure our footing, as the incline is considerable. From the Port there is a magnificent view of the Mont Perdu and Cylindre, the semi-frozen lake of Mont Perdu lying immediately at our feet. 10 minutes of rough descent from the Port brought us down to the lake, where we halted an hour. At 11.45 we left the lake; and striking due W. up the glacier, which is at a gentle slope, 40 minutes over the snow, and 40 minutes over the rocks of crystalized limestone, brought us to the top of the Col d'Astazou at 1.5 P.M.

To descend, bear S. for a few steps, and then to the W. The descent requires caution, as the rocks are very precipitous till you reach the glacier; but Passet knows every inch of the way. At 3.25 P.M. we reached the bottom of the col, and were at Gavarnie before 5. At the top of the Port de Pinéde, and also in descending from the Brèche d'Astazou, we found the *Artemisia glacialis* growing in abundance; and lower down, on the limestone rocks, are very fine specimens of the *Gnaphalium Leontopodium*, and *Oxytropis Uralensis*. Izards are almost sure to be seen in the course of this excursion. The ascent of Mont Perdu may be effected by passing on to it from the glacier above the lake; but in that case it is almost imperative that the night should be passed at the Cabane de Gaulis.

Section 26.

TAILLON AND FAUSSE BRÈCHE.
10 hours.

For those who wish to see something of a Pyrenean glacier, without too long a course, this is a most interesting excursion. A guide is necessary. It would be prudent to secure both the Passets, then there can be no danger. Take the road to the Port de Gavarnie, and 20 minutes after attaining the tableland, cross the stream coming down from the Port, and make for the foot of the glacier. If the snow is in good order, there is no difficulty in zigzagging a course up this, and 2 hours 40 minutes from Gavarnie brings you to the snow col, over which you may pass E. into the gorge of Sarradets, and the usual route to the Brèche de Roland. In spring, indeed, this approach to the Brèche is preferable to that by the cirque; and should the lower stage of the glacier be difficult of access, by mounting from the cirque and passing this col W. from Sarradets, you reach the same point. Here the difficulties begin, and a rope, axe, and guide experienced on ice are required to surmount the remaining part of the glacier interposed between you and the Fausse Brèche, which presents crevasses, seracs, and all the phenomena of a Swiss glacier, though on a less extended scale. An hour will probably serve to overcome this, and on attaining the crête, 30 minutes W. up a gentle slope of snow will suffice to give the summit of the Taillon, which is very easy of access. Thence there is a grand view. In the late summer the uppermost rocks are almost bare of snow; and here at a height of 3,146 metres, several little Alpine saxifrages and androsace's may be seen in flower. The return should be made by the Brèche de Roland. Descending on the same arête, pass S. of the **Fausse Brèche** (2,948 m.), where there is rather a mauvais pas, from the very friable nature of the schist rock. Beyond this the road is very easy: 1 hour from the Taillon places you in the Brèche de Roland, whence to Gavarnie, 3 hours.

Section 27.
SOURCE OF CASCADE AND MARBORÉ.

Another interesting and not over long glacier course is that to the source of the cascade, at a height of 2,331 metres: time about 8 hours; but it must be made before the end of summer, when a bergschrund renders the passage from the glacier on to the rocks projected from the Marboré, which form the last barrier, impossible; and even when this buttress is surmounted, the descent SW. to the source is exceedingly steep, and even dangerous, being exposed to an artillery of stones from the moraine of the upper glacier.

To those who do not mind fatigue and a long course devoid of danger, Count Russell strongly recommends the tour of the cirque on its upper level. Mount by the Brèche de Roland. From the Brèche you must descend 15 min. SSE.: then rise up a snowy col, whence there is a fine view of Mont Perdu; and continue to ascend NNE. in 1 hour 20 min. from the Brèche the arête of the cirque E. of the Tour de Marboré is gained. The slopes on the Spanish side are so gentle that a mule might be made to mount. From here the effect of looking down on the cirque some 1,200 metres below is very striking. The cabane may be well distinguished, but a man is invisible. Passing above the glacier of the cascade, in 2 hours more you may gain the summit of the **Pic du Marbore**, the NW. buttress of the Cylindre. The top of this mountain is a level plateau as large as the 'Champs de Mars,' on which a race might be ridden 3,253 metres above the sea. The return to Gavarnie must be made the same way. There is no difficulty in this magnificent course, but it is a long day's work of about 12 hours' walking.

Count Russell has truly remarked

that the mountains of the cirque of Gavarnie are much easier ascended from the south side. Passing through the portal of the Brèche de Roland, the almost level terrace above the cirque offers a very grand and by no means difficult approach to the **Cylindre**, almost a virgin summit, and in height little inferior to the Mont Perdu.

Section 28.
ASCENT OF THE PIMÉNÉ.
2,803 metres = 9,196 feet.

On foot, 8 hours. Guide advisable, but not indispensable.

The Piméné is recommended by Ramond as one of the mountains most deserving an ascent, from its fine view and comparatively easy access. It may be ascended either from Gavarnie by the Brèche d'Allanz, or from Héas, 1st by the Lac de Hosse, or 2ndly by the Val d'Estaubé.

I have many times ascended the Piméné, and strongly confirm what Ramond said of it—that there is no mountain so readily accessible that equals it as a point of view. It is an excursion less difficult, and far better worth the labour, than that to the Brèche de Roland. A few details as to the route may be useful to those who feel competent to undertake the ascent without a guide, for in settled weather there is no danger. The Piméné rises due E. of Gavarnie—a man on the summit being visible from the front of the inn, even to the naked eye. To ascend, cross the stone bridge about 100 yards above the inn, and follow up the stream coming down from the E., keeping the left bank. The path first winds among some shrubs, and in 1 hour 10 minutes emerges on a grassy plateau, on which there is a cabane occupied by shepherds. Leaving this on your right, steer E. up the gorge, to the source of the stream, which is the last good water. Thence, turning above the rocks to your right, ascend the steep grass shoulder that forms the southern buttress of the Piméné. In ascending this, keep the precipitous rocks on the left, the course being slightly to the E. of ENE. From the southern point of the Piméné the bearing of the higher peak is 13° W. of the true N. It is easily reached in 35 minutes, crossing the saddle, and climbing up the crête. The Piméné summit is 9,200 feet above the sea, 4,848 feet above the inn at Gavarnie, and 486 feet above the lower peak. The summit is composed of chloritic slate, the dip of the strata being from SW. to NE.

On the top of the Piméné are some good plants, and the bright little red flowers of the *Androrace ciliata* grow here in great abundance, side by side with the yellow tufts of the *A. vitaliana*. The summit of the Piméné is 1,457 metres above the inn at Gavarnie, from which it is distant 3,156 metres in a direct line. About 70 metres below the summit on the S. side some rocks present a tolerable *abri* for those who care to pass the night at this elevation, and contemplate the scene in all its grandeur; but there is no wood, and therefore ample *capotes* should be taken.

The following list of the magnetic bearings will enable the traveller to recognise the principal points in this magnificent panorama, remembering that 18° is the true North :—

Pic de Bergons . . .	8°	
Péne Taillade . . .	22	
Péne de Pourry . . .	27	30'
Pic Ayré . . .	38	30
Pic du Midi . . .	41	10
Coumelie . . .	45	
Neouvielle . . .	51	30
Pic Long . . .	58	
Pic Badet . . .	68	30
Col de Cambiel . . .	74	
Pic Cambiel . . .	79	30
Pic des Aiguillons . . .	98	
Tour de Trumouse . . .	101	30

Pic de la Munia, highest of the Cirque de Trumouse	}	120° 20'
Port de la Canaou	. . .	154
Port Biel	. . .	138
Pic d'Estaubé	. . .	147
Lofty Mountain in Spain	. .	151 30
Port de la Canaou d'Estaubé	.	158
Pic de Pinéde	. . .	168 30
Port de Pinéde, leading to Lac de Mt. Perdu	}	175 30
Brèche d'Allanz	. . .	183
Mont Perdu	. . .	187
Cylindre	. . .	194
Brèche d'Astazou	. . .	207
Pic de la Cascade	. . .	216
Tour de Marboré	. . .	226
Casque	. . .	233
Brèche de Roland	. . .	249
Fausse Brèche	. . .	243
Taillon	. . .	249
Gabietou	. . .	252 30
Port de Boucharo	. . .	260
Sommet Rouge	. . .	263
Pic de Tentenéra	. . .	268
Col de Tentenéra	. . .	271
Gavarnie (inn)	. . .	287
Pic Serbigliana	. . .	288
Mt. Peternelle	. . .	295
Cascade Tapou	. . .	299
Mt. Ferrand	. . .	301 30
Vignemâle (summit)	. .	303
Col de Vignemâle	. . .	307 30
Sommet Blanc	. . .	309 50
Pic Paymourou	. . .	309 to 316
Pic Baletous	. . .	312
Pic de Malle Rouge	. . .	322 30
Sommet de Malle	. . .	332 30
Pic Ardidieu	. . .	344 30
Pic Aubiste	. . .	345
Pic Barbe de Bouc (nearest)	.	345 30
Pic Viscos	. . .	358

Section 29.
GAVARNIE TO BOUCHARO BY THE PORT DE GAVARNIE.—VALLÉE D'ARRAS.

5 hours on foot or horseback.

On leaving the inn at Gavarnie, turn to the right in the direction of the valley of Ossoüe, from which a path sufficiently easy strikes south-west round the base of the Taillon, and 2 good hours from Gavarnie brings you to the Port at a height of 2,280 metres = 7,481 feet. The view from the Port is not extensive. On the Spanish side the path descends rapidly to the S.E., over the escarped rocks that form the buttresses of the Taillon, to the hamlet of **Boucharo** (*Bujaruelo*), which is at the head of the valley of Broto. The rocks on either side of the Port are limestone, but above Boucharo, they are intersected by a dyke of slaty schist. At Boucharo there is a solitary inn, which affords descent quarters for Spain.

1. From Boucharo you may reach Panticosa in 7 hours, crossing the mountain of Bendenera; or in 12 hours by the valleys of Torla and Biescas.

2. From Boucharo you may also pass on foot to Cauterets by the **Col d'Aratile** over the western shoulder of the Vignemâle. Time 10 hours.

Two hours below Boucharo (1,326 metres) is the Spanish town of **Torla** (height 996 metres) which is worth a visit as being a perfect specimen of the dirt and discomfort of a Spanish town. The house where travellers are entertained for a consideration, is the hereditary mansion of an ancient and noble family, the present mistress being the Marchesa Vio. 1 hour below Torla is the town of Bielsa, and thence to Fanlo, about 5 hour's walking through a wild country, and some extensive fir forests.

From Gavarnie an interesting and novel route may be taken to Bagnères de Luchon, on the Spanish side of the chain, passing through Torla, Fanlo, Bielsa, or Escalona, and Plan; and thence by an especially grand and beautiful course over the Port de Gistain and the Port d'Oo to Luchon. None of these days are long except the last, which is rather a fatiguing course of about 16 hours, but one of the grandest and most enjoyable to be found in any country.

The scenery between Boucharo and Torla is very magnificent, but will not compete with that of the **Vallée d'Arras**, the valley which is seen E., facing you, as you make the turn at the Echelle de Torla. Of the few persons, that have hitherto visited this valley, I have met none who have been disappointed in their expectations. The rocks are so highly coloured, so varied in form, as well as so stupendous, and the forest scenery below so magnificent, that, apart from glacier scenery, it can hardly be surpassed in any country. The upper part of the valley affords excellent shelter for a bivouac, and though it may easily be visited from Torla or Boucharo in the day, I should counsel all those who value luxurious night-quarters not to sleep at either of those places, but to encamp 'sous les etoiles' in the valley. If the party are on horseback, the journey each way must be made by the Port de Gavarnie; time as follows:— 5 hours to Boucharo, including halts, thence descending S. by the path to Torla, 50 minutes to the chapel and cascade of St. Heléne, on the right bank of the stream; 10 minutes beyond, a second cascade, that of l'Echelle, is passed. Both these cascades proceed from the melting snows of the Pic Sebouillat, and in early summer are sufficiently imposing. After passing the second cascade, the botanist may pick from under the box bushes, on the right of the path, an abundant supply of that rather local fern, the *Asplenium fontanum*. Shortly beyond this the path, which is very rough, passes over some limestone rocks containing angular fragments of silex, the **Echelle de Torla**; and at the bend, 30 minutes above Torla, descends to the stream, over which it is carried by a wooden bridge, and continued up the rocks on the N. side of the stream coming down from the Valley d'Arras. Continue to follow the path, nearly due E., and in about 1 hour leave on the right a wooden bridge, by which there is a communication with the mule-track leading down to Torla on the south bank of the stream. The scenery here is magnificent in the extreme.

On either side of the river beautiful park-like glades in the forest, which consist principally of beech and box, with heavy and almost impenetrable belts of fir above. High above these, on either side, are stupendous walls and pinnacles of rock, of the richest colouring and most picturesque form, those on the north side especially resembling towers and battlements. Half-way up the valley, at the last grange, a difficult passage leads up the rocks of the **Cotatoir** to the Brèche de Roland. The woods abound with flowers; *Pyrola rotundifolia*, and a very rare and beautiful liliaceous blue flower *Aphyllanthes monspeliensis*, in the lower part; and at the extreme top the *Pyrola uniflora*. (June). Beyond the Cotatoir the path mounts rapidly through the fir forest, leaving below on the right a very fine waterfall. In the upper part of the forest beech trees supplant the fir, and on emerging from the trees, two limestone caverns, the Cueva d'Arras (1,684 m.) are seen to the left of the path, which are quite weather-proof, and afford all that can be desired as quarters, except food, which may be procured at Torla, 4 hours distant. Here a night should be passed by any party disposed to explore this valley, which presents no ordinary attractions to the artist, sportsman, botanist, and mountaineer. In the open glades below the forest the *Gentiana ciliata*, and *Ophioglossum lusitanicum* are found; in the long grass beyond the **Cueva**, the *Gentiana cruciata*, and *G. burseri*, are very abundant. The distance of this cave from Boucharo is 5 hours, and from it the summit of the Mont Perdu may be reached in less than 6 hours.

From the '**Cueva**' you may pass to Fanlo in 4 hours by a brèche in the chain on the S. side of the valley, which is not very easy. From this cave the brèche bears SE., but to reach it the valley must be ascended for 30 min.

before crossing the stream. On the rocks before reaching the brèche, the rare *Anemone Halleri* is in flower at the end of June. The height of this brèche is 2,266 metres. There is not the slightest difficulty in the descent to Fanlo, which is in sight from the brèche. The direction is 10° W. of S.; the time 1 hour 40 min. The height of **Fanlo** is 1,360 metres above the sea. Ask for the Casa del Senor, where you will be well taken care of, and will find a clean bed.

The party, if on foot, may return by the Valley d'Arras, and enter Spain by crossing over the Brèche de Roland. Thence to the Cabane de Gaulis, 3 hours; from whence, descending by some steep rocks, down which it is not over easy to find the way, 1 hour 30 minutes brings you to the cabane at the head of the Vallée d'Arras, occupied by Spanish shepherds till Oct. 10, who will supply goats' milk and bread. About 1 hour 30 minutes from the cabanes, following down the stream on its right bank to the Cueva d'Arras, at the entrance of the forest.

Section 30.
GAVARNIE TO BATHS OF PANTICOSA.

The excursion from Gavarnie to Panticosa is a very convenient one for the traveller who wishes to see a little of the Spanish side: somewhat fuller details may therefore be acceptable. The better and shorter road to the Port de Gavarnie is to keep the road leading to the cirque till past the church; then follow the path to the right. There is not much difficulty in reaching the hospice of Boucharo without a guide in good weather; the ascent on the French side (3 hours) is uninteresting; but the descent through the forest to Boucharo, which lies hid in the bottom of the gorge, is picturesque. The Port de Boucharo is not high enough to offer a striking view; but the following bearings (magnetic*) may enable the traveller to identify some of the points. Looking to France, the Fourche de Brada, 48° 30'; the Piméné, 80° 30'; the Brèche d'Allanz, 97°. Looking towards Spain, a mountain known as the Sommet Rouge, 264° 30', then the Pic Tentenéra, or Bendenéra, 272°; to the right of this the Fourquette de Tentenéra, 279° 30', over which lies the way to Panticosa; then the Pic d'Ourdisset 2,88^c; and the Sommet de Serbigliana, 303°. The hospice at Boucharo is a station of Spanish carabiniers, consisting of a sergeant and guard of 16 men. The establishment is a curious medley; but in an upper chamber, above the cows and pigs, there are three tolerable beds. From Boucharo to the top of the Fourquette de Tentenéra in $3\frac{1}{2}$ hours. The top of this pass is generally covered with snow. The imposing mountain seen in the W. is the Péna Colorada. In descending the gorge to the W. there is some little difficulty in getting down to the lower valley, owing to the precipitous rocks in which it terminates: but after passing the waterfall, the path is easy along the left bank of the stream. To reach the baths of Panticosa, it is not necessary to descend to the town of Panticosa; but crossing the valley, you strike the main road to the baths 1 hour 30 minutes below them. This new road is an excellent piece of engineering; and from June 20th to the end of September diligences are constantly running upon it. The baths of Panticosa have been of late years much improved. The bedrooms in the Grand Etablissement are excellent for cleanliness and comfort; the eating, also

---

* The present variation of the compass in the Pyrenees is 18° W.; therefore 18° is the true N.; and to reduce any of the magnetic bearings to the true bearings, 18° must be subtracted.

good, is supplied in a separate but contiguous building. The waiters and chambermaids are mostly French. Some of my friends consider Panticosa (see Sec. 6) one of the most charming spots in the Pyrenees; but it is too confined. The **Port de Marcardou**, leading from it to Cauterets, is one of the most beautiful of the Pyrenean passes. To reach the top from the baths requires about $4\frac{1}{2}$ hours. Starting from the baths, the general direction is very nearly N. The track mounts high above the left bank of the torrent, and then winds round a series of small lakes which occupy the upper plateau, and is rather puzzling; so that this pass should hardly be attempted without a guide. It is never altogether free from snow; and on June 19th, when I last passed, the lakes were all frozen, and scarcely a spot of ground uncovered. The fine mountain ESE. of the port is Mt. Peterneille (2,904 m.), which hides the Vignemâle. The descent on the French side is ENE., first over snow and rock; always following the stream. After passing a tableland, $1\frac{1}{2}$ hour from the top, the track becomes more marked, and passes among some noble fir trees along the left bank of the stream, lower down emerging on the pasturages of Cayan, where it bears to the E., and 4 hours from the top strikes the Cauterets-road at the Pont d'Espagne. The time from the baths to Panticosa is 9 hours. Mules pass in very favourable seasons. From Panticosa baths to Huesca is 70 kilometres = 44 miles; whence to Saragossa is 2 hours' rail.

It is quite feasible to pass direct from the baths of Panticosa to Gavarnie; but the route is a very wild one, and ought not to be attempted in bad weather. It occupies about 10 hours' walking, and probably no human being will be met with on the road.

Starting from the baths of Panticosa at 6.30 A.M., rise 10° N. of E.: very steep but easy. One hour to a small circular lake. Leave this on the right, and continue to mount. Thirty minutes more to a long lake, having the shape of a boomerang, the Lac de Brassato. Leave this also on the right, and continue to mount northwards, passing by a low ridge of rock to a desolate waste with several small lakes. Skirting these, make for the col due E., which is reached at 9. From this col, the summit of the Vignemâle bears east. Descend E., and at 10.10 reach the bottom of the col, and the junction of the stream which runs N. to the Col d'Aratile. You are now in the upper Serbigliana valley, where in midsummer there is a possibility of finding some Spanish shepherds. Following down the stream which runs ESE. and then SE., a faint track leads over pleasant turf, abounding in good plants, to the foot of the Plan d'Aube, 11.30. Thence 2 hours to the top of the Plan d'Aube, which is a spacious col, and not difficult. Descend eastward, bearing rather to the left to avoid the rocks. At 3 P.M., strike the Valley d'Ossoüe, just above the cabane of Millaris; from this point it is 2 hours' walking down the valley to Gavarnie.

Section 31.
GAVARNIE TO CAUTERETS.

By the Val d'Ossoüe and Lac de Gaube.

On foot, 12 hours. Guide necessary. Hippolyte or Henri Passet.

From Gavarnie to Cauterets is an expedition strongly to be recommended, from the noble views it offers of the Vignemâle. But the weather must be perfectly fine; otherwise there is not merely useless fatigue, but absolute danger, as this is one of the wildest of the Pyrenean passes, and there is no shelter of any kind on the route. Starting W. from Gavarnie, after leaving on your left the path that leads to the Port de Boucharo, you keep the right bank of the Gave d'Ossoüe, skirting the northern base of the Mont Cardal. Two and a half hours' walking brings you to some steep rocks and a waterfall at the foot of the great glacier of the Vignemâle. The first difficulty here begins; you must pass the abrupt shoulder of rock to the left of the waterfall: and this passed, you come upon some beds of snow, over which you pass, still keeping the stream on your right. The col for which you are aiming presents itself in the NW., a lofty ridge of red rock. To reach this, you must either pass over the elevated plateau on the left, or (which is the shorter and better way), if the torrent is sufficiently bridged over by the snow, you follow up the gorge, and cross the stream by a snow bridge, in the face of some almost perpendicular calcareous rocks; these you must surmount, availing yourself of the slight ledges which they present, and after this climb there is no more difficulty. For 20 minutes you still follow up the stream, and then, leaving on the right the col to the N., which leads by a pass of some difficulty to the Lac d'Estom, you strike due W., and in about 5 hours from Gavarnie find yourself on the top of the col, of which the height, according to my computation with the barometer, is 9,148 feet —3,282 feet above the Lac de Gaube. From this point the Petit Vignemâle is easily reached, striking up the shoulder to the SW.; but the chasms that intervene between this point and the great Vignemâle (the Pic Long) appear to me impracticable. To the N. of the col is the mountain of Poey la Hun. The descent is at first over the glacier, and afterwards along the moraine. Arrived at the bottom of the moraine, you leave the stream on your right, and the path lies across a succession of open and almost level grassy basins (paramos); S. of which is the Col d'Aratile, leading into Spain. From the Cascade de Spumouse, close to which the track passes, the path winds over rocky fragments and stunted fir trees till it reaches the border of the lake. From this point the boatman on the other side is within hail, and, for the consideration of 1 franc per person, he will row you across in 10 minutes. Those who do not take the boat must pass round the western side of the lake by a tedious and stony track, occupying 40 minutes.

From Lac de Gaube to Cauterets, 2 hours. (Sec. 16.)

Section 32.
Gavarnie to Cauterets by the Valley of Lutour. On foot, 11 hours.

This route is about as long as the last, and as difficult, but equally fine, and the course to Cauterets by this route, and return by the Vignemâle is much recommended. On leaving the inn at Gavarnie, cross the Gave of Ossoüe, which descends from the W., and, turning the flanks of the Pic Cecugnac, pass the hamlet of Saugué, and descend into the valley of Aspé. From here the col is in view WNW. between the two formidable peaks of the Malle Rouge (2,969 m.) to the left, and the Som de Mâle (2,973 m.) to the right. After an hour up the valley of Aspé, when arrived almost at its extremity, strike due north; a steep but not difficult ascent over the grass, and latterly stones, of nearly 1 hour 30 minutes, places you on the Col de la Houle, 5 hours from Gavarnie. From here you must descend some hundred metres, and then a scramble for 1 hour NW. over sliding schists and steep snow to the foot of a narrow and rather difficult little brèche, the Col de Malle Rouge, 2,700 metres(?), N. of the peak of the same name. From here there is a most noble view of the Vignemâle. The descent is very easy to the W., and thence NW. over patches of snow, interspersed with half frozen little lakes. Forty minutes from the col the 1st lake of **Estom Soubiran** (2,460 m.), and 30 minutes farther the 2nd lake, Estom Soubiran. Skirt both these lakes, leaving them on your left; and then descend to the NNW. as you best can. Here is the principal and only real difficulty of the route. You are on the top of an enormous cliff of rock, whence the torrent descends in a furious cascade. On the right bank is a passage down the rocks, but precipitous, and not easy to find. It is the one I adopted. It would probably be better to cross to the left bank, and endeavour to join the slight track which descends from the Hourquette d'Araillé, and the Hourquette des Olettes, two passes communicating with the Vignemâle gorge; the former on the Lac de Gaube side, and the latter on the side of the valley d'Ossoüe. When once below this wall of rock, all difficulty is over; keep the left bank of the torrent for 30 minutes to the lake **Estom** (1,782 metres), 13 kilometres from Cauterets, at the head of the very picturesque valley of Lutour. [Just below the lake there is beautiful ground for a pic-nic; delightful turf, and firewood to any extent. The woods are full of strawberries, and horses can be taken within ½ an hour of the lake.] Below the lake d'Estom cross to the right bank of the stream, and the path, which skirts the roots of the Pic d'Ardiden, continues on this bank for 1 hour 20 minutes, when it crosses to the left bank by a stone bridge; and after passing a fine cascade, emerges on the road of La Raillére, 2 hours from the Lac d'Estom; thence 20 minutes to Cauterets.

Section 33.
GAVARNIE TO HÉAS.

Route A. By the mountain, 3½ hours, on horse or foot. Guide not necessary.

The mountain road from Gavarnie to Héas is so easy and so enjoyable, from the beautiful and varied views that it presents, as well as its treasures in the shape of mountain flowers, that I strongly recommend it.

Twenty minutes below Gavarnie, on the road to Gèdres, before reaching the Chaos, opposite the bridge of Bareilles, the path mounts the grassy slopes to the E., and following the line of granges, goes round the shoulder of the **Coumelie**, and not over it. When the upper pastures are once attained, the road lies nearly on a level. The lake of **Hosse** is not seen, being left on the right at a considerably higher elevation, 1,963 metres = 6,440 feet. These pastures in the late spring are a mass of turquoise blue, with the flowers of the *Myosotis alpina*; and I have no where seen finer specimens of the *Ranunculus amplexicaulis*, which are in full flower in the middle of June. At the granges of Gèdres, which are half way, the path turns to the S., and shortly after this the chapel of Héas is in sight. In descending to the valley of Estaubé keep close to the rocks, which are on your right. The path leads down to the stream by some granges. From this point the summit of the Mont Perdu is in view, looking over the Port de Pinède. Three minutes down the left bank brings you to a bridge; cross this, and a rough descent of 20 minutes by some picturesque rocks, **the Pas des Glouriettes**, brings you to the gave of Héas, over which there is a fragile wooden bridge. Cross this, and below the huge block known as the **caillou de l'Arraye**, you emerge on the route from Gèdres, 30 minutes along

E

which brings you to the chapel of Héas, with one or two scattered houses, one of which is a little inn (chez Paget, dit Chapelle), which affords homely refreshment and two available beds. The chapel, placed at a height of 1,547 metres = 5,075 feet above the sea, is a place of considerable sanctity, from a statue of the Virgin which is said to have perched upon a rock that stands near it. On two days in the year, the 15th of August and the 8th of September, a considerable number of pilgrims resort here to visit her shrine. According to the legend, the chapel was built by three masons, who, during the time of their pious labours, were miraculously supported by the milk of 3 goats, who with their kids daily descended to them from the mountains; till at last, tired of their milk fare, the masons having resolved to kill a kid that they might eat its flesh, neither the goats nor kids were seen any more, and the builders ran great danger of being starved.

From Gavarnie also there is a pleasant walk to Héas by the S. side of the Pimené, crossing the col just N. of the little Pimené, also by the Brèche d'Allanz; but the former is the finest. Neither of these routes presents any difficulty to an ordinary mountaineer; the time by each is about 7 hours.

Route B. By Gèdre, 3 hours 30 min., on horse or foot.

Gavarnie to Gèdre 7 kilometres.
Gèdre to Héas 8 kilometres.

In coming from Gavarnie, at Gèdre Dessous, you have to cross the stream and ascend on the other side to the group of houses known as Gèdre Dessus. In coming from St. Sauveur to Héas, it is not necessary to descend to the Lower Gèdre. From the Upper Gèdre the path leads S.E. above the right bank of the stream, and in 15 minutes brings you to a bridge over the stream that descends on the left from the Pic de Cambiel. Up this gorge is the way to Aragnouet by the Col de Cambiel. See *post*, Sec. 48. In about 30 minutes the path, which had been suspended above the torrent, approaches the level of the stream, and the gorge opens into a little basin formed by the confluence of the stream which pours down on the right from the valley of Estaubé, which is seen opening to the S. between the Coumelie and the mountain of Héas, which also goes by the name of the Montferrand. Beyond this the path passes a chaos of fallen rocks resembling that on the route to Gavarnie; 45 minutes from this (2 hours from Gèdre) you come to Héas. The road from Gèdre to Héas is as yet very rough, and the guides are not very fond of exposing their horses' feet to the wear and tear of the stones; but a new road is in progress along the left bank of the gave.

Section 34.

HÉAS TO THE CIRQUE OF TROUMOUSE.

4 to 5 hours to go and return, on foot.

A little S. of Héas, the principal valley divides into two branches. The western gorge is that which comes down from the Port de Canaou, and down which the Gave de Maillet leaps in a series of falls, the larger gorge to the left, that of Touyères, down which also a torrent pours in several fine cascades, the highest of which is that of Mataras. For those who visit the Cirque de Troumouse, the best point of view is the granitic plateau that divides the two streams. To reach this follow up the left bank of the westernmost stream, by the track that leads to the Port de Canaou. When almost under this, after passing the snowy débris of the huge avalanches, which last late into the summer, the rocks are turned, and the upper plateau is gained without difficulty in about 2 hours from Héas. Here the rocky defile suddenly opens into a widespread undulating plateau, enamelled with flowers; among which the carmine tints of the *Daphne cneorum* and the *Androsace carnea* are conspicuous, intermingled with the white ranunculus, the yellow gageas and geums, and violas, and the blue myosotis; while the snow-clad ledges of rock which hem you in, contrast strikingly with this rich and varied verdure. If time allows, the twin rocks known as the Sisters of Troumouse may be reached from this point, and the return made on the other side of the cirque by the gorge of Touyéres.

Section 35.
PORT DE LA CANAOU.
To Bielsa, 8 hours.

From Héas a very fine and not difficult passage into Spain may be made by the Port de la Canaou, 2,771 metres = 9,081 feet. Three hours to the top of the Port, which is of clay schist, with a cross cut to mark the frontier on the E. side. From the Port 1 hour of easy descent to the fountain at the foot, known as **Mount Saint**, thence 4 hours to the Spanish town of Bielsa. For this or any other excursion about Héas, Chapelle, the master of the inn, is a most competent and willing guide. He is also an experienced hunter of izards, for whom he professes to have a great affection. For the route from Héas to the Valley d'Aure, see Sec. 48.

Section 36.
CIRQUE OF ESTAUBÉ.

At the head of the desolate cirque between that of Troumouse and Gavarnie, are 4 ports or exits, about the nomenclature of which some confusion seems to have existed. The westernmost port, the **Brèche d'Allanz**,

long. 1' 54" E. of Greenwich, leads from the cirque to Gavarnie by the N. side of the Pimené: its height is 2,505 metres. The central one is a narrow gap, ascended by a small glacier at a very steep incline, the base of which is marked by a projecting rock or tus. Its longitude is 2' 24" E., and its height 2,866 metres. It is marked wrongly in the map of the Etat Major as the Port de Pinéde, to which it is not the approach. It leads directly to the Lac de Mont Perdu, and its proper designation is the **Canaou d'Estaubé**. Facing the Brèche d'Allanz, a large gap or depression is seen to the E. Its longitude is 3' 57" E., and its height 2,665 metres. This is properly the **Port de Pinéde**, as it leads down to the hospice of Pinéde and the valley of Bielsa, passing at the foot of the cascades which issue from the lake of the Mont Perdu. In descending this Port, keep to the right as much as possible. At the hospice of Pinéde even bread is scarcely to be procured. From the hospice to Bielsa it is 2½ hours' good walking; and the artist will linger to mark the gradations of colour which paint the rocks of the **Tres Sorellas**, the Spanish name of the Mont Perdu, on the S. side of the valley. The actual time of walking from Gavarnie to Bielsa by this route is 10 hours. In Bielsa the Posada is in the Plaza on the left as you enter. Meat here is a rash experiment. I should advise the traveller to be content with bread, wine, eggs, and chocolate. The name of the master is Antonio Vidaillet, who is an honest man, and his beds clean. N. of the Port de Pinéde, in long. 4' 27" E., is a fourth port, marked in the map as the **Port Biel**, at a height of 2,762 metres. By this also you may descend to the hospice of Pinéde; or, after passing this, remount to the left, and so reach the cirque of Troumouse, through the Port de la Canaou.

Section 37.
CIRQUE DE TROUMOUSE.
Upper circuit, 12 hours.

Mountain amateurs making any stay at Héas, are strongly recommended to engage Chapelle, and undertake this course. Following the course of the Touyéres, the E. branch of the torrent above the right bank, in 1 hour 20 min. you reach the tableland of the cirque, at the cabane of Lious Aoube. Thence ESE. over strata of rather slippery limestone, gradually mounting till the crête is gained. Follow this SE. till stopped by the **Pic de Gerbats**, an inaccessible rock (2,920 m.). Here turn to the W. and descend a few steps into the cirque by a very mauvais pas, where a slip would be certain death; skirt the precipices under the Gerbats, and again mount S. on to the ridge of the cirque. Continue to follow the crête, having frightful precipices on either side, that into Spain being vertical. On reaching the **Pic de Serra-Mourenne** (3,086 m.), descend a little into the cirque, passing under the summit, and beyond this cross on to the Spanish side of the ridge, traversing some very deeply-inclined beds of snow. Once more on the rock, a steep but not very difficult scramble of 40 min. up the rocks SW. places you on the highest point of the Cirque, the **Pic de la Munia**, 3,150 metres. The view of Mount Perdu is grand; the wide brèche to the E. of this mountain is the **Col de Niscle**, by which there is a passage to Fanlo. Descend the rocks NW. for 20 minutes, and then a glissade down the glacier N. will land you in front of a little cheminé, by which you may descend into the Cirque de Troumouse, close to, and just W. of the **Deux Sœurs**. This passage is not easy to find. In default of Chapelle, a son of Thérése is a capable guide.

Section 38.
LUZ TO BARÈGES.
7 kilometres = 4½ miles. Carriage-road.

From Luz there is only one escape to the eastward, the road leading up to **Barèges** (Hôtels de l'Europe and de France), thus far practicable for wheels, but from this point to cross the mountains to Gripp on the Bigorre road, the traveller must trust to saddle-horses or his own feet. The road from Luz to Barèges is 4½ miles of continual ascent, gradually passing from the shady trees and green pastures to the barren mountains, whose scarred sides, strewn with huge stones, attest the fearful ravages of the winter torrents. The town of Barèges is the sort of place we should expect from the approach, one long street of wretched houses, with a gap at intervals where an avalanche has swept them away, and a row of wooden booths has been substituted. Here they sell hardware, cotton stuffs, and some of the coarser sort of the celebrated woollen Barèges shawls; but the finer ones are manufactured and sold at Bigorre. The architecture of the town is certainly not imposing, and the climate does not more recommend it. Barèges being at an elevation of 4,084 feet (1,245 metres) above the sea, all the chilly mists of the surrounding mountains here collect, and the wind at times blows down this defile with such violence that for 5 months in the year the place is uninhabitable. The ground is covered with snow to the depth of 15 feet, and all the population emigrate, except 7 or 8 of the most hardy, who remain to take care of the houses and furniture, and are kept close prisoners for many weeks. The bath establishment is a dreary stone building. 16 gloomy little cellars, admitting neither air nor light, are set apart for those who can afford the luxury of a private bath, the price of which is 1 franc 25 cents; and for these there is such a demand, that, without being registered in the doctor's book you can hardly procure one. In addition to these smaller compartments (*baignoires*), there are 2 public baths or *piscines*, which are supplied with the water that has been used in the private baths; and, that nothing may be wasted, there is a third, gratis for the poor, which is fed with the water that comes from the two piscines. The Barèges waters are the strongest in the Pyrenees, and are reputed to be very efficacious in curing gunshot wounds and ulcers, on which account there is here a military hospital; but after inspecting these gloomy underground dungeons, with their reeking vapours and dark fœtid waters, I was disposed to acquiesce in the remark of a French gentleman who was with me, 'Il faut avoir beaucoup de santé pour y guérir.' Barèges was for a time the residence of Madame de Maintenon with her pupil the Duc de Maine, who was brought here to try the efficacy of the waters, and was at last cured by them, so much the worse for his country.

From Luz to Luchon by the Tourmalet road, 92 kil. = $51\frac{1}{4}$ miles.

Lac Escoubous (2,044 m.). **5 hours to go and return.**

At 3·84 kil. from Barèges quit the main road to the Tourmalet, and strike S.S.E. along the right bank of the stream, following the rough horse track. The lake is immediately behind a red rock, which is seen before you. Below this rock, the path crosses the stream, and mounts by zigzags on the right, i.e. W. side of the rock. The Lac Escoubous abounds in good trout, which, on a cloudy day with a breeze, afford sport to the fly-fisher.

Section 39.

ASCENT OF THE PIC DU MIDI DE BIGORRE.

On foot or horseback.

To toil up a mountain without a prospect of a view, is of course pure waste of labour, but if the sky be tolerably clear, no visitor to these parts should omit the ascent of the Pic du Midi de Bigorre. Unlike its namesake at the head of the Val d'Ossau the ascent of this mountain presents no difficulty even to ladies; and those who wish to spare themselves fatigue may ride their horses to the very top, though they will probably choose to dismount for the last thousand feet. The ascent may be made either from Barèges or Gripp, but the better plan is to take the mountain en route on leaving Barèges, sleep at the cabane on the top, and descend the next day on the other side to Gripp, as a considerable elevation *must* be attained in passing the col on leaving Barèges, and this all comes into the day's work. After leaving Barèges, the mule track should be kept for about $1\frac{1}{2}$ hour (6 kilometres), and then, turning to the left, a steepish climb leads over the shoulder of the mountain in an hour and a half more to the Lac d'Oncet, a mere mountain tarn which at no period of the year is wholly free from ice. On the north brink of this a wooden cabane has been erected, where the traveller will find very fair night quarters, and may sleep tolerably unless disturbed by one of the storms of wind (tourmentes), which are here not unfrequent and of exceeding violence. Humble as is the appearance of this '*cabaret*,' its construction cost 14,000 francs to the company who erected it in the year 1855, previous to which a cowshed was the only shelter on the mountain. The summit of the mountain is 506 metres = 1,660 feet above this cabane, and the zigzag track leading to it is so steep and stony, that if horses are brought I should advise their being left here. If the weather be fine, there is no point, except the Pic des Posets or Maladetta, so favourably situated as the Pic du Midi for a general view of the highest mountains of the Pyrenees, detached as it is from the main chain, and yet attaining an altitude of 2,877 metres = 9,439 feet. To the north is the Pic de Montaigu, 2,341 metres = 7,680 feet, and beyond this the view stretches far over the level and vine-clad plains of Languedoc, dotted with considerable towns, Toulouse on a clear day being distinguishable in the N.E. horizon. To the south the whole range of the higher Pyrenees lies before you, from the Montvallier in Ariège to the Pic du Midi d'Ossau. Those who have no experience of mountain climbing are unable to appreciate the exquisite pleasure after toiling up the rocks for several hours, of arriving at the summit, and the agreeable sensation of seeing all around you. There is, doubtless, something solemn and imposing in the aspect of a boundless horizon, whether viewed from the summits of the highest Alps, or the humbler eminences of 2,000 or 3,000 feet. 'L'esprit comprend ce que le corps domine.' The human mind thirsts after immensity, and immutability, and duration without bounds; but it needs some tangible object from which to take its flight—something present to lead to futurity—something bounded from whence to rise to the infinite. The vault of the heavens over our head, sinking all terrestrial objects into absolute nothingness, might seem best fitted to awaken this sense of expansion in the mind; but mere space is not a perceptible object to which we can apply a scale, while the Alps, seen at a glance between heaven and earth—met, as it were, on the confines of the regions of fancy and sober reality—are here like written characters traced by a divine hand, and suggesting thoughts such as human language never reached. Infinity of space, as poets in every language say, is reflected within ourselves; it is associated with ideas of a superior order; it elevates the mind which delights in the calm of solitary meditation.

It is fine
To stand upon some lofty mountain-top
And feel the spirit stretch into the view.

It is true, also, that every view of unbounded space bears a peculiar character. Seen for the hundredth time, it can never pall upon us or become monotonous. The prospect ever varies, according as the clouds reposing on the plains extend in layers, are conglomerated in groups, or present to the astonished eye through broad openings the ice-ribbed mountains, the habitations of man, the labours of agriculture, or the verdant tint of the tortuous valleys. It is a mistake, too, says Baron Humbolt, to suppose that the highest and most arduous ascents are attended with the greatest amount of pleasure to repay the toil.

Colossal mountains, such as Mont Blanc and Monte Rosa, or even the Maladetta, compose so large a mass, that the plains covered with rich vegetation are seen only in the immensity of distance, and a blue and vapoury tint is uniformly spread over the landscape, so that the views from their summits are neither so beautiful, picturesque, nor so varied as those from heights of from 6,000 to 10,000 feet.

There is but little snow on the Pic du Midi in summer, owing to its isolated position; but it unfortunately stands a prominent buttress to attract the mists that accumulate on the plain. Cloud and storm is, I am afraid, the normal state of this mountain; but by sleeping at the cabane above Lac Oncet, and ascending to the top for sunrise and sunset, there is a double chance of a view. Once, out of 3 ascents (which is, I believe, the average), I have been rewarded with a most magnificent panorama. The sunset effect seen from the top of the Pic du Midi is especially fine; and it is curious to watch the huge conical shadow projected by the mountain, with its point sharply defined, as it gradually encroaches on the still glowing mountains beyond Arreau, till finally lost in the sky, over the Col de Pierrefitte.

From the cabane on the Pic du Midi, there is no difficulty in passing the col to the W. of the Lac d'Oncet; and thence over another little col, the **Col d'Aoube**, and so to the Lac Bleu; time 2 hours 30 minutes. You may also pass into the valley of l'Esponne by the lake Peyrelade; but the descent upon the lake is by a rather difficult cheminé.

Section 40.
FROM BARÈGES TO BAGNÈRES DE BIGORRE.

Route A. By the Tourmalet; 8 hours, walking, in carriage, on horseback, or on foot. 39 kilometres = 24½ miles.

The new carriage road is now complete; and if that over the Stelvio pass is allowed to fall into disuse, I believe this is the highest carriage pass in Europe. It is, however, very uninteresting; and except as the most direct communication scarcely to be recommended. The road is early blocked by snow, and is always damaged by the winter avalanches; so that it is traversable in carriages for only about 3 months. Those who are on foot will find it shorter and better to keep the N. side of the valley, avoiding the long sinuous sweeps made by the carriage road; 10 kilometres = 2½ hours, along the left bank of the Bastan River, brings you to the top of the **Col Tourmalet**, the road being sunk by a cutting 15 metres below the col. Though this elevation is 2,122 metres (6,962 feet), there is scarcely anything of a view, which is shut out on either side by bare rocks, consisting principally of clay slate mingled with quartz, and presenting singularly contorted strata. On the east of this col, the **River Adour** has its rise. 9 kilometers from the col brings you down to the hamlet of Tramesaigues, where you meet the track that leads down from the Pic du Midi. From this point, 4 kilometres, to the little village of **Gripp**, 17 kilometres from the col, where the Hôtel des Voyageurs is a very good half-way house for obtaining refreshment. The trout are especially commended. From Gripp to Bagnères de Bigorre the distance is 16 kilometres (10¼ miles). At Ste. Marie (4 kilometres beyond Gripp), the road enters the route départementale;* and 12 kilometres (8 miles) along this, passing through Campan (6 kilometres), brings you to Bigorre.

Route B. By the Lac Bleu.

From Barèges you may also cross over to Bigorre by the **Lac Bleu**, descending by the Val de l'Esponne. This is a more interesting route than the Tourmalet or the Hourque de Cinq'Ours, but can only be made on

---

* Nothing can be more admirable than the system of roads in France, and the perfect repair in which they are kept without any tolls. First come the Routes Impériales, the great highways, which are kept up at the expense of the Government, and along which stones are set to mark the distance of each hectometre, or every 100 metres. Next in order to these come the Routes Départementales, which are under the control of the Prefect of the Department, and which are more than equal to our best turnpike-roads. Thirdly come the Routes Communales, which are kept up at the expense of the adjoining commune. None of the carriage-roads in France are allowed to be constructed with a greater inclination in any part than an angle of seven degrees.

foot—time about 8 hours. Starting from Barèges, the Lac Bleu is most easily reached by the Col d'Aoube. In descending to the Lac Bleu, after passing the small lake called Le Laquet keep rather to the right; descending upon the zigzags by the **Pas de Chevre**. The **Pas de l'Ours**, some 300 metres more to the W. is more difficult. At the Lac Bleu, a sheet of water 1,968 metres = 6,457 feet above the sea, at the foot of the Pic d'Asblancs (2,630 metres = 8,629 feet), the western buttress of the mass of the Pic du Midi, the annual fall of rain and snow has been computed by M. Colomé, the chief engineer of the department of the Hautes Pyrénées, at more than 2 metres, or more than double that of the plain to the north.

The Lac Bleu is one of the largest of the Pyrenean lakes, its surface comprising 49 hectares, while the Lac d'Oo has only 39·16 hectares, and the Lac de Gaube only 16·4 hectares; but the Lac Bleu is very inferior in beauty to the other two.

If starting from Luz, the most direct route to Bigorre is by the **Pène Taillade**; or rather by the **Pène Pourry**, to the E. of this route, as follows: at 2·76 kilometres from Luz cross the Bastan by a little bridge, and ascend the rocks to the village of Serts. From thence continue up the valley, crossing to the left bank of the stream in about 30 minutes. Bear always rather to the right; direction NNE. The Pène Taillade is a V shaped cleft in the rock; but it is a passage of considerable difficulty, and it is much better to make for the gap a little more to the east, known as the Pène Pourry. There are three contiguous passages, any one of which is practicable; but the scramble up the rocks, or probably snow, is sufficiently steep. The centre passage is rather the easiest, height 2,475 metres. The descent upon the Lac Bleu is steep, but not difficult; keep the W. side of the lake, 2 hours from the col brings you to the cabane. From the cabane 1 hour down the zigzags to the head of the Valley of Lesponne, and 3½ hours more to Bagnères de Bigorre.

Section 41.

Bagnères de Bigorre is a lively and fashionable watering-place, 553 metres = 1,814 feet above the sea, with pretty country walks, shady boulevards, and plenty of society. The indigenous population of the town is 8,000, and during the season it receives as many more visitors, for whom, and upon whom, the natives live. The hotels (de Paris, Londres, and France) are in general very good, well served, good cuisine, and not exorbitant in charge. The surrounding scenery, though pretty, owing to its distance from the higher mountains, is not nearly so grand as that of its rival, Bagnères de Luchon, so that the tourist, if at all pressed for time, may be content to omit visiting this place; but those whose sole object is to while away a few days indolently and agreeably, and who are not what the French call 'passionné pour les montagnes sauvages,' should by all means come to Bagnères de Bigorre.

The principal season at Bagnères de Bigorre is from June to October, but those who would then prefer the sultry climate and fashionable promenade of the Coustous to the invigorating air of the mountains, will hardly derive useful information from this work. As a winter station, however, Bigorre has considerable *agrémens*, and for all persons but those whose lungs are affected, is in my judgment preferable to Pau. Being 349 metres higher, the climate is much more bracing, and also much drier, and does not exercise that depressing effect upon healthy persons for which Pau is notorious. At Bigorre during the winter there are generally about 100 English. These, and some of the French residents, form a very pleasant and most sociable society, meeting without any ceremony. There is a very pretty little English church, established principally by the exertions of Mr. Lyte, the well-known photographer and chemist, who has been residing at Bigorre for many years. Mr. Lyte is also the proprietor of a mineral spring at Mondang, strongly recommended for its combination of iron and sulphur, and for this water he has established a buvette at Bigorre, where it may be procured in bottles fresh from the source.

Bigorre is also the head-quarters of the *Société Ramond*, lately established, for promoting exploration in the Pyrenees. The *Bulletin* of the society, published quarterly, may be procured of M. Caznave, who is also the proprietor of the *Petite Gazette*. His office is opposite to the Hotel de France.

On the right of the road, half-way between Ste. Marie and Bigorre, the marble quarries of Campan are worth a visit, as also the marble works of M. Aimé Geruzet at Bigorre, where specimens of every variety of the Pyrenean marbles may be seen. Many of these, as the Sarrancolin, Œil de Perdrix, Le Roncé de Bise, and La Griotte, are exceedingly beautiful, surpassed in colour by none of the Italian marbles, but their texture is not quite so fine. The bathing establishment at Bigorre, a handsome stone building, with marble corridors, billiard and reading rooms, and ornamental gardens, is a great contrast to that at Barèges. Here there is no primary rock near the surface, and the waters consequently are not sulphurous, as is usually the case in the Pyrenees. The sources are mostly saline and chalybeate, of no great efficacy; but the amenities of the place are such that they attract a host of soi-disant valetudinarians of the higher classes, dyspeptic lawyers, literary men, and merchants, and with them many of that numerous class who are ennuyés because they have too much money and too little care. All these you meet at Bigorre. They drink and bathe a little, and the rest of the day they devote to promenading on foot or on horseback. The evening they spend at balls or conversaziones at the Casino. For self-constituted invalids of this kind, Bagnères de Bigorre is just the place: 'Peu à peu ils s'égayent, et se guérissent en s'amusant.' Montaigne, who was, by the way, as sceptical in medicine as in other things, thus spoke of these baths: 'Toutesfois aussi n'ai je vue guères de personnes que ces eaux aient empiré; et ne leur peut on sans malice refuser cela, qu'elles n'esveillent l'appétit, facilitent la digestion, et nous prestent quelque nouvelle alaigresse, si on n'y va pas trop abbattu de forces; ce que je desconseille de faire; elles ne sont pas pour relever une poisante ruyne; elles peuvent appuyer une inclination legiere, ou prouveoir à la ménace de quelque altération. Qui n'y apporte assez d'alaigresse pour pouvoir jouyr le plaisir des compaignies qui s'y trouvent, et des promenades et exercises à quoy nous convie la beauté des lieux ou sont communement assises ces eaux, ils perdent sans doubte la meilleure pièce et plus assurée de leur effect. A cette cause j'ai choisi jusques à cette heure à m'arrester, et à me servir de celles ou il y avoit plus d'amœnité de lieu, commodité de logis, de vivres, et de compaignies, comme sont en France les bains de Banières.'

Section 42.
L'Heris.—Puits de la Pindorie.—Grotte de Judios.

The **Bedat**, 881 metres; the **Monné**, 1,258 metres; and the

Castel Mouly, 1,142 metres, are three eminences close to Bagnères, presenting fine points of view in the course of a morning or afternoon walk. At the foot of the Bédat is an extensive grotto, or rather a series of subterranean passages, of which a very elaborate model, executed by M. Vaussenat, civil engineer, may be seen at the Mairie. This gentleman and M. Pambrun have lately published a very correct map of the country round Bigorre, which will be found useful by those making walks or rides in the neighbourhood.

The **Péne de l'Héris** (1,593 metres = 5,226 feet), composed of changed limestone rock, is also a favourable point of view to the SE. of Bigorre, which may be ascended in about 3½ hours, and the same to return.

A new road has lately been made, and it is easy to ride on horseback; passing the village of Asté, and up the picturesque gorge southwards. The path keeps on the N. side of the gorge, and after passing through the wood, emerges on a grassy plateau, which in spring-time is perfectly covered with wild flowers, the prevailing tint of which is blue, the *Polygala alpestris*, *Horminum pyrenaicum* and *Gentiana acaulis*, being the dominant species; the gentians I have nowhere seen so fine. On this plateau there is a cabane, where refreshments may be obtained; thus far from Bigorre is a ride of less than 3 hours. To reach the top of the Pène, is a steep but not difficult climb up the passage known as the **Pas du Chat**, 35 minutes from the cabane to the top. The rather rare *Ranunculus thora* is pretty abundant on the "Pas du Chat."

South of the Pène de l'Héris, runs the parallel crête d'Ordincéde, which rather exceeds it in height, and is preferred by many as a point of view. It is easily reached from the cabane in about the same time. From the top of this crête you may descend S.E. to the **Cabanes d'Ordincède**, whence a path leading down the mountain side strikes the Bigorre road, at 9·5 kilometres from that place.

The return may be agreeably varied and the distance scarcely increased, by continuing along the crête on the N. side of the gorge as far as the Palombières, from which there is a fine view; but the apparatus arranged in the tree tops for the bird catching is of no great interest, and the birds seem now to have become wiser, so that but few are taken. The season for catching the pigeons is in the first days of October.

There is one excursion to be made from Bagnères which I must not omit to mention, as unique in its kind in the Pyrenees—that to the Ice-cave, or natural glacière, known as **Le Puits de la Pindorle**. It is a not very fatiguing day, but an early start should be made; and a good rope of 25 metres, and 2 or 3 strong guides are indispensable. I can especially recommend Pedelhez, the garde forestier of the Duc de Grammont, living at Medous. On foot, the shortest route is along the high road to within 100 metres of Ste. Marie; thence across the stream, and mount the gorge leading up to Ordincède by a rough path keeping the south side. From the top of the plateau, nearly 2 hours through the wood to the Puits, which is situate among the trees at a height of 1,512 metres. There are three entrances; that in the NE. corner being a mere well with jagged sides; the other two entrances, one on the S. and the other in the SW. corner, are very steep, but not absolutely perpendicular; that on the S. is the largest and easiest, if provided with a good rope. The depth of the cave is 20 metres, and the floor always of ice, with ice columns prostrate or upright, and varying in size with the season of the year. The return may be well varied by mounting the arête to the SW. (height of col 1,764 metres); whence 1 hour 40 minutes easy descent to the Bigorre road, 2 kilometres above Ste. Marie (height 902·2 metres), crossing the stream below a huge rock, on which is placed a singular little hanging garden. If ladies are of the party, the Puits is reached with least fatigue by this way, taking a carriage from Bigorre to the 14th kil. stone.*

Another interesting excursion may also be made to the **Grotte de Judios,** a picturesque cavern on the N. flank of the Péne de l'Héris. The time to go and return is about 6 hours, passing by the village of Gerde. If, with the intention of exploring the cavern, candles should be taken; with these it is easy to penetrate about 200 metres in the cave, in which there are some fine stalactites. At the end of this first cave is a well, which can only be descended with the aid of a rope. You then find yourself in a second cave, also of considerable dimensions. At the upper end of this second cave, there is a small chink, through which a man can barely squeeze his body. Having done so, he finds himself in a third cave, of smaller dimensions, but very rich in stalactites, and of a high temperature even in mid-winter. Beyond this there is no further opening. It is not very easy to find the entrance to this cave, without some one who knows the place.

In going or returning, this excursion may be varied by passing through the village of Asté instead of Gerde. The distance is the same; and those who are more charmed by the historical associations than the natural beauties

* For a fuller notice of this ice cave, which is the only one of which I am aware in the Pyrenees, see the 2nd number of the bulletin of the Société Ramond.

of a place, may take an interest in the ruined château of Asté, where Henri IV. used to visit the fair Corizandre de Grammont. The fountain close by is dignified with the name of the Lac du Bourbon.

Section 43.

THE PIC DE MONTAIGU TO PIERREFITTE.

The Pic de Montaigu 2,341 metres =7,681 feet), 11 hours to go and return to Bigorre, and about the same time to descend to Villelongue, in the valley of Pierrefitte, on foot, though, with guide, practicable on horseback. Leaving Bigorre by the SW., 1 hour 30 minutes to the summit of the col between the Monné and the Castel Mouly. Mount by the gorge, and on reaching the foot of the Monné, cross the stream and mount by the path on the flank of Castel Mouly. From this col the summit of the Montaigu is just visible behind La Peyre. (1 hour 30 minutes.) Make across the same level SW. for a little stony brèche, and thence through a wood to La Peyre. (3 hours 30 minutes.) Skirting this mountain, 1 hour farther to the actual base of Montaigu, whence 1 hour to the summit, passing on the left a little tarn. The actual peak is attacked from the N., the least precipitous side. It may also be ascended from the valley of Lesponne, in which case, a carriage may be taken for 17 kilometres as far as the cabane at the foot of the ascent to the Lac Bleu. Thence at once strike W. up the shoulder of the mountain. Though steep, there is no difficulty, till the last 10 minutes, the summit being approached from the W. by a very narrow arête, with a precipice on either side.

You may also descend on the western side in the valley of Pierrefitte. Descending to the SSW., in 1 hour 30 minutes you come to a desolate plateau, with a small lake, into which falls the Cascade de Paspiche, the source of the stream of the Isaby. Following down the course of this stream for 1 hour 40 min., you come to the ruined abbey of St. Orens. Below this, crossing the stream by a little bridge, 15 minutes brings you to the hamlet of Ortiac, where there is a ferruginous source, and 20 minutes more to the bridge of Villelongue, whence to Pierrefitte, 1 kil. = half a mile.

Section 44.
BIGORRE TO LOURDES.

A. By the road 20 kil. At 6·5 kil. from Bigorre, 1·5 kil. short of Montgaillard, take the turn to the left, which mounts the côte. At Loucrup, 10 kil., there is a magnificent view of the high chain of the valley of Azun and the Pic de Baletous. The best point for the view is the far end of the village, just before the descent. One seldom sees a happier combination of rich foreground tints, and mountain distance, than this presents in the late autumn.

B. On foot or horseback this course may be made over the hills by Labassere, the **Croix Blanche** (746 m.), and Juncalas. Though the route is more varied, as a picture, the view from the Croix Blanche is not equal to that from Loucrup. To Lourdes the distance is rather longer.

	kilometres
CROIX BLANCHE	11
Junction with road in Valley of Juncalas or Castelloubon	4·5
Pont Neuf, junction of main road to Argeles	8
Lourdes	2·5
	26 kil. = 16m.

If going to Argeles, the distance is much the same by the two roads; as from the Pont Neuf, the bit between that and Lourdes is gained, and it is only 10 kilometres to Argeles. To Argeles you may also take the road along the right bank of the river, turning to the left at the mill before reaching the Pont Neuf. From the mill to Pont de Tillos 9 kil.; thence to Argeles 1·5 kilometres.

The best hotel at Lourdes is the Hôtel de France. The three lions of the place are the lake, the *château*, and the miraculous *grotte*, which, at least, have the merit of being easily accessible.

Section 45.
BAGNÈRES DE BIGORRE TO BAGNÈRES DU LUCHON.

Route A. By Lannemezan and Montrejeau. 82 kilometres = 51 miles. Post road.
Route B. By Labarthe and St. Bertrand de Comminges. 77 kil. = 48 miles. Post road.
Route C. By Arreau and the Col de Peyresourde; wheelroad. 70 kil. = 43½ miles. 37 kil. to Arreau by the Hourquette d'Aspin.
Route D. By Arreau and the Col de Pierrefitte; on foot or horse.

36 kil. to Arreau.
31 kil. to Luchon.
———
67 kil. = 42 miles.

This last is the route most recommended, as shortest and most beautiful. If route C is taken, the Lac d'Oo may be visited by the way.

Route A. By Montrejeau and the plain, as follows:—

Bigorre to Escaladieu	12 kil.	= 7½ miles
Mauvezin	3	1¾
Capvern	4	2½
Lannemezan	18	11
Montrejeau	10	6¼
Croix de Bazert	4	2½
Estenos	18	11¼
Luchon	21	13
	90 kil.	= 55¾ miles

Good carriage-road; diligences morning and evening, doing the distance in about 10 hours. Two kils. beyond Capvern the road joins the road to Tarbes, distant from here 27 kil. = 16¾ miles, and at the **Croix de Bazert** the road enters the main route to Toulouse, distant from here 97 kils. = 60 miles. This route is not particularly interesting, but the railway now in progress from Tarbes to Toulouse is to pass through Montrejeau, distant from Luchon only 39 kil. = 24 miles, and this will then be the shortest approach to Luchon. At Montrejeau (*mons regalis*) there is no good inn; but there is a tolerable buffet for refreshment at the station.

Route B. By Labarthe and St. Bertrand de Comminges, as follows:—

Capvern	19 kil.	= 11¾ miles.
Labarthe	7	4¼
St. Bertrand de Comminges	18	11¼
Junction of Luchon road	4	2½
Luchon	29	18
	77 kil.	= 48 miles.

There are diligences daily along the road as far as Labarthe; from thence to the town of Montrejeau 20 kil.; 19 kil. to the railway station—if on foot, when it is not required to mount into the town; but the river can be crossed by the railway bridge below it.

Route C. By Arreau and the Col de Peyresourde.

Ste. Marie	12 kil.	= 7½ miles.
Paillole	7	4¼
Hourquette d'Aspin	6	3¾
Arreau	12	7½
Louderville	12	8
Col de Peyresourde	7	4¼
Garin	5	3
Cazaux	2	1
Luchon	7	4½
	70 kil.	= 44¼ miles.

Route D. By Arreau and the Col de Peyresourde; wheel-road. 70 kil. = 44 miles.

From Bagnères de Bigorre to Bagnères de Luchon there is a practicable carriage-road, passing over the Hourquette d'Aspin by Arreau. By this, which is somewhat hyperbolically called the route 'par la montagne' there are diligences every day in summer; fare in coupé 20 francs. The hire of a carriage and horses should not exceed 60 francs. The Hôtel d'Angleterre at Arreau (the best) affords a very tolerable night's lodging, and I should advise every one to divide the distance into two easy stages by sleeping here, as, by taking it in one day, they will have scarcely leisure to enjoy the very beautiful and varied scenery over which the road passes. The whole distance from Bigorre to Luchon is 70 kilometres = 44 miles, Arreau being 1 kilometre beyond the half-way; the whole of the road is very hilly. In passing through the forest after Paillole, remark how profusely the trees are festooned with lichens; Usnea barbata, and Usnea florida, the latter distinguished by its little circular discs. The view of the mountains from the top of the col beyond Gripp, called the Hourquette d'Aspin (1,497 metres), is especially beautiful. Looking back, the bold form of the Pic du Midi de Bigorre is seen to great advantage; the view to the S.E. is yet more attractive, where the lovely Val d'Aure stretches southward among the pine-clad spurs of the mountains, which are blended imperceptibly with the snowy peaks above and beyond Luchon. To feast on the view in all its completeness, mount 20 minutes to the crête N.W. of the col; you will not regret the labour. The Pic des Posets is well seen, the monarch of the group in the S.S.E.; for the Maladetta, a little more to the E., is scarcely visible.

Conspicuous above the fir forests in the W.S.W. is the imposing mass of the **Pic d'Arbizon** (2,831 metres). The ascent of this peak may be made from Arreau, but not directly. Ascend the Valley d'Aure as far as Guchen on the left bank of the Neste. The semi-circuit of the peak must be made so as to attack the summit from the W. A guide may be procured at the village of Aulon. The pedestrian in descending to Arreau from this col, may cut off about 5 kilometres, by avoiding the zigzags. Five minutes after passing the col, strike S., and descend the gorge upon the village of Aspin; 30 minutes beyond this the path emerges on the main road, 2·1 kilometres from Arreau. From Paillole (1110 metres), where there is a decent auberge, there are two other passages into the Valley d'Aure for the pedestrian, the Hourquette d'Arreau (1,517 metres), and the Hourquette d'Ancizan, which passes still nearer to the Pic d'Arbizon, and descends upon the village of Ancizan, 4 kilometres above Arreau.

A very pleasant excursion of three days' moderate walking may be made from Luz to Luchon, taking the Pic du Midi en route. The portmanteaux and principal luggage should be sent from Luz by *poste*, reserving to carry only what is absolutely necessary for the comfort of three days. The following extract from my note-book will show the time and length of the different stages:—

'*Aug.* 22.—Booked and paid for our luggage to Luchon, the charge for the united kit of the three, weighing 43 kilogrammes, being 17 francs—not exorbitant, considering the long detour which the post-road makes, passing round by Tarbes and Montrejeau, and so to Luchon up the valley of the Garonne. Having paid our bill at the Hôtel des Pyrenees, we left Luz at 10.30, and having halted at Barèges for *déjeuner*, and taken a *bain sulfureux* while it was preparing, we continued our course up the mountain, and reached the inn on the Pic du Midi at 4 P.M. Having secured our beds—a business of the first importance at all these mountain inns—and ordered dinner, which is the next essential, we ascended to the top, to see the sun set. It was very cold on the summit, and the view was a good deal spoilt by a storm of snow and hail which came on, preventing that glorious prospect of mountain and plain which, from the recollections of my first ascent, 4 years before, I knew lay at our feet; but the glimpses of the cloud-capped peaks in front were a sufficient reward for the toil. In the night a tremendous storm of wind burst upon us, breaking in the windows, and carrying away a great part of the wooden outhouse where the horses were stabled. The appearance of the lake beneath us (Lac d'Oncet) sufficiently attested the violence of the wind. Overnight, its surface had been covered with a thick and uniform sheet of ice; and in the morning this was completely broken up into fragments, which were piled like icebergs one upon another.

'*Aug.* 23.—In the morning rain had succeeded to the wind; but, taking advantage of a lull, we left the auberge on the Pic du Midi at 9.30, and descended the valley to Gripp. Here we refreshed, and again at Paillole, and, thence passing over the Hourquette d'Aspin, reached the Hôtel d'Angleterre (the best) at Arreau at 6 P.M.

'*Aug.* 24.—Left Arreau at 8.30 A.M., and, passing by the lower road the Port de Peyresourde, reached Louderville at 11. Here is a post-house, where bread, cheese, and beer may be procured. From Louderville the road is not particularly interesting. Seven kilometres of zigzag ascent, which may be shortened by the pedestrian, brings you to the top of the port (1,545 metres), but not remarkable as a point of view. On the NE. side of the port, there is a small lead mine. As you descend towards Luchon from the 4th to the 6th kilometre stone below the port there is a fine view on the right of the mountains of the Port d'Oo and the glacier of the Ceil de la Vache. At exactly 2·2 kilometres from Luchon on a rock to the left of the road, from which it is distant 20 metres, the maiden hair fern, and the hart's tongue are seen growing very abundantly. We reached Luchon at 4 P.M., one of our party being very tired, though each of the days' journeys had been but about 23 miles. In entering Luchon by this route, many persons turn off at the village of Oo to visit the lac of the same name, which is about 7 miles from the village; but it is far better to devote an entire day to this excursion from Luchon, as, if called upon to name the most beautiful spot in the Pyrenees, I think I should name the Lac d'Oo and its cascade, and it is a pity that the sense of fatigue and the desire to hasten onward should in any way interfere with the enjoyment of such a scene.' (See *post*, Sec. 55.)

Route D. By Arreau and the Col de Pierrefitte. 67 kil. = 42 miles, as follows:

Arreau	36 kil.	= 22½ miles.
Jezeau	2	1¼
Bareilles	4	2½
Col de Pierrefitte	7	4½
Bourg d'Oueil	3	2
Mayregne	4	2½
Junction of Peyresourde	6¼	4
Luchon	4½	2¾
	67 kil.	= 42 miles.

For the route as far as Arreau, see Route C. From Arreau the remainder of the journey must be made either on horse or foot. From Arreau to the top of the col 3 hours, 8¼ miles. A carriage-road is in progress, but as yet far from completion. The ascent of the Pic de Monné may be made from the top of the col; but for this and the descent down the valley d'Oueil into Luchon, see *post* (Sec. 85). For those who can walk or ride, this last is the route most recommended. There is no difficulty in finding the road, though the path for the last 30 minutes is rather indistinct; in mounting the col, keep the stream on your right.

Section 46.
VAL D'AURE.
The Val d'Aure; on foot.

Those who have already, en route to Arreau, passed over the Hourquette d'Aspin, will do well, on returning, to explore the beauties of the Valley d'Aure, of which there is such a glorious view from that col. The luggage must, however, be sent round, as the mountains at the head of this valley can only be traversed on foot. The great drawback is the want of any tolerable sleeping quarters in this valley. The inn at Aragnouet is execrable; and those who are compelled to make their night quarters in this region are recommended to throw themselves on the hospitality of the garde forestier Fourgat; who will give them supper and a clean bed, of course taking care to pay for their lodging. The house of the 'garde,' is near 2 kilometres below Aragnouet, being the first house in Castets, and on the left side as you ascend the valley. By starting early it is possible to reach Héas the same evening, where there is a very decent little inn; but this makes a very long day, with all the hardest work at the close; and really to enjoy this magnificent scenery in going from Arreau, I believe there is nothing to be done but sleep at Aragnouet. In coming the other way it certainly would be advisable to push on down the valley and reach Arreau. The lower part of the Val d'Aure is very populous and highly cultivated. There is a good wheel-road on either side of the Neste d'Aure (*neste* being a Celtic word for gave or stream) as far as the village of **St. Lary**. The principal road is along the left bank of the stream, a mile shorter than that along the right bank; the distance by this road being 5¾ miles to **Vielle Aure**, the principal town of the valley, where the road crosses the stream, and 11 kil. = 7 miles to St. Lary. The road along the right bank does not pass Vielle Aure, but continues on the same side of the river as far as St. Lary. At **Cadéac,** 2 kils. from Arreau, where there is an establishment of baths and a comfortable hotel, there is a bridge across the gave connecting the two roads. At St. Lary there is a French douane; and at the last house in the village, on the left, very good bread, cheese, and other refreshment may be procured. The meadows about St. Lary are quite beautiful and fragrant in early May with the white flowers of the *Narcissus poeticus.* The carriage-road is continued up the gorge along the right bank of the Neste, and after leaving on the left the path that strikes southward to the valley of Rioumajou, and the Port de Plan, passes by the ruined tower and village of Tramesaigues, 2 miles (3 kil.) beyond St. Lary. Just beyond Tramesaigues you pass under a ruined gateway, the remains of an old fort made as a defence for the valley against the Spaniards. High on the left bank of the stream are perched the church and village of Get. Just beyond this the stream is crossed by a stone bridge, and the path from this to Aragnouet is along the left bank. A mile after crossing the Neste, the stream and gorge of the Moudang are seen striking southward to the port of the same name (2,487 metres = 8,159 feet) leading to Bielsa. A mile and a half farther you come upon the hamlet of Castets, where a stone bridge crosses the torrent that descends from the valley of Couplan. By this gorge you may cross over to Barèges in about 9 hours, passing by Lakes Doredon and D'Aubert, and descending by the valley of Escoubous. About a mile farther brings you to **Aragnouet,** 12 kilometres = 7½ miles from St. Lary, and 23 kilometres = 14½ miles from Arreau. Aragnouet, so called from the Aragonese emigrants who founded the place in the middle ages, though a wretched village, at a height of 1,210 metres (3,970 feet), is the *chef lieu* of the upper valley, its principal inhabitant being a French *douanier*, who is well pleased to see any stranger, and relieve by a little gossip with him the dull monotony of his daily life. With the exception, however, of a few Spanish muleteers, there are few strangers who pass this way; there is nothing like an inn, but there are two houses in the place where the traveller may get something to eat, though the fare is not very dainty, and the fleas will take care that he is ready for an early start in the morning. The charge for board and lodging will probably be exorbitant; but I found that this was merely experimental, and the good lady was quite content on receiving about half her demand.

In descending the valley d'Aure, the pedestrian who has slept at Castets may, if pressed for time, pass at once by a mountain path to Luchon without descending to Arreau, and so making an angle. Below Tramesaigues take the upper road on the right; this soon ceases to be a wheel-road, but there is always a good path. Leaving the village of St. Lary considerably below, on the left, as also the villages of Ens and Azet, turn the shoulder of the mountain, and make ESE. for the low grassy col forming the watershed, the Col de Passis (1,576 m.) This col is reached in 2 hours from Tramesaigues; and 1 hour from the top descending WNW. leads to the village of **Genos,** where humble refreshment and a night's lodging may be had. It is better, however, to push on to Luchon. From Genos descend 15 minutes to the river; over which there is a bridge; cross this, and continue the path 30 minutes up the other side of the valley, when you emerge at Loudervielle on the main road, thence to Luchon over the Port de Peyresourde 21 kilometres = 13 miles.

Section 47.
VALLÉE D'AURE INTO SPAIN.

There are several passages into Spain by the long lateral gorges running S. from the Valley d'Aure: those to the E. of the Pic d'Ourdisset leading to the Valley of Gistain, and those to the W. into the valley of Bielsa. The first, the **Port de Plan**, running S. from Tramesaigues through the forest, is practicable for a carriage some distance above Tramesaigues. The scenery is very beautiful: uninterrupted woods for 2 hours, when the path emerges in a verdant little basin with cabanes, whence the gorge of Peguère on the E. leads up to the port of the same name. From here the Pic Lustou (3,025 m.) presents an imposing aspect in the E., and more to the S. is seen the rival mountain, the Pic de Batoua (3,035 metres). Leaving this valley on the left, continue S. 1 hour more brings you to the long wretched barrack called the hospice of Rioumayou. Here the road, which now is only a mule path, bifurcates: the W. branch being carried to the right over the **Col d'Ourdisset**, W. of the Pic of the same name, and thence to Bielsa. [A wheel-road is partly completed on the French side of this port, but it is as yet undecided whether this port or the next westward, the Port de Moudang, is to have the preference as the grand route of communication with Spain. I am inclined to think the road might be more easily constructed by Gavarnie. At each of these three ports the road is in progress on the French side, but on the Spanish side nothing has been done.] The path to the left from Rioumayou takes a S.E. direction up a series of zigzags, and, leaving more to the E. the path which leads to the Col de Caouarére, then passes nearly S. over undulating mounds, half pasture, half rock, to the summit of the dividing ridge (2,457 m.) The passage is so open as hardly to entitle it to the name of a port. From the port the Pic de Posets is in view a little S. of E., but the most striking mountains on the Spanish side are the twin peaks of Los Libones, close at hand in the S.W., connected with the frontier chain by an easy col running N. and S., called the Passo de los Caballos. To the E. an eminence of red schist well deserves the attention of any botanist. It is of very easy access, and not distant half an hour. It is one of the few stations for the *Papaver pyrenaicum*, with red flowers, which I found here in abundance in the middle of August; also some fine specimens of the *Petrocallis pyrenaica*, the *Viola valderia*, and other good plants, From the Port de Plan descend SSE. to the hospice of Plan, which is reached in 2 hours; but do not let the traveller imagine that he can make this hospice his night-quarters, or procure here any refreshment. Though called a hospice, it is only tenanted by surly carabiniers, who give nothing for love or money; and the traveller must continue his course 1½ hour farther down the valley, first on the left, and then on the right, bank of the stream to San Juan, Plan, or Gistain. Of these three, the last affords the best quarters. Inquire for the house of Don Pedro.

From Plan (1,127 m.) in a very long day it is possible to reach Luchon, or, at all events, the cabane at the lake, passing over the Port d'Oo. Start as early as possible, as there are two elevated ports to cross. Proceeding N. from Plan, 15 min. to the village of San Juan, where possibly there are better quarters. 10 minutes from this the path descends to the stream, which it crosses to the left bank. The road mounting from the right bank leads to Gistain. Continue along the left bank, and above the hospice, 1 hour 10 min. from San Juan, cross to the right bank. An hour above the hospice, the path opens out into a grassy plateau; and, leaving on the left the track ascending to the **Port de la Pez**, mount ENE. the hill side, which at first presents a semblance of culture, gradually changing into a series of stony ravines. Remark the *Gentiana cruciata* here abundant in August. Continue E. above the right bank of the stream, which you must again cross on reaching the foot of the col, 1 hour from the top. Here you must make a steep mount to gain the table land above the left bank of the torrent, which is in fact the N. shoulder of the Posets; and after passing a delicious fountain, you reach the col, 6 hours from Plan. The descent lies partly over snow, partly over slaty rock. Make a little N. of east for a red mamelon of rock, with a few stunted trees, at the SE. foot of which is the cabane, 1 hour from the col. A very white bit of rock to the N. marks the commencement of the ascent to the Port d'Oo, which is neither difficult nor dangerous, but very rough. The time from the Paoules to the port is about 2½ hours; thence to the Lac d'Oo, 4 hours.

The gorge of **Moudang**, which opens southwards, 5 kil. above Tramesaigues, presents some grand scenery, especially near its entrance, where it is thickly wooded, opening out after about 2 hours into a wild cirque. Here the chase of the izard and bear may be prosecuted with some chance of success. The two most conspicuous peaks rising from this cirque are the Pic de Lasloue (2,383 m.) and the Pic de Lia (2,775 m.). At the foot of this last rises a remarkable chalybeate source. This spring has lately become the property of an English gentleman, Mr. Maxwell Lyte, a well-known chemist and photographer, who has resided at Bigorre for many years. It is both ferruginous and sulphurous, and gushes in immense abundance from the slaty rock. It is remarkable for its superior medicinal qualities, and especially by the fact that the waters on keeping in bottle retain their virtue and limpidity. It is now imported in large quantities, and is found to succeed in many cases where every other known preparation of iron has failed. In the cirque are some granges and a little house of Mr. Lyte, at a height of 1,556 m. To pass the Port de Moudang (2,487 m.), mount along the left bank of the Chourrious, the stream which comes down from the SSE. The Moudang stream descends on the SW. from the Port de Héchempy.

A third port, still more to the W., is that of **Bielsa**, 2,465 m. = 8,087 feet, one of the easiest and most frequented in this part of the chain leading from Aragnouet to Bielsa, in 7 hours. On quitting Aragnouet, strike due S. There is a mule-track all the way, though in some places faintly marked. Always keep the eastern branch of the stream; that to the W. comes down from the mountains of Baroude. From the port, on the Spanish side, the path is well marked. About 1 hour from the port a fine cascade falling sheer over the rock is passed on the right, and 1½ hours beyond this the path strikes the main valley at the ruined hospice of Bielsa, crossing the stream just above the ruin. A little below this the path from the Port de Moudang converges on the left, beyond which remark, close to the stream, a huge chaotic block, the largest I have ever noticed; then the humble village of Persa, with a post of carabiniers, and 40 minutes beyond **Bielsa.** Hotel in the Plaza Major, with accommodation of the roughest description, but an honest landlord.

From Bielsa, Fanlo on the W. may be reached in 6 hours; or even Gavarnie in a long summer's day, by the Canaou d'Estaubé and the Brèche d'Allanz.

Section 48.
ARAGNOUET TO HÉAS AND THE PIC DES AIGUILLONS.

From Aragnouet, the passage over the mountains may be made either to Gèdre by the Col de Cambiel, or to Héas by the Col des Aiguillons; but if the weather be fine the latter is much to be preferred, on account of the superiority of the view.

A. By the Col des Aiguillons, 8 hours on foot.

On leaving Aragnouet, you cross the stream, and half an hour brings you to the little hamlet of **Le Plan**. Leaving this on your right, ascend the wooded hill on the right bank of the stream, and a hardish tug up this brings you in about an hour to the open pastures on the top. On reaching the corner of the gorge, where the stream turns southwards, you see before you, in the WNW. corner, a snowy depression in the ridge, by which the track leads under **Mont Cambiel,** down to Gèdre. But instead of crossing to this, strike due S. down the valley, and, passing on your right a small basin of water under a rocky buttress, you soon reach the top of the col that forms the extremity of the gorge. From this a climb up the rocks to the WSW., crossing some patches of snow, brings you to the **Col des Aiguillons**. The pic lies a little farther to the south, but it may be reached by passing along the rocky ridge. Keep rather to the western side of the ridge, as the pic is most easily scaled from that side. The height of the Pic des Aiguillons is 9,741 feet (2,969 metres), and from it you have a glorious view of the Mont Perdu, Cylindre, Tour de Marboré, Vignemale, Piméné, and a host of other mountains. I know no spot from which the Mont Perdu especially may be seen to such advantage. From Aragnouet to the Pic des Aiguillons, the time occupied is 4 hours of pretty good walking. In descending from the col, you must first cross over to the other side of the valley, as an abrupt chasm in the rocks prevents a direct descent. In descending to Héas, keep always on the left bank of the stream.

[N.B. This ridge is usually crossed at a spot more to the north of the Col des Aiguillons, called the Hourquette d'Héas. Two small cairns, visible only on the Héas side, mark the passage. If this pass is taken, you must still cross over to the south side of the valley in descending to Héas, so as to have the stream always on *your right*, as the rocks on its right bank are nearly perpendicular.]

3 hours' walking from the Col des Aiguillons will bring you down to the **Chapel of Héas**. As you descend, you may get a peep at the desolate rocks of the Cirque de Troumouse, which, though rather larger, is less imposing than that of Gavarnie. To visit this from Héas, and return, will take about 4 hours. (Section 34.) At Héas there is a little auberge, where you may get a very tolerable supper and a clean bed, if you choose to rest there for the night. From Héas Luz may be reached in 3½ hours, as the road is all the way downhill, and after Gèdre, which is nearly half way, it is a perfectly smooth and good road.

The path as far as Gèdre is very rocky, winding amongst débris fallen from the mountains, and in one place there is a chaos somewhat similar to that on the road to Gavarnie. For the road to Luz see Sec. 23; to Gavarnie, Sec. 33.

The easiest and most direct route from the Valley d'Aure to Gèdre is by the **Col de Cambiel,** which during the late summer is just practicable on horseback; time 8 hours. On leaving Aragnouet, you cross the stream, and half an hour brings you to the little hamlet of Le Plan. 20 minutes beyond again cross the stream, and a hardish tug up the wooded hill on the right bank of the river Badet brings you in about an hour to the open pastures on the top. On these in May there is an abundance of the *Erithronium dens canis*, or dog-tooth violet.

After skirting a little lake, the path descends NW., and recrosses the river Badet by a rough bridge; here there are 2 shepherd's cabanes, the last in the valley. The path winds up the rocks to the W., and thence up a schisty débris to the top of the col, which is rarely free from snow upon the east side.

The descent to Gèdre is long, but presents no difficulty. At the granges of Cambiel, 2 hours below the col (height 1,700 m.), where there is much verdure, but as yet no trees, you have choice of two paths; that on the right bank is a little shorter, but, if on horseback, you must cross the stream, and take the left bank.

At Gèdre, the Hotel des Voyageurs, chez Perissère, affords a clean bed, and comfortable quarters, to any pedestrian who here finds he has had enough. Botanists passing Gèdre will miss an opportunity if they do not pay M. Bordère, the schoolmaster, a visit.

In ascending the Col de Cambiel from Gèdre, be careful to keep to the right as much as possible; the gorges to the left lead up to the Pic de Cambiel, which, though 3,175 m., is very easy of access. The col is not in view till above the cabane of Saoucet, 50 min. above the granges, where there is an excellent source.

From the summit of the Col de Cambiel (2,600 m.) if the weather be fine, and time permits, I strongly recommend the ascent of the peak on the S., the Pic de Salette (2,960 m.) Follow the arête, the ascent presents no difficulty, 1 hour to the top. On the rocks, which are calcareous, intermingled with slaty schist, are some good plants—*Thalictrum alpinum, Saussurea depressa, Papaver alpinum* with white flowers, *Saxifraga grœnlandica, Androsace ciliata,* and *Androsace helvetica.* From the top there is a magnificent view, especially of the Mont Perdu. S. of this mountain, over the Col de Niscle, there is a view over Spain as it were cased in a frame. The rocks are too precipitous to admit of descending to Héas, and a fissure in the arête excludes the possibility of pursuing it westward, so that the return must be made the same way to the top of the col.

Section 49.

COL BADET.

From Castets to Gèdre a very beautiful walk, though not the most direct course, may be made as follows:— Castets to Lac Doredom, up the gorge of Couplan, 3 hours by a new good horse road almost practicable for wheels. One hour above Castets a fine waterfall is seen on the right coming down from the Col de Portet. At the Lac Doredom there are no shepherds before August or after September 15. From Lac Doredom to Lac de Cap Long 1 hour 30 min. Starting from the SW. end of Lac Doredom where there is a cabane, cross the stream coming down from the upper lake; and ascend by a rough path on its left bank to the almost circular lake of Loustallat. On reaching this, cross the stream, and skirt the N. shore of the lake Loustallat. In 10 minutes you come upon the large and long lake of **Cap Long**, the property of M. de Campassens. Coast the S. shore of this at an elevation of about 30 metres above the lake on the granite rocks, and, on reaching the W. end, bear southward, crossing the stream and ascending the rocks, which are very easy and not steep. In 1 hour 30 min. from the lake you come in sight of the glacier E. of the Pic Long, which is remarkable as the only glacier off the main chain, on the N. side of the Pyrenees. The Pic Long (3,194 m.) is the highest peak seen in the SW.; at a little distance SE. from that is the Pic Badet (3,164 m.) almost equally precipitous; and still further SE. is the Pic Cambiel (3,175 m.), very easy of ascent, and commanding a magnificent view. Between the peaks Cambiel and Badet is a col leading over to Gèdre, and communicating with the path of the Port de Cambiel. The height of this col is full 2,800 metres; there is a little lake near it on the east side. The ascent of this col from the E. is very easy; leave the lake on your right, and make for the lowest point. In descending to the W., make a little to the N. at first, taking the rocks 'en traverse.' From Castets to Gèdre by the Lac Doredom and this col is a walk of about 11 hours. N. of the Pic Long, between that and the Pic Néouvielle, is another pass, the **Col de Bugarret** (2,711 m.) leading over to Pragnères.

Section 50.
PIC LONG OR VIERGE.
On foot.

From Gèdre, the mountaineer who loves an unusual and difficult ascent may climb the Pic Long or Vierge, N. of the Port de Cambiel. Martin of Luz, and Marc Sesquet of Gèdre know the route; the latter made the first ascent in 1846 with the Duc de Nemours. Follow the road to the Col de Cambiel as far as 50 min. above the granges. Then turn NE. and traverse the schisty talus to attain the Col de Badet, between the pic of that name and the Pic Long, which, though very high, is not difficult. On the other side the view is limited—a desert of slate and snow—in which is cradled a small tarn. Leave this on the right, and, steering NW., attack the eastern glacier of the Pic Long, aiming at the brèche to the left of the highest pyramid. As you mount, the glacier is more steeply inclined, and towards the top it will probably be necessary to cut steps. On attaining the brèche, you must again pass to the western side of the arête, and here the difficulties are considerable; but a perilous escalade NNW. for 20 minutes, using hands and feet, will place you on the summit, 5 hours from Gèdre. The Pic Long is a pointed granite pyramid, on the west, north, and south-east guarded by precipices, which spring from small but veritable glaciers. On the N. the little lake of Tourrat glitters in the sun some 700 metres below, almost at your feet; and in the NNE. at the foot of the Pic d'Aubert, is seen the opaque-green basin of the Lac de Cap Long. Count Russell descended on the western side of the peak, returning for the first 20 minutes by the same route as far as the brèche, and then descending SW., leaving to the right the torrent of Estibère-Male and the Pic d'Araillé. This route is shorter, but requires, as M. Russell says, '*des pieds sans peur et sans reproche.*'

Section 51.
VALLÉE D'AURE TO BARÈGES.

There are two very beautiful and not over long excursions to be made from the Valley d'Aure to the Valley of the Bastan; and as a guide is not always at hand, a few notes of the routes may be useful.

A. Starting from Aragnouet or Castets; time 8 hours.

Ascend the gorge of the Couplan, following the new road 2 hours 40 min. to **Lac Doredom** (1,870 m.) Cross the dam below this, and skirting the eastern margin of the lake, mount through the fir-wood at its NE. corner. In 30 minutes emerging from the wood, strike N. across a grassy plateau to the **Lac d'Aumar** (2,202 m.), skirt the west shore of this, and you will presently see the **Lac d'Aubert** (2,168 m.) below you on the left. Leaving this below, mount over some granite rocks with stunted firs, rough, but by no means difficult, and then incline to the west, making for the col which is in view. In 5 hours from Aragnouet you attain this, the **Col d'Aubert** (2,500 m.), NE. of the Néouvielle. In ascending this col, you may see fine specimens of the *Anemone verna, Ranunculus amplexicaulis*, and *Erithronium dens canis*, at the end of May. In descending amid snow and rock, the path is scarcely discernible, but steer N., leaving the Lac Nègre on the right, and the Lac Blanc below you on the left. 1 hour 15 min. from the col to the Lake Escoubous; and thence to Barèges, 1 hour 50 min. The path descends from the left or W. corner of the lake; the rocks on the E. side are precipitous.

B. Starting from Vielle; time 8 hours 30 min., 10 hours from Arreau.

From Vielle to Vignec 10 min., whence the ascent lies over grassy hills to the Col de Portet (2,213 m.), 2 hours from Vielle. The shorter path keeps along and above the right bank of the stream, but those who require a guide (quite necessary if early in the year) must cross to the village of Soulam on the left bank, 30 min. above Vignec, where a peasant, by name Aspé, will guide them to the top of the col for 5 francs.

From the Col de Portet, the route descends W. to the stream, and then turns NNW. through the fir-forest of Bastanet, which presents some striking scenery. There is a path all the way, but in the early season it is covered with snow. At the upper end of the gorge you come upon two small lakes, frozen in May, to the right of which rises the Pic de Port Biel. [On the N. side of this peak is a rugged granitic basin containing some 15 lakes of different sizes. This region may be well visited from Gripp or Paillole.] From the two small lakes mounting over the granite boulders by their side, you attain a first col, whence a second is seen a little farther. This, the highest point, is the **Port de Barèges** (2,470 m.) It is reached in 5½ hours from the town of Vielle.

In descending from the col, you pass by the two small lakes of Aigues Cluses, frozen in early May; and beyond these a chaos of granitic rocks presents a fatiguing obstacle, which may be in some measure avoided, by keeping above them. The path then crosses to the left bank of the stream, till at last it falls into that leading down from the Lake d'Escoubous. From the top of the col Barèges is easily reached in 3 hours.

The system of mountains separating the valleys of Barèges and Aure is somewhat complicated; and if there is the slightest chance of the weather being overcast, it is prudent to take a guide or a shepherd who knows the way.

Section 52.

PIC DE NÉOUVIELLE.

The ascent of the Pic de Néouvielle is seldom made; most people contenting themselves with the comparatively tame course of the Pic du Midi. It was first climbed by M. de Chausenque in 1847; and the natural starting-point is Barèges; but by allowing a long day, it may be well combined with the passage of the Col d'Aubert (Section 51). From Barèges 4 hours to the Col d'Aubert. Pass this, and strike due S. to the base of the large snow-field, which always here exists. Without absolutely amounting to a glacier, at the end of August, this snow is partly converted into hard blue ice; and as near the top the inclination is nearly 60°, steps must be cut, or the rocks hugged, as much as possible, to avoid the ice. Having passed the snow, make for the brèche on the W., communicating with the valley of La Glaire; and thence climb the rocks, which are of sound granite, to the summit. The return may be made the same way, traversing the snow in a long glissade, or the route may be varied by passing the brèche in the arête N. of the peak, and leading to the gorge of La Glaire, the Brèche de Chausenque. Thence a rough descent of 1 hour 30 minutes over huge poised blocks, interspersed with snow, conducts to the lake of La Glaire (2,185 metres) at the head of a chaotic valley; where sterility always reigns. This is the last step of a chain of lakes which succeed one another on the N. of the Néouvielle, evincing the bed of some ancient glacier. From the Lac de la Glaire to Barèges, a short 2 hours, principally along the left bank of the stream. This course is not one of danger, or even difficulty, in fine weather; but it is long. 12 hours at least should be allowed, if starting from Barèges, to return to the same point, and 14 if the ascent is to be made in the course of the passage of the Col d'Aubert.

Section 53.

Of all the watering-places of the Pyrenees, **Bagnères de Luchon** is the one possessing most attractions, fully equalling Bagnères de Bigorre in the comforts and appliances of Parisian life, and far surpassing it in the grandeur and beauty of the surrounding scenery. The tourist, although pressed for time, must make Luchon his head-quarters for at least 10 days, a period which, with fine weather, may suffice for some of the principal excursions; but in a stay of 6 weeks he will not tire of it, and will not have seen all. Luchon is divided into two distinct parts, the town proper and the Allée d'Étigny, a fine boulevard, with hotels and lodging-houses on either side, extending half a mile from the town to the bathing establishment, which is a handsome building of white stone at the southern extremity. At most of these hotels and lodging-houses in the Allée d'Étigny comfortable quarters are to be had; but I can especially recommend the Hôtel d'Angleterre, of which M. Seveilhac is the proprietor. The prices of rooms are from 2 to 5 francs per day; and board—that is, two full meals at the table d'hôte, at 10 A.M. and 5 P.M.—is 6 francs per day. M. Seveilhac is exceedingly obliging to English travellers, and is always willing to give them the fair exchange for English money and bank and circular notes. At a place like Luchon this will be found a great convenience. The Hôtel de Bonnemaison (M. Vidal), opposite the gardens of the thermal establishment, is also a large and excellent hotel. I can also strongly recommend the Hôtel des Bains (Madame Merans), which perhaps has the best table d'hôte of any; the Hôtel du Parc, des Princes, and Hôtel Sacarron. These are the best; but there are many others fairly good, and plenty of very comfortable apartments. Those of Bertrand Estrujo, in the Rue de la Cité, are especially clean and comfortable; and Estrujo lets the best horses for a regular ride along the road of any one in Luchon. The Allée d'Étigny takes its name from M. d'Étigny, the Intendant of the Provinces in 1761, who not only constructed this and otherwise embellished the place, but by opening the roads over the Col de Peyresourde and up the valley of the Garonne, made it accessible; and it soon became a favourite resort both for pleasure and health.

The bathing establishment in size and completeness is the first in the Pyrenees. It contains 4 *piscines* or public baths, 100 *baignoires* or separate baths, and one large swimming bath, all of marble, besides douches and shower-baths, *douches écossaises*—' on se soufre de toutes façons.' The reputation of the Luchon waters is increasing every year; the diseases for which they are especially held specific being paralysis, rheumatism, cutaneous affections, sprained limbs, and all kinds of wounds and ulcers. There are upwards of 50 sources, all strongly sulphurous, and varying in temperature from 82° to 150° Fahrenheit. Like all sulphurous waters, the waters of Luchon give out an odour like that of rotten eggs, more or less strong according to the quantity of sulphur contained, which is generally most abundant in those of the highest temperature. The waters of many of the sources present a greasy appearance, which comes from the small flakes or filaments of *barégine* and *sulfuraire* which one sees suspended in the water, and which stick to the body of the bather. At their point of emergence from the rock, the waters of most of the sources are pure and colourless; but some of them, on contact with the air, begin to change, and take a greenish yellow appearance. Others, at first yellowish, gradually become white and milky after they have been exposed for some time in the bath, and again become clear and transparent on the waters of certain other sources being mixed with them. This remarkable phenomenon of whitening (*blanchiment*), which is peculiar to some of the sources of Luchon, is supposed to be owing to the precipitation of sulphur under the influence of the exterior air, and depends in a great measure on the state of the atmosphere, on some days taking place more rapidly than on others. The principal sources are those of the La Reine, La Blanche (which are generally used mixed), and Le Pré, those of the last being the most sulphurous of all, and of a temperature of 140° Fahrenheit. The most favourite time for bathing, as well as for drinking the waters, is before breakfast; and the day is divided into convenient and inconvenient hours, and the tariff of prices arranged accordingly, varying from 70 centimes to 1 franc 20 centimes for a bath, including linen.

It is not easy to assign a date to the first discovery of the waters of Luchon, but it is probable that in very early times the inhabitants of this district found out their virtues, and formed for themselves rude baths by cutting holes in the living rock, for in many places among the Pyrenees holes may yet be seen near the sources, which would seem to have been formed with this object. Certain it is that the waters of Luchon, as also of Bigorre, were known to and frequented by that bath-loving people, the Romans. The 'distractions' of these two places seem to have made them favourite resorts of a people equally civilised, and consequently equally *ennuyée*, with the modern French; and the *thermes* and *piscines* which have lately been constructed are only the renewal of those ancient *thermæ* and *piscinæ* which existed, probably on the same site, under the Roman emperors. Ample remains of these were found in fragmentary inscriptions, marble tablets, and broken columns, when Luchon and Bigorre were re-established by modern luxury; and though most of these have been removed to the museum at Toulouse, there is still left on the spot some classical evidence to encourage the confidence of the modern bather. Just within the principal entry of the thermal establishment stands a votive altar, supporting a Roman vase, with this inscription:—

NYMPHIS AUG. SACRUM.
Dedicated to the holy nymphs;

and on either side of this are two tablets with the following inscription:

NYMPHIS	LIXONI DEO
T. CLAUDIUS	FABIA FESTA
RUFUS	V. S. L. M.
V. S. L. M.	
T. Claudius pays his vow to the nymphs, being cured of his disease.	Fabia Festa pays her vow to the god Lixon, being cured of her disease.

The last inscription is remarkable, as, from the name of the god Lixon (some Celtic deity, to whose tutelage the place was especially commended), the name of the town, Luchon, seems to have been derived. Other marble tablets there are, bearing similar inscriptions, such as—

NYMPHIS	MONTI
AUG......	BUS. QG.
VALERIA	AMŒNIS
HELLAS	V. S. L. M.
V. S. L. M.	

the last being the only one in which allusion is made to the mountains, the nymphs being the beneficent deities to whom the cure is generally attributed. All of these conclude with the same four letters, V. S. L. M., which may stand either for *votum solvit libens meritum*, gladly pays the well-earned offering; or for *votum solvit liberatus morbo*, pay his offering, being freed from his disease.

Of all the Pyrenean waters, those of Luchon are the strongest next after those of Barèges; and certainly, if any one is disposed to make a trial of the sulphur cure, I should recommend this place beyond any other, and have little doubt, if the patient has faith in the waters, he will derive at least as much benefit from them as from any of the German springs. For sprains and external wounds they certainly have real healing powers.

The baths at Luchon, however, are mere adjuncts to its many attractions. The main charms of the place are its scenery, its delicious climate, and the many and beautiful excursions which are so easily made from this spot, situate in the very heart of the highest mountains of the Pyrenees. The excursions from Luchon may be made on foot, on horseback, by *chaise à porteurs*, and some few in carriages. Of horses and carriages, as well as guides, there is an ample supply at Luchon, at reasonable charges; tariffs are affixed at the bathing establishment and at the principal hotels, with the prices for the different courses, and by the day. If the guide, in the course of an excursion, as is sometimes the case, attempt imposition, and strike work for a higher price, it is always advisable to accede to his demands, and report him, on returning, to the Commissioner of Police, M. Jourdain, or the Juge de Paix, who are strictly impartial magistrates, and will take care that he makes restitution. The charge for a horse is 5 francs a day, and for a guide the same, exclusive of refreshment for man and horse; but the guide expects also to have a horse provided for him: a guide on foot is entitled to 6 francs. If the excursion is made on horseback, a guide is indispensable to look to the horses; but if the tourist is sufficiently hale and hearty to trust to his own feet, there is scarcely an excursion from Luchon, with the exception of the ascent of the Maladetta, requiring other guide than a good local map and compass; and the interest of pioneering one's own way will be found not a little to increase the pleasure. 'Vous aurez les jouissances de l'imprévue et vous ferez la découverte du pays. Le moyen de s'ennuyer est de savoir où l'on va, et par où l'on passe: l'imagination déflore d'avance le paysage.' The map of the environs of Luchon, by M. Lézat, copied from the relief plan in the museum, is one that may be depended on. To all the ordinary points of interest there is a well-marked track, quite practicable for horses, and this constitutes one of the principal drawbacks to the enjoyment of this exquisite scenery; for a noisy cavalcade of frivolous Frenchmen, who come galloping down the road, cracking their whips and hallooing, ill harmonises with the solitude and grandeur of the mountains. Of guides there are more than enough at Luchon, and among them some serviceable fellows, with a really good knowledge of the mountains, though seldom beyond their own district. Among these I can recommend the following, who will not grumble at being asked to use their own legs and carry your knapsack:—

Barrau (Pierre), Rue de Pique (also a chasseur).

Haurillon (Jean), active and obliging.

Redonnet (Jean), dit *Mitchot*, Rue de Courtat.

Redonnet (Pierre), dit *Nate*, Rue Miègevielle (somewhat too old, but thoroughly trustworthy).

Lafont (Bernard).

Sors (Jacques).

The above are specially qualified as *guides des sommets*, and not only may be trusted as pioneers over the mountains, but when the day's work is done, will be found cheerful companions and useful aids in preparing the bivouac *al monte*.

I can also recommend Jean Estrujo, Jean Courrège, and Firmin Barrau, the last especially as a porter where a second guide is wanted. Most of the Pyrenean guides have been spoilt by being allowed too much of their own way, and are very apt to assume the position of master rather than of servant; but with an Englishman, especially when they find he is not a novice among the mountains, they are soon brought into their proper place. The charge for all the more ordinary excursions is regulated by a tariff, a copy of which is affixed at the entrance of the *Établissement des Bains*. The usual charge is 5, and in some cases 6 francs for each man and each horse per day; but those courses not named in the tariff are subject to a special arrangement between the employer and the guide. The price, therefore, should be arranged beforehand. For the Maladetta excursion, 30 francs per guide is the established charge; and for any mountain excursion where the guide is required to sleep *sub Jove*, 10 francs a day seems to be required as remuneration, exclusive of provisions. For a course extending over more than 4 days, I should consider 8 francs per day quite as much as ought to be given. As keepers of the best horses, I can recommend especially Bertrand Estrujo, and after him Jean Marie Lafont and Joseph Abadie. At Luchon there are three seasons. The first, from the 15th of July till the 15th of August, is the fullest and most aristocratic. During this time the town presents the gayest possible appearance, many will think too much so. Fine toilettes are all the mode, and the fanciful and elaborate costumes of the male sex, especially the *cavaliers!* are scarcely less conspicuous than those of the women. During this season hotel prices, and especially lodging-houses, are proportionately dear. To this succeeds the season of the French bourgeois, which lasts about another month; and then comes the season of the English; which sometimes with a fine autumn is not the least enjoyable; but the year is too far advanced for appreciating the unrivalled and diversified flora, extending almost to the mountain tops, both on the French and Spanish side, which form one of the main attractions of the Pyrenees. The tourist interested in Alpine botany should explore the mountains about the beginning of June. Later, in August, he can form no idea of the brilliant and beautiful colouring which then enamels the mountain sides. The botanist or geologist making Luchon his head-quarters will do well to seek the acquaintance of M. Fourcade, who lives in the Place de la Croix de la Mission. He is very zealous in the cause of science, and is well up in the botany and geology of the district. He is also in every way an intelligent and agreeable companion, and one who does not mind hard work: anyone bent on a botanising or geologising excursion to Castaneza or elsewhere will find him a great addition to the party, if he can be induced to join. M. Fourcade has also a boutique at the head of the Allée d'Étigny, where he exposes for sale a promiscuous but very interesting collection of dried plants, antiquarian remains, objets d'art, &c.

Those who require literature will find a fair library of books, English and French, at the shop of M. Lafont, 36 Allée d'Étigny, where there is also a good collection of maps, views, and photographs.

Section 54.
Climate of Luchon.

In the tables at the end of this little book, certain facts are given connected with the general climate of the Pyrenees; but each of the Pyrenean ba-gnères has special meteorological phenomena, dependent on its elevation above the sea, the direction of its valley, and other conditions: and as Luchon is not only the principal and gayest of these bagnères, but also the best quarters for seeing the beauties of the Pyrenees, whether we prefer landscape on the mountains, or figure groups in the Allée d'Étigny, a brief notice of its climate may not be uninteresting.

Bagnères de Luchon is situate in $42°\ 45'$ north latitude, and longitude $35'$ east of London, or $1°\ 45'$ west of Paris.

The town of Luchon, though one of the lowest of Pyrenean baths, is placed at a height of 629 metres $= 2,063$ feet above the sea, an elevation rather exceeding that of Madrid; and its mean temperature is in consequence some 6 degrees Fahrenheit lower than that of the plain, the average mean temperature during the summer months being about 1 degree Fahrenheit higher than that of Paris, but not so uniform.

The changes of season from winter to summer, and again back to winter, are very sudden; scarcely any spring or autumn intervenes, the periods of change being the first days of June and the last days of October. The official bathing season comprises five months, from the 1st of June to the 1st of November, of which July is the most preferred, and then August. The greatest heat generally takes place between the 1st of July and the 22nd of August, and these 52 days comprise the finest weather of the year. The midday sun during this period is sometimes very powerful, but it is not oppressive, and the nights are always tolerably cool and fresh. By the table at the end, the exact moment of sunrise and sunset at Luchon may be calculated; but by reason of its position among the mountains the actual continuance of sunshine upon the town of Luchon is much shortened. The chain above Montauban and Juzet intercepts the morning rays, and the mountain of Superbagnères on the west soon shuts out the afternoon sun; so that during the months of June and July the sun is upon Luchon for no more than 11 hours, 10 hours during August, 9 in September, 7 in October, and during the short days of the winter solstice the sun gladdens Luchon only for 3 hours, rising at 10·30 and setting at 1·30.

The mean height of the barometer at Luchon is 27·94 inches (709·9 millimetres), and its extreme range between 27·47 and 28·41 inches (or 698 and 722 millimetres). The elevation of Luchon above the sea—629 metres = 2,064 feet—interferes in some measure with the double diurnal oscillation of the barometer observable on the plain. The maximum attained by the barometer is generally about 11 P.M., and those observations will be found most to be depended upon which are taken about 2 hours after sunrise or shortly after sunset.*

Its mean daily variation, as observed by M. Lézat, is as follows:—

Mean daily height of barometer at Luchon during the months of June, July, August, and September:—

	Inches	Millim.
Morning at sunrise	27·943	=710
Midday	27·927	=709·6
Evening (11 o'clock)	27·947	=710·2

The diurnal variations of the hygrometer generally coincide with those of the barometer; the moisture of the air being greatest at sunrise, and least at midday.

The most prevalent wind at Luchon during the summer months is that from the west, and it is this wind unfortunately which brings the rain and clouds as it sweeps along the chain, saturated with the moisture of the Atlantic Ocean. Now this wind, in its progress eastward, is broken by obstructing points of the chain, and gra-

---

* Sir John Herschel gives the *maxima* of the barometric oscillation at 9 A.M and 10½ P.M.; the *minima* at 4 P.M. and 4 A.M. But he remarks that on mountain stations there is a diurnal oscillation in the barometric pressure of quite a different nature from that above considered. 'The whole vertical column of air from the sea level to the top of the atmosphere being dilated by the increase of diurnal temperature, it is evident that in the hotter portion of the 24 hours a certain portion of the air below the cistern of the barometer must be lifted above it, and vice versâ for the colder. The actual weight of air incumbent on the mercury at a fixed altitude above the sea level must thus be alternately varied in excess and defect of its mean amount. The effect of this cause on the mercurial column is easily calculable for any given elevation and diurnal change of temperature. At a height of 26,000 feet, the total diurnal fluctuation due to a change of temperature of 30° Fahr. would amount to the very considerable quantity of 0·672 inches. The effect at inferior elevations for 10° of diurnal oscillation of temperature is calculated in the following table:—

Altitude in Feet	Diurnal Oscillation in 10° Fahr.	Altitude in Feet	Diurnal Oscillation in 10° Fahr.	Altitude in Feet	Diurnal Oscillation in 10° Fahr.
	Inches		Inches		Inches
1,000	0·022	6,000	0·111	11,000	0·168
2,000	0·043	7,000	0·125	12,000	0·176
3,000	0·062	8,000	0·137	13,000	0·184
4,000	0·080	9,000	0·148	14,000	0·191
5,000	0·096	10,000	0·159	15,000	0·197

These quantities, it will be seen, are quite large enough, at any considerable elevation, to overlay and mask the real diurnal oscillation.'—Herschel, *Meteorology*, ed. 1861, p. 168.

dually becomes dissipated; hence the eastern parts of the chain are much less rainy than the western; and at Mont Louis, the last of the Atlantic clouds are attracted northward along the chain that stretches to the Cevennes, so that, in the Pyrénées Orientales, the only rain that falls is brought by the south and east winds from the Mediterranean.

The quantity of rain that falls annually at four of the principal places on the north side of the Pyrenees is given from observations by Doctor Lambron as follows:—

	Days on which rain falls	Mean quantity that falls in the year	
		Metres	Inches
Pau	129	1·085	42·71
Toulouse	118	0·561	22·08
Carcassone	51	0·728	28·66
Perpignan	70	0·492	19·36

I have not sufficient data for giving the annual amount of the rain-fall at Luchon; but the mean quantity of rain that falls at Luchon during the five months of June, July, August, September, and October, is 13·167 inches = 0·334 metres.

MM. Lambron and Lézat thus give the result of their observations, extending over six successive years, during the summer months:—that of 109 days, from the 13th of June to the 30th of September, there are on an average 38 days completely fine, 52 days more or less covered by cloud or mist, or the two combined, and 19 days completely enveloped in cloud, when the sun is not at all seen; that out of these 109 days there are 38 on which rain will fall, and that on 16 out of these 38 there will be thunderstorms; that of the three summer months, July is on an average the least rainy, September the most rainy, and August the month most subject to thunderstorms; that in the course of this period snow falls on the summits exceeding a height of 6,000 feet 3 or 4 times, and this may happen even in July, the month in which least snow falls on the mountains.

Although so small a proportion of days, only one-third, is absolutely and entirely free from cloud and mist, it must not be supposed that the days suitable for mountain excursions are thus limited. It is only in settled bad weather that these cloudy vapours hang on the mountains during the entire day; ordinarily they begin to form towards 11, 12, or even as late as 1 o'clock, at which hour they begin to gather round and envelop the peaks and ridges of the chain; from these, as they increase in volume, they gradually ascend towards the valleys, though never, even in the most violent storms, do the clouds descend so low as the town of Luchon itself, but are suspended over it at a height of some 300 or 400 feet, where they are condensed.

This periodical formation of cloud towards midday is probably thus to be accounted for. The mists are due to the watery vapour contained in the atmosphere, which becomes condensed when two air strata of different temperatures come in contact; for the mean temperature of the two becoming lower than that of the hottest, the mixture is super-saturated with more vapour than it can contain, and forms mist. Thus we find that on fine days, when the morning sun has made the air of the valleys warmer than that of the peaks, the strata of warm air, becoming lighter, gradually float up the mountain sides, and there encounter the colder strata; the consequence of which is that the vapour with which each was charged exceeds the capacity of the mean temperature of the two, and mists and clouds are formed. which prevail ordinarily during the afternoon portion of the day. The practical result to the excursionist of this periodical formation of midday cloud is to recommend an early start on any expedition among the mountains, especially with a view to the ascent of the higher peaks. He ought always so to time his departure as to arrive on the summit by 11 o'clock at the latest; otherwise there is every chance of finding a portion at least of the mountains intercepted by a curtain of cloud, which is as likely as not to descend in a drenching rain before he gets home, for of the rain that falls at Luchon in the course of a day, at least two-thirds fall in the afternoon.

In the neighbourhood of Luchon there is a peak which forms a pretty good indicator of the probable state of the weather, the Cap de Moncaupé or Signac, situate at the north extremity of the chain which forms the valley of Luchon. Although this mountain is only 1,910 metres = 6,266 feet in height, it stands well advanced from the east, and may be plainly seen from the promenade in front of the bathing establishment. Now, when in the morning there is mist on the summit of Signac, or, as the Luchonnais say, when he has his hat on, you may be sure that before long the mountains of the main chain will be also completely covered, however clear may be the sky, or bright the sunrise. The only exception is when the wind comes from the east, or, still better, if from the south-east (the vent d'Espagne), for this last wind not only has a remarkable avidity for absorbing the moisture, but drives the clouds from the mountains towards the plain. If the mists on Signac are thick and black, they are the precursors of rain as well as of fog on the main chain.

The popular adage on the weather may also generally be trusted:—

    Montagnes clairo,   Mountain clear,
    Bibero escuro,   Plain misty,
    Plujo per segur,   Rain for sure.

July, August, and September are the three months most favourable for seeing the Pyrenees; in the first of these there is the least rain; but for excursions in the high mountains the two latter are to be preferred, as less liable to cloud and mist. This no doubt must be mainly attributed to the snows being then mostly melted, and the rocks thoroughly warmed, and so having less effect in cooling the warm air as it rises from the valleys, and condensing its excess of vapour.

The weather in the Pyrenees is not generally liable to any very sudden changes; after a certain number of days settled fine, there comes a break-up in the weather, and then we may expect some continuous days of storm and rain. The winds which bring the finest weather at Luchon are the north-east and the south-east; it is the latter especially that brings the finest and hottest weather during the summer months. When the wind is south or south-west, the air is peculiarly enervating and oppressive even to the inhabitants; but it never continues many days in this quarter without bringing a thunder-storm, which restores the natural temperature and freshness. Thunderstorms among the mountains are not unfrequent, occurring much oftener than upon the plains, but the peaks which attract the clouds attract also the electricity; so that the valleys are generally pretty safe from the lightning. There is no instance of it having fallen in the town of Luchon; though of course the effect of the storm is heightened by the rolling echoes of the thunder among the mountains. For the same cause also, hail rarely falls in the Pyrenean valleys, though in the plains of the south of France it is often so violent as completely to devastate the agriculture, grievous in its effects as the seventh plague of Egypt.

Section 55.
LAC D'OO.

17 kil. = 11 miles. 4 hours to go, 3 hours to return. Excursion to be made on foot or on horseback, or carriage as far as the Cabanes d'Astau.

	Distance Kil.	Height above sea	
		Met.	Ft.
Luchon		629	2,064
Pont de Mousquères	1·698		
Chapel of St. Aventin	2·762	855	2,805
Junction of the road to the Valley of d'Oueil	·208		
Village of St. Aventin	1·190	950	3,116
Village of Cazaux	1·350	979	3,212
Village of Oo	2·135	930	3,051
Cabanes d'Astau	3·700		
Cabane du Lac d'Oo	3·581	1,497	4,911

16·624 kil. = 10½ miles

One of the first excursions to be made from Luchon is that to the Lac d'Oo on Seculejo, which may conveniently be reached on horseback. Quitting Luchon to the west by the Allée des Soupirs, which leads into the valley of Larboust, *some* 50 *minutes* brings you to a small chapel on the road, dedicated to **St. Aventin**, who on this spot is said to have suffered martyrdom by decapitation, after having been imprisoned in the tower of Castel Blanca which stands on the hill to the right. His body, buried, according to the legend, on this spot, was miraculously discovered by the lowing of an ox many centuries afterwards, in perfect preservation; and in the village of St. Aventin, a little farther on, the rudely carved effigy of an ox may be seen sculptured on the church wall. All mountainous countries are essentially favourable to the growth of superstition, and signally does this appear in the district about Luchon, especially in this valley of Larboust, the population of which is one of the most priest-ridden and superstitious in the whole of France.

The number of crosses and shrines to the Virgin that are seen in every village, and at every turn of the road, may be the effect of genuine religious feeling; but such a chapel and inscription as this to St. Aventin can only be tolerated by gross credulity.

From the Chapel of St. Aventin 25 min. to the village of Cazaux. Half a mile beyond this, on the Bigorre road, is a church of which the walls are covered with some very curious and grotesque paintings. The clock-tower of the church in each of these villages in the Val de Larboust is of a peculiar and picturesque construction.

At Cazaux, 6,412 metres from Luchon, the route quits the main road to Bigorre, and descends to the left through a double hedgerow over a very rough road, across some meadows, which are strewn with rounded blocks of a peculiar kind of porphyritic granite that have probably been carried down by some ancient glacier from the mountains of the Port d'Oo.

From Cazaux to Oo, 35 min. The village of Oo is 9 kilometres (6 miles) from Luchon. Here there is a bridge across the stream, over which the road passes.

From Oo to the Granges d'Astau at the foot of the mountain, where the steep ascent begins, occupies a full hour if on foot. The path is up the left side of a spacious and dreary valley, strewn with boulder-stones, the bed of an ancient glacier, in which there is little to attract the eye. A carriage *may* just be taken, by those ladies who cannot ride, as far as the Cabanes d'Astau, leaving one hour of ascent on foot to reach the lake. The master of this cabane, Lasalle, is an obliging and civil man, supplying refreshment on much more reasonable terms than the fermier at the Lac d'Oo; and in the summer he can generally supply horses and saddles;

so that those who come thus far in a carriage, need not be at the expense of sending on horses, even if they are too indolent to reach the lake on foot. A little above these cabanes, on the opposite side of the valley, may be seen the cascade known as the Chevelure de la Madeleine, and the gorge of Esquierry leading up to the Pas de Couret and Val de Louron. (See *post*, Sec. 61.)

At the huts of Astau a wall of slate rock extends across the valley, damming in the waters of the lake above. Up this the road mounts by a series of zigzags, very steep, but practicable for horses; and in 50 minutes more you emerge upon the **Lac d'Oo**. Just before reaching the top, remark the striated and polished rocks to the left of the path, as a further evidence of the vast glacier which once swept down this valley, and deposited the boulders we noticed on leaving Cazaux. Crossing the stream by which the waters of the lake find their outlet at the northern extremity, you reach the cabane, and pause to admire the beauties of this exquisite scene.

Before you lies the Lac d'Oo, 1,497 metres = 4,911 feet above the sea, a deep dark basin of most cold clear water, fed by the ice streams from the mountains, and shut in on all sides except the north—that where you are standing—by precipitous rocks. Only at midday does the hot sun kiss its waters; and only at midnight does the pale moon glimmer on their bosom. Quite near to the foot of the mountains there is a slight appearance of vegetation, a dull line of grey and brown, then a patch of pale green; but above this the everlasting snow lies summer and winter alike. The only sound is the roar of the very fine cascade, at the far end of the lake, which forms the centre of the picture, and is again reflected in a white streak on the dark mirror of the waters. The circumference of the lake is about 2 miles, its superficial surface 39·16 hectares, and its average depth 75 metres = 246 feet. The cascade causes its current to flow along the bottom of the lake, there being no stream perceptible on the surface. The temperature varies at different depths, being colder at the bottom than at the top of the water, From observations made when the temperature of the air was 57° Fahr. it was 44° at the bottom of the lake, 47° at half its depth (123 feet), and 52° at the surface. The distance across the lake is about ½ mile (740 metres), and it is well worth while to cross in the boat *to the foot of the fall, occupying* 15 *minutes*, as by so doing you form a juster estimate of its magnitude. The whole height of the fall, including the talus of débris at the foot, which prevents its falling into the lake, is 250 metres = 820 feet. Specimens of the rocks—chiefly mica-schist and granite, with tourmaline—may be picked up at the foot of the fall; but those who cross to it must be prepared for a ducking from the clouds of the falling spray. On returning from the fall, the next object will be the repast, which in the meantime has been got ready for you at the cabane, probably lake trout and a well-cooked omelet, the staple fare at all these kind of places; and what more can you desire? The cabane here was built by the commune of Oo. It is let for terms of 3 years to the highest bidder, and the fermier who tenants it for 3 or 4 months pays a sum something exceeding 3,000 francs annually for rent, and makes what he can. In addition to his charges for refreshment, he is empowered to demand certain dues,—25 centimes for every horse, and 25 centimes for every person. The system is a vile and niggardly one; and the commune would consult every one's interest, by securing a tenant who would strictly attend to his duties, rather than surrendering it to the highest bidder. The result is that the object of the fermier is to screw his rent out of the passing travellers, rather than to study their comforts; and especially is this felt to be the case during the third year of his tenancy; for it is a losing concern, and the same man never renews his term.

At this cabane there are two very tolerable bed-rooms, one of them containing two beds; which at this elevation ought to be exempt from the plague of fleas, which infest them. Still all this may be amended under a new regime; and the situation is so charming in itself, as well as convenient for exploring some of the very grandest scenery in the Pyrenees, that the mountaineer will submit to a little imposition, and be grateful for accommodation of any kind at such a spot.

The gorges and mountains around the Lac d'Oo especially deserve the attention of the botanist. In rising from the village of Oo to the frontier chain, the plants present almost a complete epitome of the flora of the central Pyrenees, especially if you include the two lateral valleys of Medassoles and Esquierry, which are justly celebrated as the *Jardin des Pyrénées*. Each plant is confined to its peculiar zone with such regularity that, from a few specimens, it would not be hard to estimate the elevation within a thousand feet.

The following is a list of some of the principal plants which I found in the neighbourhood of the Lac d'Oo :—

3,000 ft. to 5,000 ft. above the sea.
Aconitum Napellus (purple-flowered Monkshood).
Antirrhinum sempervirens.
Aquilegia pyrenaica.
Aspidium fragile.
Astrantia major.
Cistus Helianthemum.
Digitalis ochroleuca.
Euphrasia lutea.
Galeopsis Ladanum.
Geranium phœum.
Helleborus viridis.
Hyoscyamus niger.
Hypericum nummuloïdes.
Meconopsis cambrica.
Scrophularia Scorodonia.
Solanum Dulcamara
Spiræa Ulmaria.
Teucrium Scorodonium.
Valeriana pyrenaica.
Vicia pyrenaica.

5,000 ft. to 6,000 ft.
Bupleurum angulosum.
Centaurea montana.
Eriophorum Scheuzeri.
Gentiana lutea.
Hepatica triloba.
Lilium Martagon.
Meum athamanticum.
Myrrhis odorata.
Oxalis acetosella.
Pinguicula flavescens.
Polygonum viviparum.
Potentilla alba.
          alchemilloïdes.
Ramondia pyrenaica.
Reseda glauca.
    lutea.
Saxifraga adscendens.
    aizoon.
    cæspitosa.
    pyramidalis.
    umbrosa.
Scabiosa columbaria.
Thalictrum Candollei.
Veratrum album.
Viola biflora.
    cornuta.

6,000 ft. to 7,000 ft.
Aconitum lycoctonum (white-flowered Monkshood).
Alchemilla pyrenaica.
Allium Schœnoprasum.
Allosorus crispus (Parsley Fern).
Anemone alpina.
Angelica pyrenaica.
Arnica montana.
Artemisia.
Aspidium Lonchitis (Holly Fern).
Asplenium septentrionale.
Astrantia minor.
Colchicum autumnale.
Doronicum scorpioïdes.
Erica Tetralix.
Eriophorum capitatum.
Geum montanum.
Gnaphalium dioicum.
Helianthemum alpestre.
Homogyne alpina.
Juncus arcticus.

Lonicera pyrenaica.
Orchis nigra.
Parnassia montana.
Phyteuma hemisphæricum.
Polygala angustifolia.
Rhododendron ferrugineum.
Rosa pyrenaica.
Salix pyrenaica.
 reticulata.
Saxifraga aïzoïdes.
 stellaris.
Scabiosa succisa.
Silene ciliata.
Tofieldia palustris.
Trifolium alpinum.

7,000 ft. to 8,000 ft.

Azalea procumbens.
Dryas octopetalon.
Empetrum nigrum.
Erigeron alpinum.
Gentiana verna.
Gnaphalium Leontopodium.
Luzula spicata.
Pedicularis pyrenaica.
 rostrata.
Poa alpina.
Primula integrifolia.
Ranunculus alpestris.
 gracilis.
 gramineus.
 Thora.
Scutellaria alpina.
Seseli montanum.
Soldanella alpina.

8,000 ft. to 9,000 ft.

Androsace carnea.
Arenaria ciliata.
Aster pyrenaicus.
Cardamine alpina.
Gentiana acaulis.
 nivalis.
Leucanthemum alpinum.

9,000 ft. to 10,000 ft.

Cerastium latifolium.
Draba Wahlenbergii.
Hutchinsia alpina.
Linaria alpina.
Oxyria reniformis.
Ranunculus glacialis.
Saxifraga nervosa.
 oppositifolia.
 grœnlandica.
Statice Armeria.
Veronica alpina.

If the weather should not prove favourable, by no means leave Luchon without making a second attempt. Under sunlight or cloud, if seen at all, a visit to the Lac d'Oo will always repay; but the most favourable time for appreciating this scene is on a fine summer's night, when the mountain tops gleam cold in the moonlight, and the twinkling stars are mirrored in the tremulous waters.

> Reflected in the lake, I love
> To see the stars of evening glow,
> So tranquil in the heaven above,
> So restless in the wave below.
>
> Thus heavenly hope is all serene,
> But earthly hope, how bright soe'er,
> Still fluctuates o'er this changing scene,
> As false and fleeting as 't is fair.
> *Heber.*

Section 56.

PORT D'OO AND LAC GLACÉ.

On foot — to Venasque, 11 hours; 8 hours from Lac d'Oo to go and return to Port d'Oo.

Above the Lac d'Oo lies a series of lakes amid very wild and snowy scenery, which are well worth visiting. This expedition is sometimes made in one day from Luchon, taking horses as far as the Lac d'Oo; but as it occupies 16 hours, the start must be made not later than 4 A.M., and it is much better to put up with the accommodation of the cabane at the Lac d'Oo, and sleep there for the night, starting at daybreak.

Mounting a track that is carried up the mountains on the left (E.) side of the Lac d'Oo, about 1 hour of ascent brings you to the **Col d'Espingo**, from which there is rather a fine view of the higher mountains. By traversing the Lac d'Oo, in the boat, and landing to the left of the cascade, you may ascend by a steep path which leads up to the track that skirts the brink of the lake. By this course you save about 15 minutes of walking, but the path is not very easy to find. It lies a little to the left, i. e. N. of a small waterfall which comes down into the NE. corner of the lake. From this col 15 minutes' gentle descent in a NNW. direction brings you to the

Lonicera pyrenaica.
Orchis nigra.
Parnassia montana.
Phyteuma hemisphæricum.
Polygala angustifolia.
Rhododendron ferrugineum.
Rosa pyrenaica.
Salix pyrenaica.
    reticulata.
Saxifraga aïzoïdes.
    stellaris.
Scabiosa succisa.
Silene ciliata.
Tofieldia palustris.
Trifolium alpinum.

      7,000 ft. to 8,000 ft.

Azalea procumbens.
Dryas octopetalon.
Empetrum nigrum.
Erigeron alpinum.
Gentiana verna.
Gnaphalium Leontopodium.
Luzula spicata.
Pedicularis pyrenaica.
    rostrata.
Poa alpina.
Primula integrifolia.
Ranunculus alpestris.
    gracilis.
    gramineus.
    Thora.
Scutellaria alpina.
Seseli montanum.
Soldanella alpina.

      8,000 ft. to 9,000 ft.

Androsace carnea.
Arenaria ciliata.
Aster pyrenaicus.
Cardamine alpina.
Gentiana acaulis.
    nivalis.
Leucanthemum alpinum.

      9,000 ft. to 10,000 ft.

Cerastium latifolium.
Draba Wahlenbergii.
Hutchinsia alpina.
Linaria alpina.
Oxyria reniformis.
Ranunculus glacialis.
Saxifraga nervosa.
    oppositifolia.
    grœnlandica.
Statice Armeria.
Veronica alpina.

If the weather should not prove favourable, by no means leave Luchon without making a second attempt. Under sunlight or cloud, if seen at all, a visit to the Lac d'Oo will always repay; but the most favourable time for appreciating this scene is on a fine summer's night, when the mountain tops gleam cold in the moonlight, and the twinkling stars are mirrored in the tremulous waters.

> Reflected in the lake, I love
>   To see the stars of evening glow,
> So tranquil in the heaven above,
>   So restless in the wave below.
>
> Thus heavenly hope is all serene,
>   But earthly hope, how bright soe'er,
> Still fluctuates o'er this changing scene,
>   As false and fleeting as 't is fair.
>   *Heber.*

Section 56.

PORT D'OO AND LAC GLACÉ.

On foot — to Venasque, 11 hours; 8 hours from Lac d'Oo to go and return to Port d'Oo.

Above the Lac d'Oo lies a series of lakes amid very wild and snowy scenery, which are well worth visiting. This expedition is sometimes made in one day from Luchon, taking horses as far as the Lac d'Oo; but as it occupies 16 hours, the start must be made not later than 4 A.M., and it is much better to put up with the accommodation of the cabane at the Lac d'Oo, and sleep there for the night, starting at daybreak.

Mounting a track that is carried up the mountains on the left (E.) side of the Lac d'Oo, about 1 hour of ascent brings you to the **Col d'Espingo**, from which there is rather a fine view of the higher mountains. By traversing the Lac d'Oo, in the boat, and landing to the left of the cascade, you may ascend by a steep path which leads up to the track that skirts the brink of the lake. By this course you save about 15 minutes of walking, but the path is not very easy to find. It lies a little to the left, i. e. N. of a small waterfall which comes down into the NE. corner of the lake. From this col 15 minutes' gentle descent in a NNW. direction brings you to the

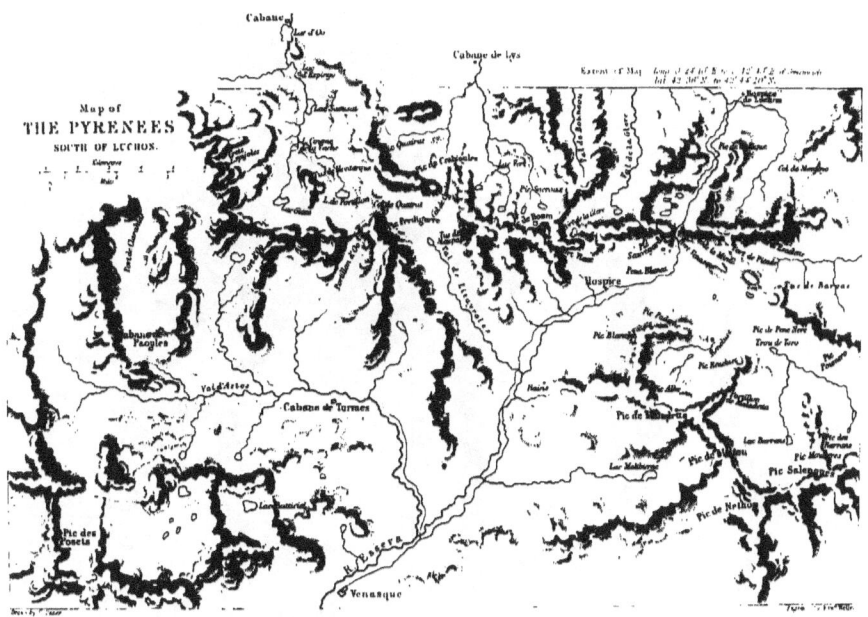

Map of THE PYRENEES SOUTH OF LUCHON.

second lake, the **Lac d'Espingo**, 1,875 metres = 6,152 feet above the sea, a considerable-sized basin of water in a grassy pasturage, which is occupied by goat-herds and cow-herds, who have here some wretched-looking huts. In these huts cheesemaking to a large extent is carried on during the months of July and August, though all the rest of the year they are quite deserted.* It is said that a bed may be procured in one of these huts at a pinch, but it would probably only be a share of one occupied by one or more of these herdsmen, without counting other occupants whose company it would be still more difficult to shake off. In the Lac d'Espingo there are some good trout, but it is the last lake containing any fish, the waters of those higher than this being too cold. Passing the Lac d'Espingo on your right, you come to a rude bridge over the stream; cross this, and 15 minutes over some large glacier-worn rocks on the left bank of the stream brings you to the third lake, on nearly the same level, the **Lac Saousat**, 1,960 metres = 6,433 feet. The mountains above

you on the left are the Crête de Spijoles. Leaving this third lake on the left, a steep ascent over the rocks brings you in an hour to a fourth lake, **Coume de la Baque**, 2,090 metres = 6,857 feet. It is a mere mountain tarn, of less size and depth than the preceding.

Immediately below this fourth lake the stream is divided into two branches, the left-hand branch leading up to the Lac de Portillon, in which it has its rise, and the branch to the right leading to the Lac Glacé, properly so called, though both lakes have always some portion of their surface frozen.

After leaving the fourth lake, mount the steep rocks on the right. A tiny streamlet coming down from the W., confluent with the main stream just below the lake, serves to mark the way; follow the bed of the stream, and mount by a little cheminée on to the upper plateau; whence there is a rough track, not always very discernible, mounting SSW., over snow and bare rock to the ridge overhanging the fifth lake, the **Lac Glacé**, at a height of 2,670 metres = 8,760 feet. The only difficulty is in surmounting the wall of rocks from the fourth lake. In descending especially it requires care to strike the right cheminée.

Descending to the edge of the lake, on its north brink may be seen a vein of argentiferous lead, the working of which has been discontinued for some time. The rock generally is partly of granite, and partly of mica-schist. From the Lac Glacé you may pass over the glacier of the Ceil de la Vache on its eastern border, and so reach the Lac de Portillon in 2 hours. (See Sec. 57.) In order to reach the **Port d'Oo**, you must leave the Lac Glacé on your left, and a toilsome scramble of 50 minutes, partly over glacier and partly over rocks half buried in snow, leads up to the ridge, where there is a slight depression.

---

* The manner in which these peasants combine to carry on the cheesemaking deserves to be noted as a striking instance of the economy of the division of labour and capital. The owners of the cows get credit each of them entered in a daily book for the quantity of milk given by each cow; and at the end of the season, that is about the last day of August, there is a 'distribution des fromages,' when each owner receives the weight of cheese proportionable to the quantity of milk his cows have yielded. By this co-operative plan, instead of the small-sized, unmarketable cheeses only which each could produce out of his three or four cows' milk, he has the same weight in large marketable cheese, superior in quality because made by people who attend to no other business. The cheeseman and his assistant are paid so much per head of the cows in money or in cheese; or sometimes they hire the cows and pay the owners in money or in cheese. One of the most remarkable points in this system of combination is the confidence which it supposes, and which experience must justify, in the integrity of the persons employed.

The barrenness of this icy waste is only redeemed by the green tufts and purple flowers of the *Ranunculus glacialis*, which here grows in profusion. The Port d'Oo is the highest of the Pyrenean passes, being 3,001 metres = 9,846 feet above the sea level. The view from the port is exceedingly wild, but not extensive. Towards Spain the rugged mass of the Pic des Posets, 3,367 metres = 11,046 feet in height, and second only to the Pic de Nethou, blocks in the view; and on the French side the prospect is hemmed in on all sides by the mountains, in one narrow point only extending to the plain. From the Port d'Oo the return to the Lac d'Oo occupies about 4 hours; but those who choose may descend on the Spanish side by the Val d'Astos to the town of Venasque, which may be reached in 5 hours 30 minutes, and, there sleeping, return the next day over the Port de Venasque. (Section 74.) The descent on the Spanish side is not difficult; though the first few steps involve a little climbing. 40 minutes over the rocks in a SSW. direction brings you to a small lake. Skirting this on your right, follow the course of the stream downwards, and 1 hour 30 minutes will bring you to the **Cabane de Turmes**, 1,680 metres = 5,511 feet. At the Cabane de Turmes you must cross to the right bank of the stream. From this spot a rude path carried along the stream brings you to Venasque in 2 hours. The Port d'Oo, though more traversed than the Portillon, described in the next section, is one of the loftiest and most difficult of the Pyrenean passes, and so little frequented that there is no station for the douane either on the French or Spanish frontier, so that provisions must be taken with you on this expedition; and, indeed, on any of the mountain excursions, it is advisable to have with you a supply of bread at all events.

From the Port d'Oo you may also descend upon the **Cabane of Paoules** (2,022 metres = 6,635 feet), higher up the valley of Astos, at the foot of the Port de Gistain. From the lake S. of the Port d'Oo strike SW. across the rough granite boulders, 40 minutes, to a depression in a lateral ridge. Cross this, and continue in the same direction for 30 minutes. A faint track marks the descent from the rocks, and thence 15 minutes over the loose stones to the cabanes, where there are always some Spanish shepherds during the summer months. From the cabanes de Paoules to that of Turmes, 1 hour 30 minutes along the right bank of the stream.

Section 57.
PORTILLON D'OO.

To Venasque 11 hours on foot; and Lac de Portillon 8 hours to go and return from Lac d'Oo.

From the Lac d'Oo to the fourth lake the road is the same as that to the Lac Glacé, occupying 2 hours 30 minutes (Sec. 56). In order to reach the Lac de Portillon, just before reaching the fourth lake, follow up the branch coming down on the left. Always keeping this stream on your left hand, a roughish climb of 1 hour 30 minutes brings you to the **Lac de Portillon**, 2,650 metres = 8,694 feet above the sea. The scene here is magnificent, the glaciers of the Ceil de la Vache streaming into the lake on the S. and SW. sides, and floating on its surface in broken icebergs.

From this spot the **Tus de Montarqué**, 2,933 metres = 9,623 feet high, may be easily ascended in about an hour. The Tus de Montarqué forms the northern buttress of the snowy arête that separates the Lacs Glacé and Portillon, and affords a magnificent view of these two lakes and the adjoining glaciers, on which izards may generally be seen. Not being one of the frontier peaks, it affords no extended view to the S.; but northward the prospect comprises the Monné and mountains on the far side of the valley of Larboust.

In order to reach the Portillon d'Oo, when arrived at the northern outlet of the Lac de Portillon, you must cross the stream, passing over the snow bridge, beneath which the waters are thundering. A very steep little snow slope then has to be surmounted; and when you have gained the rocks at the top of this, the direction is southward over some loose débris of schistose rocks which keep sliding down into the lake. These passed, you come to a moraine; and up this it is a rough scramble till you emerge on the comparatively level and snow-covered plateau which forms the top of the glacier; 10 minutes over this brings you to the **Portillon d'Oo**, 3,044 metres = 9,987 feet above the sea.

To the E. of this gap rises the **Pic de Perdiguère**, height 3,220 metres = 10,564 feet. This pic, though not quite 600 feet higher than the port, is from it nearly inaccessible. On the SE. side it may be ascended without difficulty from the Val de Litayrolles. The Pic de Perdiguère is a noble point of view, and the ascent strongly recommended. Starting from the cabane of the Lac d'Oo, 3½ hours to the Lac de Portillon. Thence mount SSE. over rolling stones, moraine, and snow, and 1 hour 30 minutes of a fatiguing though not dangerous ascent, leaving to the right the glacier and port of the Portillon, brings you to the northern base of the Perdiguère, at the foot of the Col de Perdiguère, called by M. Russell the Col Supérieur de Litayrolles. Here the ascent for an hour is over eternal snow, to the top of the col between the Pic Perdiguère and the Pic Royo. From the top of this col, which is one of the highest in the Pyrenees, (about 3,100 m.) mount the crête to the SSW., 30 minutes suffice to place you on the highest point of the Perdiguère. The summit may also be reached by the Col de Litayrolles, properly so called, more to the N., which is somewhat lower. 5 hours from the Lac d'Oo to this col. Thence 2 hours to the summit of the Perdiguère, leaving the Pic Royo to the right. There is no danger in this ascent, and those who undertake it will be surprised at the extent of the surrounding glaciers. Remark in the SE., cradled in the most savage gorge of the Maladetta, the little-known lake Gregonio, one of the most extensive in the Pyrenees. If too late to return to the Lac d'Oo, or to descend to Venasque, a cabane at the foot of the valley of Litayrolles will afford shelter for the night.

A wine-bottle placed under a rock on the N. side of the port de Portillon contains the barometrical and thermometrical readings of two separate occasions on which the writer passed this way; but it is scarcely ever visited. The vegetation here is confined to four species of plants, the *Hutchinsia alpina*, *Saxifraga grœnlandica*, *S. oppositifolia*, and the *Ranunculus glacialis*, of which both varieties, the white-flowered and the purple, are here to be found.

Looking northwards, the prospect reaches as far as the Monné, on the other side of the valley of Larboust; but its greatest beauty consists in the striking contrasts of form and colour presented by the foreground; the sharp serrated peaks of the warm ruddy rocks, alternating with the undulating sweep of the white snow cols, are each of them relieved against the deep blue of the sky. Turning towards the S. you look full upon the Pic des Posets, with its red arching crest set up in defiance above the snows. The descent on the Spanish side is steeper than towards France, but there are no glaciers, and after traversing some snow beds, too steep for a glissade, you come to a waterfall and stream. Follow this, keeping it on your right hand, and 2 hours from the top of the Portillon brings you to the **Cabane de Turmes,** 5,511 feet above the sea. From the Cabane de Turmes to Venasque, 2 hours. (See preceding section.)

Those who do not care to mount the whole way to Port d'Oo or the Portillon, should on no account, if the weather is fine, omit an excursion to one of these frozen lakes, as there is no spot in the Pyrenees which gives you a better idea of the glacial regions of these mountains. The Lac de Portillon is the one most recommended; though inferior in size, to my mind it surpasses in beauty the far-famed Marjelen-See.

Section 58.
PIC DE CRABIOULES.

Descending to Luchon by Val de Lys;—on foot; 12 hours should be allowed from the Lac d'Oo.

Luchon to Lac d'Oo. (Sec. 55.) Lac d'Oo to Lac de Portillon, 4 hours. (Sec. 57.)

From the Lac de Portillon, for 15 minutes the route is the same as that to the Port de Portillon, but when the snow col is surmounted, instead of striking across to the S., you must continue upwards to the SE., and shortly come to a smooth col of snow, extending from the Pic de Crabioules to the Pic Rouge, the northern buttress of the Perdiguère. This has to be surmounted, and about 1 hour 30 minutes from the Lac de Portillon will place you on the top of this, the Col de Litayrolles. From this point a very steep climb up the rocks on your left will place you in 30 minutes on the summit of the **Pic de Crabioules,** 3,219 metres = 10,560 feet. From the top there is a magnificent view; but on the occasion of my ascent, foul weather and storm in the W. gathering over the Vignemale, the stormiest of all the Pyrenean mountains, warned us not to loiter, and we descended by exactly the same steps to the spot whence we had mounted. From here, after coasting across the snow on the Spanish side above Lac Litayrolles, 30 minutes, we re-entered France, by a depression between the Pic de Crabioules and the Tus de Maupas. Before this the storm had burst upon us in a heavy snow, which passed into rain and a severe thunderstorm as we descended. The descent on the French side is first a series of glissades over the snow; and at the foot of this you come to the source of the stream that forms the series of cascades terminating in the Cascade d'Enfer. From this spot you may descend directly through the forest of Lys, along the right bank of the stream, leaving on your left the series of cascades. But this route is difficult and even perilous, owing to the steepness of the rocks. The better course, though somewhat longer, is, leaving the stream on your left, to pass over to the wooded plateau on the E. side of the valley, which brings you down to the **Cascade de Solage** below the Lac Vert. From this spot the Cabane de Lys is easily reached in less than an hour: the whole time from the Col de Crabioules to the Cabane de Lys being about 3 hours 30 minutes. From the Cabane de Lys to Luchon, 1 hour 50 minutes (Sec. 65).

The descent from the Col de Crabioules into the Valley de Lys is simple and easy enough if you keep the right route, but unites all kinds of dangers if you once get out of it, and I should advise no one to undertake it without a guide.

If, in making this, or the preceding, excursion, the quarters at the Lac d'Oo should be already occupied, or the weather so fine as to make a bivouac 'sous les étoiles' desirable, there is a sort of cave on the mountain near the Col d'Espingo, which will afford very tolerable night-quarters. It is about half a mile to the E. of the pathway up the side of the Lac d'Oo; and to reach it, you must strike off to the left when the path has gained the upper level of the cascade. A horse with provisions may be taken as far as this point, if the expedition is made from Luchon.

Section 59.

LAC D'OO TO VAL DE LYS, AND SO TO LUCHON.

Ten hours on foot.

From the Lac d'Oo another very delightful walk—an excursion somewhat less fatiguing than that by the Pic de Crabioules (Sec. 58)—may be made over the Pic Quairat, returning to Luchon by the Val de Lys; but for this excursion also it is necessary to sleep at the Cabane d'Oo, and fine weather is indispensable. 1 hour 30 minutes to the Lac d'Espingo.

Thence, striking up the mountain to the east, 2 hours of steep but not dangerous climbing brings you to the cairn on the top, which is the point to make for. At this spot halt to contemplate the view over the **Val de Lys**, which lies stretched at your feet, and by descending to it due east, it may be reached in about 3 hours by the **Cascade d'Enfer**. To the NE. is the Pic de Céciré (height 2,397 metres = 7,864 feet), and to the south the still loftier peak Quairat (3,059 metres = 10,036 feet). The peak on which you stand is the Pic Montaroy (2,803 metres = 9,196 feet). This is a very fine point of view, higher than the Céciré, and a favourite haunt of the eagles. This excursion is not much known to the guides, but there is no difficulty if you descend from the col straight to the Val de Lys, passing north of the Tus Rouge. By keeping along this lateral ridge to the southward for a mile, you come to a small lake at the foot of the Pic Quairat, and then, skirting the Glaciers de Crabioules close under the main chain, the excursion may be prolonged, and some very wild scenery explored; but there is considerable difficulty in effecting the descent to the Val de Lys from the Glaciers de Crabioules, which is very abrupt, and there are deep fissures in the mountain, which require to be out-flanked.

The note I made when I passed this way may be useful to any one who is inclined to attempt it:—

'*August* 29.—Having slept and breakfasted at the Lac d'Oo, started from the cabane at 7·10 A.M., Lac d'Espingo at 8·30, the cairn on the top of Montaroy at 10·30. Here halted to rest for 30 minutes; three eagles whirling round me within easy gun-shot. The Val de Lys at my feet seemed so easy to reach that I decided not to descend at once, as too easy a day's work; so, crossing the low col to the east of Pic Quairat, I came down on the south side of a huge red mountain spur, keeping close under the Glaciers de Crabioules, and following downwards the course of the stream, at 1 o'clock reached a wild sort of cirque, resembling that of Gavarnie, with a considerable cascade pouring down from the snow-capped ledges of rock. To follow the course of this stream was impossible, as it is one continuous fall through a series of deep gullies, and, to get round the precipitous fissures which barred my descent, I was obliged to reascend nearly to the top of the mountain on my left. From this it was by no means a "*facilis descensus*" till I dropped down upon the birch woods, where the work became easy, and, keeping to the left of the Cascade d'Enfer, I reached the Cabane de Lys at 5·30, where having dined, I arrived at Luchon (7½ miles along a good road) a little after 8, much tired.'

The **Pic Quairat** may be well ascended by this route, scaling the peak Montaroy from the Lac d'Espingo, and thus attacking the Quairat from the back, on the eastern side, which is more practicable than that overhanging the cirque of Oo. The southern of the two peaks of the Quairat is the loftiest, 3,059 metres = 10,036 feet. It is almost a perfect pyramid in form, hence the name 'Quairat' (*carré*). It is composed principally of that variety of granite, with large oblong crystals of felspar, known as '*actif*' by the French geologists, from its more recent eruption, having bands of schistose and metamorphic rock intermingled. From the summit of the Quairat there is a very fine view of the glaciers and lakes of the frontier chain, over which the crest of the Maladetta can just be seen. The descent can be made by the Valley de Lys, either by skirting the little lake and the Glacier de Crabioules, or more easily by retracing your steps as far as the Montaroy, and passing to the north of the **Tus Rouge**.

Section 60.
PORT DE CLARABIDE AND PORT D'ENFER.
3 Days.

The rocks on the south side of the Port d'Oo and the upper part of the Val d'Astos well deserve a visit from the izard hunter or the botanist, even if the Pic des Posets be not ascended. Passing over the Port d'Oo, or one of the ports to the east of that, an encampment for one night may be well made at the Cabane de Paoules. Thence you may return the next day over the Port de Clarabide by the following route. Making up the col to the westward, the Port de Gistaou, the top of this ridge is gained in about 2 hours without difficulty. A lofty, red, and very savage-looking pic that rears itself in the middle distance intercepts the Mont Perdu; but the well-known shape of the Cylindre is visible due west, and 10° farther north the Pic des Aiguillons. From this spot the Pic des Posets bears 10° E. of S., and the Pic de Nethou 10° S. of E. A narrow ridge is continued northwards from the Port de Gistaou, and, keeping along this, you descend eastward to a source issuing from some beds of snow, where you pass from the red schistose rocks, 'gravier rouge,' the prevailing rocks of these mountains, to the granite boulders of the Clarabide. A very rough transverse scramble over these will bring you to the foot of the Port de Clarabide, and continuing up the west side of this, still over huge granite boulders, in about an hour more you reach the top.

From the top of the Port de Clarabide, by keeping along the 'gravier' at the same level to the east of the gorge, you may reach the Port d'Oo without descending. From the top of the Clarabide the rock of Paoules, the best spot for a bivouac in this locality, is visible at the bottom of the gorge, bearing SSE. To descend from this port to the Val de Louron, keep to the left, i.e. west side of the gorge, where there is a slight track: by so doing you avoid the very rough and wearisome scramble over the granite boulders, which fill the pass from the Lac Poussioles upwards. The Lac Poussioles is situate NNE. of the Port de Clarabide, but lies hid by an outjutting buttress of precipitous rock. To reach this lake, on emerging from the granite boulders, the path winds to the west, and, after making a steep and rather precipitous descent into the head of the valley, continues for a short space along the right bank of the stream, and then mounts to the east under a steep buttress of rock.

On the plateau above the Lac Poussioles the encampment may be made for the second night; but scarcely a stick is to be found for fuel. From the Lake Poussioles an hour of ascent NNW. brings you to a col, from which you descend on a second lake, the Lac Caillaouas, or Cazoas (2,165 m.). Both these lakes abound in fine trout, which readily take almost any bait. They are too distant from a human habitation to have learnt to fear the wiles of the fisherman. At the Lac de Caillaouas, crossing the stream, you again mount NNW. along the west side of the gorge, and after passing from the granite of the Clarabide on to rocks of clay schist, attain the top of a second and higher col. From this point the view is magnificent. Looking south, the Port d'Oo is visible, bearing 10° N. of SE., under this the fine glaciers of the Clarabide, divided by the arête of the Gours Blancs. These two frozen lakes are not quite visible, but bear due SE. From this col to the Port d'Enfer, 50 minutes. For the first quarter of an hour keep N. and nearly on the same level; then a steep climb up the rocks on the right brings you to a gap, which, from its peculiarly savage character, has been named the Port d'Enfer. The view is very fine, but in some respects inferior to that from the last col, some 500 feet lower. Looking S. the crested summit of the Pic des Posets (bearing 10° E. of S.) is very striking, though partly hidden by a peak in the foreground.

Looking N., there is a magnificent prospect over the plain, extending 60° on either side of the true north. The Pic du Gar bears 30° E. of N.; the Monné 4° W. of N.; the Pic Arbizon 42° W. of N.; the Pic du Midi 48° W. of N.; the Pic Badet 10° N. of W. Behind the mountains the Valley d'Aure is well traced in its extent; and at your feet there is a fine view up the Valley de Louron, nearly as far as Arreau. From the Port d'Enfer either Arreau or Luchon may be reached in 6 hours of fair walking, without halts. The descent into the valley of Schourtigas occupies 2 hours. Leaving below, on the right, the little frozen lake of Hourgade, descend along the left bank of the stream. From the cabane of Schourtigas to Loudervielle, 1 hour 45 minutes; and thence to Arreau, 2 hours 15 minutes. To reach Luchon from the Port d'Enfer, at the cabane of Schourtigas, cross to the eastern side of the gorge, and, skirting the hill sides by a not very easy path, you emerge upon the road at the top of the Col de Peyresourde, or, better still, you may mount from the cabane to the Col de Couret on the E., and thence without difficulty descend by Esquierry and the Valley of Astau to Oo and Luchon. The time will be about the same.

Section 61.
LES GOURS BLANCS AND LAC CAILLAOUAS.
On foot, 2 days; guide necessary.

An interesting excursion among the mountains west of the Lac d'Oo, and the upper part of the Valley of Louron, may be made as follows:—
As far as the Lac Caillaouas by the Porte d'Enfer (see preceding route), 8½ hours from Luchon. [In making for the Port d'Enfer, be careful to pass the little frozen lake, and take the gap in the arête, which opens S. An opening a little N. of this conducts down NW. into a gorge barred by precipices, where M. Russell and I once passed a wretched night without food, shelter, wood, or even water through this mistake.] At the NW. corner of

the Lac Caillaouas there is a cabane, where you must pass the night. A scanty supply of juniper and rhododendron for fuel. For the return to Luchon, you have choice of two new roads.

A. By the Val d'Arrougé.

This is the shortest and most easy. A steep ascent up the rocks NE. of the Lake Caillaouas in about 1 h. 30 min. places the climber on the Col d'Arrougé at the head of the valley of the same name. Three serrated teeth on the right (the S. side) mark the passage. From the col descend E. two hours down the gorge to the Lac d'Espingo, thence to Luchon 4½ hours. It is not necessary to descend to the Lac d'Espingo; but when well above it, leave it to the right, and strike across some Alpine pasture, making for the west side of the Lac d'Oo, along which you may descend by a faintly marked track with a rather 'mauvais pas.' A little above the lake, there is a hole by the path in which *Cystopteris montana* grows plentifully.

B. By the Lacs des Gours Blancs.

From the cabane of Caillaouas skirt the south side of the lake, and mount SE. to the first and largest lake, containing 2 islets, 1 hour 15 minutes. 30 minutes farther brings you to the second lake. Beyond this there is eternal snow. You have now due S., dominating the glacier, the triple peaked mountain of **Hermittans** (3,144 m.), and just E. of this, the Pic du Port d'Oo. [To ascend the Pic Hermittans, mount S. from the second Gours Blanc, aiming for the crête between the western and centre peak. Having gained the crête, keep the southern side, and 40 minutes of a rough scramble over the granite rocks (direction E.) will place you on the summit.] Those not intending to mount the Pic Hermittans should steer ESE. across the glacier, which presents no difficulty. An hour will bring them to the col at its eastern extremity, which is a little higher than the Port d'Oo, 3,042 metres. An almost perpendicular wall of snow many metres in height here bounds the glacier. The descent of this is effected, not without some difficulty, on to the glacier of the Port d'Oo, whence the return is made to the Lac d'Oo by the usual route.

Disbelievers in Pyrenean glaciers should arm themselves with cord and axe, and make the course from the Lac des Gours Blancs, eastward. Passing over the glaciers of the Ceil de la Vache, Portillon, Litayrolles, and Graöues, in a course of 14 kilometres, they will scarcely once quit the glacier.

Section 62.
MONSÉGU, ÉCHO DE NÉRÉ, AND ESQUIERRY.
11 hours, on foot or on horseback.

This is a very fine excursion to the lateral chain of mountains dividing the valley of Oo from the valley of Louron, and one that ladies may make without difficulty. Horses may be taken all the way; there is no cabane, therefore it is necessary to take your own provisions.

From Luchon to the village of Cazaux, 1 hour 15 minutes (Sec. 55).

At Cazaux, instead of turning to the left, as for the Lac d'Oo, continue straight on, passing the church with the curious paintings; which stop to see if you have not already done so.

From Cazaux to **Garen**, 20 minutes; 1,250 metres.

The height of the village of Garen is 1,170 metres. Here stop to remark the Moraine de Garen, which rises behind the village, and is the product of that ancient glacier which once stretched down the valley from the mountains of Oo. On examining it, you will find this moraine is composed of rounded boulders and rocks, consisting of a peculiar kind of porphyritic granite, with large crystals of felspar—the same rock of which the mountains of the Port d'Oo are composed, and which is found *in situ* nowhere else in the Pyrenees.

From the moraine to **Gouaoux**, 40 minutes; 2,680 metres = 2 miles.

After the halt made on the moraine of Garen, you resume the main road to the Port de Peyresourde, but shortly quit it by the first turn you see on the left, by the small chapel of San Tritous. In the cemetery of this chapel Roman antiquities have lately been dug up, which may be seen in the emporium of M. Fourcade at Luchon. This path, which has lately been cut out of the rock, brings you in 20 minutes to Gouaoux de Larboust. This village, though perched on the flank of a mountain, is surrounded by fruit-trees; and in summer the valley has a fertile appearance, in spite of its elevation—that of Gouaoux being 1,306 metres = 4,285 feet—very nearly as high as the hospice of Luchon.

From Gouaoux to the slate quarry, 1 hour; 2,800 metres = 2¼ miles.

After leaving the village, follow the course of the stream, passing over some meadows and through a little wood of beech and fir, from which you emerge on a grassy down, where you will shortly see a slate quarry. 15 minutes beyond this you come upon a shepherd's cabane and a little lake, the **Lac de Soubiron**; 450 metres beyond this lake, on the other side of a little col, there is a spring of excellent water, where you will do well to halt and refresh. From this fountain, 1 hour and 15 minutes up the mountain side, over short and slippery turf, brings you to the **Summit of Monségu**, an elevation of 2,403 metres = 7,884 feet. The ascent is not difficult, and horses can reach the very top. From the summit, which is marked by a pile of stones indicating the boundary of the Haute-Garonne and the Hautes-Pyrénées, there is an extensive prospect of mountains. Most conspicuous among them are the savage-looking peaks to the south, from which a deep ravine separates you. These are the **Pics Nérés**, or Noires, of which the word Pyrenees is said to be a corruption. The deep gorge riven in the centre of this group is called the **Coumo de Néré**. Here, amid the eternal snows, are cradled two lakes; and from these pour two fine cascades, which are well seen from Monségu by descending a few steps towards the valley of Louron, the valley to the west. Of the near group of mountains, the summit on the left, on which a pyramid of stones may be seen, is the **Pic Néré Nord**; that to the right, more retired to the central chain, is the **Pic Néré Sud**. From the Pic Néré Nord, the buttress projecting to the left, into the Val d'Esquierry, is known as the **Pointe Arnaud**. Between the north and south Nérés several other summits are seen, the most conspicuous among which is the **Pic Rouget**. To the right of the Pic Néré Sud rises the grand Pic de Fourcade, more to the right the Pic d'Estiouère, and more still to the right the Pic de Soulan. Yet more to the right are the Pic de Clarabide and its glaciers, and the lateral chain of mountains which separates the Val de Louron from the Val d'Aure. To the left of the Pics Nérés appear the still grander mountains of Oo, and the Portillon hiding the Maladetta, which is behind them; then come the Pic de Crabioules and the Pic Quairat. On the other side, conspicuous among the lower eminences, rises the Pic de Monné, due north.

From Monségu to the Écho de Néré, 25 minutes.

Following the acclivity of the mountain due south, along the boundary of the two departments, in 20 minutes you reach the limestone rocks that overhang the Val d'Esquierry, that lies below you carpeted with all manner of rare plants and flowers. From these heights you see two little lakes without any outlet: one on the right, on the side of the mountain opposite; the other to the left, at the foot of the Monségu. In the midst of these crags, at a point where the bleak schist rocks rest on the white limestone, and just where a massive vein of quartz is interposed between the two, there is a remarkable echo, which will repeat a loud cry eight or nine times! to produce the greatest effect, the voice should be directed towards the Pic de Néré Nord. From this spot it takes 5 hours to reach Luchon, whether you return by the same way or, as is preferable, descend by the Val d'Esquierry, which, from the number of curious plants that grow here, has been justly called the Jardin Botanique des Pyrénées. Among the treasures of Esquierry the *Aster pyrenaicus* deserves especial notice, both from its size and rarity. I know no other spot where it is found. By the Pas de Couret, the col at the head of the Val d'Esquierry (2,131 metres = 6,992 feet), you may pass over into the upper valley of Louron.

From the Echo to the Cabanes d'Esquierry, 1 hour.

In descending from Monségu to the valley, it is advisable to descend from the horses, if mounted, and let the man lead them *in front*. At the bottom you will see a spot where a stream that comes down from the Pic Néré Nord disappears beneath the surface, and only shows again at the lower end of the valley, which owes its fertility to this underground irrigation. The soil seems to be composed of a blackish clay marl, very dirty and slippery.

From the Cabanes d'Esquierry to the Pont Ste. Cathérine, 1 hour 30 minutes.

Continuing down the valley, you cross the stream at this bridge, and find yourself in the road that leads up to the Lac d'Oo at the PLAN D'ASTAU.

From the Pont Ste. Cathérine to Oo, 1 hour.

From Oo to Luchon, 1 hour 45 minutes (Sec. 55).

Section 63.
SUPERBAGNÈRES.

7 hours easy, to go and return, allowing 1 hour on the top; horses may be taken.

A very beautiful view may be had on a clear day from this inferior summit, which may be reached in 3 hours, passing through the hamlet of **Gouron**; and the ordinary route is to continue along the road as far as the village of St. Aventin, but a much shorter route is as follows:—In ascending Superbagnères, through the hamlet of Gouron, leave Luchon by the Allée des Soupirs. Just before reaching the Pont des Mousquères, take a path to the left that leads across some irrigated and rather too moist meadows to the hamlet of Gouron. From the Pont de Mousquères to Gouron, 2,200 metres. From the hamlet of Gouron the path makes a sharp turn to the SE., and mounts through the forest of Artigues Ardonne, whence it emerges on the grassy pasturage forming the summit of Superbagnères.

It is also quite practicable, but only on foot, by ascending direct from Luchon, from the garden of the thermal establishment; by this route you may reach the top in little more than 2 hours. Follow the path leading up to the little chalet known as the Fontaine d'Amour (763 metres), leaving on your left the rival establishment of the Chaumière (755 metres), which is a little more to the south. From either of these two stations, which may be reached in 20 minutes, there is a charming view of the valley of Luchon.

The summit of Superbagnères is 1,797 metres = 5,895 feet above the sea, being the eastern buttress of the Pic Céciré, which may be easily reached from Superbagnères, but the excursion will be lengthened a good two hours. None of the little excursions around Luchon affords so fine a point of view as this; and if mists interfere, it is well worth making a second trial. The *return* is best made by the **Val de Lys**, in 3 hours. In some seasons a cabane is established on the top of Superbagnères, where refreshment may be procured; but this should be ascertained beforehand.

Section 64.
PIC DE CÉCIRÉ.

11 hours to go and return; horses may be taken, though in places they must be left.

If the Pic de Céciré is not reached from Superbagnères, it may be made the object of a separate expedition by the following route:—

From Luchon to Cazaux, 1 hour 15 minutes. (Sec. 55.)

From Cazaux to Medassoles, 3 hours.

After passing the village of Cazaux, cross the stream by the bridge, and a steep zigzag path leads up some meadows to the hamlet of Labach Cazaux; thence strike due south, keeping the watercourse that flows down the gorge on your left hand. There is a track marked, which conducts you over the mountains, at the upper end of the ravine, into the **Gorge of Medassoles,** which faces the Val d'Esquierry, and, though less known, is not less rich in botanical treasures.

From the top of the gorge of Medassoles 1 hour will place you on the summit of the **Pic Céciré**, 2,400 metres = 7,864 feet above the sea. The summit is most accessible on the southern flank, passing by an excellent spring (La Coume de Bourg), 20 minutes below the pic. If with horses, they must be left here. If the sky is clear, you may rest assured of a view that will repay you for the labour of the ascent. The mountain that you dominate is Superbagnères, and you have nearly the same prospect. To descend, you have choice of three routes in the Val de Lys, 3 kil. above the Pont Ravi:—A. By keeping along the arête to Superbagnères, which is most recommended; B. By leaving this arête on your left, and following the stream down to the hamlet of Castillon (this is the easiest route); C. By crossing the ridge south of the fountain of the Coume de Bourg, and following the stream down this gorge, which brings you through the woods to the Cascade d'Enfer, at the head of the Val de Lys.

From the Pic de Céciré to the Cascade d'Enfer, 3 hours.

From the Cascade d'Enfer to Luchon, 2 hours. (Sec. 65.)

Section 65.
VALLEY DU LYS AND CASCADES D'ENFER.

6 to 7 hours to go and return; good road. Horse or carriage may be taken; prices fixed by tariff.

	Distance Kil.	Height Met.	Ft.
Luchon		629	2,064
Tour de Castel Viel	2·112	772	2,532
Pont Lapadé	1·250		
Pont Ravi	1·600	834	2,736
Gouffre de Bouncou	} 1·514		
Cascade de Barrié			
Cabane de Lys	4·470	1,104	3,622
Foot of Cascade d'Enfer	0·390	1,112	3,648
	11·336 kil.=7 miles		

From Luchon to Castel Viel, 35 minutes.

Castel Viel to Pont Ravi or Arrabi, 25 minutes.

Pont Ravi to Cabane du Lys, 1 hour 20 minutes.

Leaving Luchon by the route to Venasque up the valley of La Pique, half an hour brings you to the ancient and picturesque tower of Castel Viel on the right of the road; near this spot are some manganese mines wrought by an English company. 1½ mile beyond this the road crosses the stream, and after traversing its right bank for 1¼ mile recrosses it at the Pont Ravi, where it turns sharp to the right up the valley of the Lys, the road straight on leading up the valley of the Pique to Venasque. From this point there is a good view to the SE. of the mountains overhanging the Port de Venasque, the most prominent being the Pic de la Pique. 15 minutes from the bridge brings you to the Cascade Barié, where the stream falls into a deep gorge, and farther on a path to the left conducts you to a second fall, *the Cascade Richard*. For the first half-hour the road up the Val de Lys passes under the foliage of beech-trees, chestnuts, and hazels, which, with their roots clinging to the huge blocks of rock, overhang the road with a pleasing shade. All these masses of rock consist of varieties of eurite, a species of granite composed almost wholly of quartz and felspar; and in many of them the polished surfaces and horizontal striæ may be seen, denoting the action of glaciers, as in the valley leading up to the Lac d'Oo. After leaving the Cascade Richard, the stream becomes less turbulent, and the valley opens into a green and pastoral basin, overlooked on all sides by lofty mountains, which at the head of the valley present an appearance especially imposing. From a gloomy abyss in the very heart of the mountains, amid scattered trunks of trees and riven rocks, bursts a torrent that well deserves its name—the **Cascade d'Enfer.** Above this, and on either side, rise the gloomy fir forests, tier upon tier, higher and higher, till lost in the clouds which descend to interrupt, but not to close, the picture, for high above all, through a break in the sky, gleam the white glaciers of the Crabioules. They seem so far raised above the solid earth that it takes some moments to be convinced it is not illusion; and on my first visit here, I could scarcely credit that it was down these precipices I had descended when I crossed from the Lac d'Oo three days before. The height of the Cabane de Lys above the sea is 3,622 feet, so that the descent from the Crabioules glaciers is some 5,000 feet.

From the Cabane to the Cascade d'Enfer, 10 minutes; to the Pont Nadie and back, 2 hours.

Above the Cascade d'Enfer are a series of cascades and bridges which may be visited on foot in from 2 to 3 hours, and are well worth seeing, and may be visited while your horses are baiting or your own déjeuner preparing at the cabane, which affords very tolerable fare, without great imposition. The route is as follows:—After admiring the Cascade d'Enfer from below, ascend to the right by a zigzag path through the trees and brushwood, and in about 25 minutes you reach the second bridge. Do *not* pass this, but descend for 5 minutes by the path on the left bank of the ravine to the first bridge, the Pont d'Arrougé, thrown over the Cascade d'Enfer. Here pass over to the right bank of the ravine, and, again ascending, you leave the second bridge on your right; in 10 minutes from this you reach a sort of little tower, from which you have an admirable view of the Gouffre d'Enfer, the second cascade.

In order to get a view of this from the top, continue to mount by the slight track just discernible, still keeping on the right brink of the gorge, as far as the third bridge, the Pont Nadie, whence you will have a view that will quite repay for the toil. The left (i.e. W.) bank of the stream is here impracticable; but a little track, carried up through the forest, very easy to miss, leads to the Parc, and **Rue d'Enfer**, a frightful chasm, resembling a Cumberland ghyll, but choked with unmelting snow. To descend, you must retrace your steps along the right bank as far as the second bridge, which cross, and so return to the cabane, along the right bank, by the same zigzag path by which you ascended.

From the Cabane de Lys to the Cascade de Cœur, 40 minutes to go and return.

To reach this cascade is a much less fatiguing but a less interesting walk than the preceding one. Cross the stream by the wooden bridge, and follow the path mounting through the forest to the Lac Vert. The Cascade de Cœur is formed of two streams, whose waters, crossing each other, described the shape of a heart. An avalanche of late years has destroyed this combination, though the name still remains.

From the Cabane de Lys to return to Luchon, 2 hours on foot.

Section 66.
LAC VERT, OR LAC D'ILE.
On foot or on horse ; 10 hours.

To Cabane de Lys (Sec. 65) 2 hours
    Lac Vert . . . 2 „ 30 min.
                                    4 hours 30 min.

From the Cabane de Lys the path crosses the stream by the wooden bridge and winds up the forest, passing the Cascades de Cœur, Solage, and other fine waterfalls, the last being the Cascade Tregons, which issues from the Lac Vert.

One hour from the Cabane de Lys to the Cabane d'Artigues, tenanted by some bergers, where the path crosses the stream, and continues its direction over stony débris to the head of the valley, known as the Cabane de Quiou. Here it turns sharply to the right to cross the rocky ridge on its S. side, from which it descends by the side of a little rush-covered pool into an open basin strewn with gravel boulders, the Cirque de Graouès. Cross the stream, and, striking W. up another little col, you see at your feet the **Lac Vert**, which may be reached on horseback by this route.

The Lac Vert, 1,960 metres = 6,431 feet, is a pretty basin of green water, on the N. flank of the Tus de Maupas. From the little island on its surface, it is also known as the Lac d'Ile. The cascade that pours down into it on the south side issues from an upper lake known as the **Lac Bleu**. To reach this second lake, you must scramble over the loose schists and beds of snow to the left, i.e. the E. side of the Lac Vert.

Before reaching the Lac Vert, a cascade is seen pouring down over the southern brink of the **Cirque de Graouès**. This is the overflowing of an upper lake, the **Lac Charles**, situate at the foot of the glacier of the Pic de Boum. It is a very wild spot, and may easily be reached in a climb of 25 minutes up the flank of the mountain shutting in the Lac Vert on the E. From this you may pass to some still higher, semi-frozen lakes at the foot of the Pics Mal Pintat and Mal Barré—the Lacs Pichis des Graouès, Glacé du Port Vieux, and Graouès—and so return by the flanks of the Pics Estaouas and Sacroux ; but this is a rough scramble, and a guide who knows the locality ought to be taken.

Return from Lac Vert to Luchon may be made by three routes.

A. By the route already described; the easiest and the only one practicable for horses.

B. By the Col and **Val de Bouncou**, as follows:—After recrossing the ridge that shuts in the basin of the Cirque de Graouès, mount the col to the E. 30 minutes, and then, striking down the valley NNE., 2 hours 30 minutes will bring you down to the lower part of the Val de Lys opposite the Trou de Bouncou, 10 minutes above the Pont Ravi, and only 1 hour from Luchon.

C. By the **Parc** and Cascade d'Enfer. Of the 3 routes this is the most difficult, and requires 3 hours to reach the Cabane de Lys, as follows:—Skirting the base of the Tus de Maupas, 30 min. brings you to the Lac Noir, a small piece of water on the table-land, known as Prat Lounge, deriving its name and colour from the freshwater algæ on its surface. 40 minutes from this, by the path which strikes through the forest, and not by that leading upward to the glaciers of Crabioules, brings you to the **Parc d'Enfer**, with its *rue* and cascade; which comes direct from the glaciers of the Crabioules. This *rue* is a deep gorge which the stream has channelled in the rocks, and of which the sides are as precipitous as the houses of a street. From this, 50 minutes of steep descent through the wood, always keeping on the right bank of the stream, brings you to the Pont Nadie, the uppermost bridge; whence to the Cabane de Lys 1 hour of comparatively easy descent. (Sec. 65.)

A little way above the Pont Nadie you pass the Pas de l'Ours; the only practicable spot for the bears, who sometimes cross the gorge from the forest on one side to that on the other. And here the chasseurs often lie in ambuscade for them, and sometimes succeed in killing one.

Section 67.
TUS DE MAUPAS.
3,110 metres = 10,203 feet.

On foot; 14 hours to go and return.

Lac Vert (Sec. 66)	4 h.
Lac Bleu	40 min.
Summit of Maupas	3 h.
	7 h. 40 min.

For this excursion as far as the Lac Bleu, see *ante*, Section 66. From the Lac Bleu mount the arête extending from the Maupas; and thence on to the glacier, which presents no difficulties, being of a gentle slope, and not much crevassed. The only difficulty is presented by the last arête after leaving the glacier, which is a steep bit of rock. The summit is of granite and clay schist intermingled. The return must be made to the Lac Vert, and thence to Luchon by one of the routes described in Section 66.

Section 68.
PIC DU BOUM.
3,060 metres = 10,039 feet.

On foot; 14 hours to go and return.

This is another conspicuous peak of the frontier chain, to the east of the Tus de Maupas, between that peak and the Pic Mal Barrat. It is one of the most difficult of the Pyrenean peaks, and was first ascended in 1858 by M. Lézat and three others, with the guides Pierre Barrau, Redonnet, *alias* Michot, Bernard Lafont, and Jean Estrujo. The rock forming the summit is accessible only on the eastern side; but on that side the glacier below is so steeply inclined and crevassed that the best route is as follows:—

To the Lac Bleu the route is the same as that to the Maupas. (Sec. 66.)

From the Lac Bleu strike up the rocks towards the W., and 1 hour from this lake places you on the glacier.

Traversing the glacier from the NW. to the SE., in 30 min. you reach the foot of the peak de Boum on the E. side. The difficulty consists in getting up these rocks, especially at the finish, but 30 minutes' perseverance will overcome these, and place you on the top. In crossing the glacier of the Boum, M. Lézat observed the phenomenon of red snow (*Protococcus nivalis*).*

Section 69.
PIC SACROUX.
2,678 metres = 8,785 feet.
On foot; 12 hours.

This peak, also on the frontier chain, though advanced upon the French side, is of easier access than the two preceding ones, and affords one of the finest summit views in the Pyrenees. It may be ascended either from the Lac Vert, or the port de la Glère, going by one route and returning by the other. From Luchon to the Port de la Glère (Sec. 72). Arrived at the top of the Port de la Glère, turn to the right, and mount, over some gently sloping beds of snow, to the col separating the two valleys of la Glère and Graouès, the port d'Estaou, height about 2,500 metres; 20 minutes to top of col, and from this point a climb of 30 minutes up the schistose rocks of the Pic Sacroux will place you on the summit. It is no great effort, but the schists are rather treacherous and slippery, and require care.

The view is magnificent, but the town of Luchon is just hid by the projecting shoulder of Superbagnères, the Mail de Soulans. To descend, you must return to the Col de Graouès,

* M. Lézat computes the height of the Boum only at 2,910 metres=9,548 feet. His data for this computation are the following barometrical observations which he made: Luchon, barometer 711 mm.=27·993 inches, thermometer 27° Cent. Summit of Boum, barometer 547 mm.=21·538 inches, thermometer 13° Cent. But I think the trigonometrical observation by MM. Reboul and Vidal is more correct.

and then, continuing your course SW., wind obliquely round the flank of the Pic Estaouas; there is a faint track, and the road is not at all dangerous, though at first sight it would appear so. [By continuing in this direction, you may reach the **Port Vieux**, a difficult and rarely used passage into the Val d'Essera, situate between the Pics de Port Vieux and Mal Pintat.] In order to descend to the Lac Vert, after skirting the flank of Estaouas, you again turn towards the N., leaving beneath you on the left the Lac Charles, and work a zigzac course down the shoulder of the mountain towards the Crête de la Serre des Cabales, which separates the valleys of Graouès and Bouncou. Here you have choice of two routes, either to descend into the Val de Bouncou and so to the Val de Lys, or, descending from the col to the W., 20 min. brings you into the head of the valley just below the Lac Vert, at the Cabane de Quiou. From this spot to Luchon by the Cabane de Lys 3 hours (Sec. 66).

Section 70.
LES QUINZE LACS.
On foot; 2 hard days; guide necessary.

Though this expedition is always called Les Quinze Lacs, I cannot make out more than 14; and of these, two, though seen, are not actually approached. The tour among the 15 lakes leads among the most savage scenery of the high mountains of the frontier. It is in fact the uniting of the ascents of the Pics Sacroux and Quairat into one excursion, extending over 2 days. It is somewhat fatiguing, and entails a night bivouac in a wretched cabane at the foot of the Glaciers de Crabioules, which affords but poor and comfortless sleeping quarters; a sleeping bag on this expedition will be especially appreciated. You may commence either on the side of the Lac d'Oo, or of the Port de la Glère; but the latter is recommended, as doing the hardest work the first day. The chief supply of provisions, sleeping bag, &c., should be sent on to the Cabane de Crabioules by the Val de Lys, to await your arrival in the evening. The course is as follows:—

First Day.

	A.M.
Departure from Luchon (on foot) . at	5
Fountain at foot of the Glère (breakfast)	8.30
Port de La Glère	10
Lac de Gourgouttes . . .	10.25
	P.M.
Pic Sacroux	1
Lac de Graouès	1.45
Lac Glacé de Port Vieux . .	2.25
Lac Charles	2.45
Glaciers at foot of Le Mal Plané	3
Pas des Crabes (le Pas des Chèvres) (difficult passage) . . .	4
Lac Bleu	4.15

	P.M.
Lac Vert and glaciers of Tus de Maupas . . . . at	4.30
Lac Noir	5
Glacier de Crabioules . .	6.35
Cabane de Crabioules (night bivouac)	7

Second Day.

	A.M.
Departure from Cabane de Crabioules	1
Lac Glacé de Crabioules . .	2
Col du Quairat (separating the Pic Quairat from the Tus de Crabioules, view of Lac Litayrolles)	3.30
Summit of Quairat (sunrise, view of Lac Glacé) . . .	4.30
Descent from Pic Quairat (after breakfast)	6
Lac Glacé de Portillon (view of) .	6.30
Fourth Lake	8
Lake Saousat	9
Lac Espingo	9.30
Lac d'Oo	10.30
	P.M.
Departure from Lac d'Oo . .	2
Luchon	5

Section 71.

CASCADE DES DEMOISELLES, AND CASCADE DES PARISIENNES.

Return by the Hospice de Luchon; 6 to 8 hours on horse, or foot. Guide not necessary.

A pretty little after-breakfast excursion, to be made without any fatigue, and one that admits of '*temps couvert.*'

From Luchon to Pont Ravi (Sec. 65) . . .	4·962 kil.	1 h.	
Pont Ravi to diverging path	1·315	20 min.	
Diverging path to Cascade des Demoiselles .	1·650	30 min.	
Cascade des Demoiselles to Cascade des Parisiennes	3·000	50 min.	

10·927 kil. = 7 miles 2 h. 40 min.

At the Pont Ravi do not cross the bridge, but continue straight on, along the Venasque road, passing on the right the Granges de la Bach, in about 20 minutes. A little short of the entrance to the wood, a road turns off to the right; follow this, and after descending to the stream of La Pique, which you cross by a bridge, you traverse a fertile meadow (Le Prat de Jomon), and, steering S. among the trees, you shortly reach the Cascade des Demoiselles, 7.927 kil. = 5 miles from Luchon. Here the shady trees and moss-grown rocks present a delicious retreat on a hot summer's day; and the numerous plants here to be found are a further attraction to the botanist.

From Cascade des Demoiselles to Cascade des Parisiennes, 3 kil. = 2 miles; 50 min.

Returning from the Cascade des Demoiselles to the stream of La Pique, follow its course upwards, keeping along the left bank, and a delightful shady walk, under the trees of the forest of St. Just, will bring you to a series of little falls, 10 in number, most charmingly disposed among the rocks, the Cascade des Parisiennes.

From Cascade des Parisiennes to Hospice (1,360 metres above the sea), 1 kil.; 15 minutes.

Follow up the stream of La Pique, by a good horse path, newly constructed along its left bank.

From Hospice to Luchon by Pont Ravi, 10·422 kil. = 6½ miles; 1 hour 50 minutes.

The road from the hospice to Luchon is an excellent carriage-road, carried first among the noble beech and fir trees of the forest of Charuga, and, 3·260 kil. = 2¼ miles from the hospice, passing over the bridge of La Cascade Courrège.

Section 72.
PORT DE LA GLÈRE.

On foot or horseback. 9 hours to go and return. 9 hours to Venasque. Guide not necessary.

The road is the same as that to the Cascade des Demoiselles, till the last 15 min. (Sec. 71). After crossing the stream of La Pique, the path bifurcates: that to your left leads up to the cascade; to the right is the one to be taken for the Port de la Glère.

From Luchon to this point at the entrance of the forest 7·797 kil. = 4¾ miles, 1h. 40m.
From this point to the Port de la Glère . 8·318   5¼   2h. 40m.

16·115 kil. = 10·01 mil. 4h. 20m.

Five minutes after passing the stream of the Pique, the path winds to the right, and, after passing over a stony waste, devastated by this torrent, mounts among the trees. From the foot of the forest, the path keeps on the left bank of the stream for 40 minutes, when it crosses the stream, and in 20 minutes more you emerge from the forest on an open pasture, where there is a solitary chalet. By the side of the path, after crossing the stream, notice the *Circæa alpina* (Alpine enchanter's nightshade). [In ascending, you will probably hear the tinkling bells of the cows which are pastured in the woods to browse on the thick under-growth. It is astonishing to see to what a height these not very agile creatures will wander up the steep mountain sides; but every now and then one kills itself by a fall. Last time I passed here, I found one still alive that had broken its spine.] Arrived at the chalet, you have before you a sort of cirque, and after a few minutes of level walking the pathway is carried up the rock by a series of zigzags in a SW. direction to the summit of the port, height 2,323 metres = 7,622 feet, between the Pic Sacroux on the W. and the Pic de la Glère on the E.

N.B.—The pedestrian, on reaching the foot of the cirque, will find it shorter to avoid the zigzags, and to follow the abrupt and narrow path to the left.

From the top of this port it is about 4½ hours' walk to Venasque. By descending a very little into Spain, you come to the Lac de Gourgouttes, a desolate sheet of water, and from the arête forming the S. brink of this you have a fine view over the Val d'Essera and the town of Venasque, which you do not see from the port. As a route, the port de la Glère is at present very little used, all the communication between the two countries being carried over the Port de Venasque, where the road has been much improved, so as to be traversable throughout for horses and mules, while that of La Glère, on the Spanish side, is only practicable on foot. It is, however, the most direct route between the towns of Luchon and Venasque, and a communication

H 2

of France and Spain, by means of a diligence road, and even a railway, has been projected through this port, though as yet (1867) the works have progressed no further than marking out the site for the tunnel upon the walls of rock. The length of this tunnel will only be 2,360 metres = nearly 1½ miles, a mere trifle when compared with that of nearly 6 miles which is being carried under Mont Cenis.

The return may be made:—

A. By the same road, the shortest, 4 hours to Luchon.

B. By the new horse track which leads down from the cirque of La Glère, nearly on the same level, to the Hospice du Port de Venasque, above the right bank of the stream.

This bridle-road through the forest is one of the loveliest in the Pyrenees; not because it commands a view, but from the intrinsic beauty of the forest, which is quite overgrown with flowers and ferns. Among the ferns are some fine specimens of the *Polypodium alpestre*, and the rather rare *Lilium pyrenaicum* also grows here. A rock close to the path, which in July is completely covered with the lovely purple of the *Ramondia pyrenaica*, would make a study for the flower painter. From the cirque to the hospice by this track is a short two hours. The Cascade des Parisiennes is hidden a little below the track, but may be visited before crossing the stream to the hospice. At this cascade the *Hymenophyllum Wilsoni* is said to have been found.

C. By mounting the col to the W. of the Port de la Glère, between the Pic Sacroux and the Pic Estaous, 20 minutes over some gently inclined snow slopes brings you to the top of the col, thence to Luchon by the Valley de Lys or the Valley de Bouncou, 5 hours. (Sec. 69.)

Section 73.
PORT DE VENASQUE AND PORT DE PICADE.

On foot, horseback; or chaise à porteurs, 60 francs. To go and return 10 hours, not including halts; horses and guides 5 francs each. Without horses or ladies, guide not required. Whole distance, 37·6 kilometres = 23½ miles, as follows:—

	DISTANCE	TIME	HEIGHT	
Luchon to Hospice	10·422 kil. = 6½ miles.	2 h.	1,360 met.	4,462 ft.
Hospice to Port de Venasque	6·024 3¾	2 h. 30 m.	2,417	7,930
Port de Venasque to Fountain of Peña Blanca	350 ¼	10 m.		
Fountain of Peña Blanca to Port de Picade	1·360 1	30 m.	2,424	7,953
Picade to Pas de l'Escalette	0·860 ½	20 m.	2,420	7,940
Escalette to Pas de Monjoyo	2·100 1¼	40 m.	2,078	6,817
Monjoyo to Hospice	6·100 3¾	1 h. 20 m.	1,360	4,462
Hospice to Luchon	10·422 6½	1 h. 50 m.	619	2,064
	37·638 kil. = 23·39 mil.	9 h. 20 m.		

This is one of the first expeditions to be taken from Luchon, but a fine day should be selected. The start should not be made later than 6 or at any rate 7 A.M., as before noon there is best chance of a clear view. Ladies who wish to avoid fatigue may take a carriage as far as the Hospice de Luchon, to await them on their return. At the **Hospice de Luchon,** wine of many varieties, good coffee and bread and butter, and very tolerable meat for those who are not over nice in the cooking, may be procured. A bed, also, may be procured on an occasion, but there are few who would not prefer to make Luchon their sleeping quarters. Starting from the hospice, the path crosses the stream by a wooden bridge, and, winding upwards by a series of very regular zigzags, continues on the left bank as far as a spot known as the Culets, where the water comes pouring over the perpendicular rock from the upper plateau on the left. It here crosses the stream, and, continuing up its right bank, passes the cavity known as the **Trou des Chaudronniers**, from the fate of nine tinkers who were there buried by an avalanche many years ago. Many others have, from time to time, perished in this passage; and the tourist who makes the excursion on a warm day of August, with ample stock of provisions, can scarcely realise the dangers and rigours of this journey when undertaken in the winter or in the early spring. Shortly after this, you come in view of the Man Rock, an upright block which serves for a mark for travellers when snow covers the path. Passing this on your right, and skirting to the left or E. side of 5 upper lakes or tarns of an intensely blue colour, which occupy the rock basin at the foot of the Sauvegarde, the path bends a little to the left, and a sharp pull of 20 minutes up the last and steepest zigzags places you in the gap which forms the port. It is a mere wedge-shaped niche in the rocky wall between the Pic de Sauvegarde on the W. and the Pic de la Mine on the E., and, according to the legend, was first cut by order of one of the Counts of Comminges. An iron cross here marks the boundary of the two kingdoms of France and Spain. The first burst of the Maladetta is the striking point; but it is better to descend a little to the fountain on the Peña Blanca, whence the view is no whit inferior, and you are more sheltered from the wind. Here there are generally a couple of Spanish carabiniers, a detachment from the hospice of Venasque below. They will not allow horses to descend into Spain without a permit, but make no objection to travellers passing over the Port de Picade. The Port de Venasque shares with the Cirque de Gavarnie the reputation of being the grandest scene in the Pyrenees; but do not raise your expectations too high, or you will be disappointed. The Maladetta and its glaciers, the most extensive in the Pyrenees, lies full before you, a savage-looking dome-shaped mountain, but scarcely peaked enough for the picturesque; to me it has the appearance of a squat edition of the Jungfrau, as seen from the Wengern Alp. It must be admitted, however, that the aspect of the mountain, as seen from the Port de Venasque, well earns its name of 'the accursed.' Forests of giant pines—some of them still standing, but far the greater portion uprooted by the whirlwind, or overthrown by the avalanche—form the lowest belt round the mountain. Then comes a waste and arid band of dark rock, denuded by the water streams, which have swept away every trace of vegetation; and above this sparkle the bright glacier fields, furrowed as by a network with deep crevasses.

[The valley stretching to the SE. is the **Plan des Estangs**, and here, at the foot of the Glacier de Nethou, the eastern and highest peak of the Maladetta, there is an oval rock basin called the **Trou de Toro**, at a height of 2,024 metres = 6,631 feet, whose waters disappear underground without visible outlet, and, reappearing on the other side of the mountain to the E. (the Pic de Poumero) in a copious spring, form one of the principal sources of the Garonne, the Œil de Joucou, 1,430 metres = 4,692 feet above the sea. Particles of the rocks of the Maladetta are borne onward by the stream in its subterranean course; and the arenaceous deposit at the Trou de Toro and the Œil de Joucou will be found to be identical, thus establishing their connection. This has been further put to the proof by wood shavings thrown into the upper source, which have reissued at the lower.* To descend into the valley and visit the Trou de Toro, and thence reascend to the Port de Picade, will add 3 hours to the time.

PIC DE SAUVEGARDE,
2,786 metres = 9,139 feet.

From the Port de Venasque, if the day is fine, it is well worth while to make the ascent of the Pic de Sauvegarde—the peak to the W. of the port—as the view from this point is very much more extended than that from the port itself. Not only does it comprise a much larger portion of the mountains but the town of Luchon is visible from this point. The ascent used to be of some difficulty; but some enterprising Spaniard has lately constructed a zigzag horsepath to the top; for which he has a commission from the Spanish government to levy a toll of 1 franc on every person who ascends, whether on horse or foot. I am not sure that mountains are always improved by these artificial aids; but as the large slabs of argillaceous schist are what the French call très-satinés, and afford small hold for the foot, there are probably now many who enjoy this fine point of view who would not otherwise have ventured. It was on the rocks of the Spanish side, about half-way down, that the late Archdeacon Hardwicke fell, and was killed on August 17, 1859. His body was recovered with difficulty, and is buried in the cemetery of Luchon.

The return from the Port de Venasque should be made by the **Port de la Picade,** another pass in the chain about 1½ mile to the E. of that of Venasque. To reach this, after having descended a little from the ridge as far as the spring of the Peña Blanca, the spot generally chosen for a halt for refreshment, turn to the left (E.), crossing the lateral ridge of the Poumero over an ascent of rough stones. This is the Port de Picade. The main chain is then recrossed a little farther on, turning sharply to the N. through the Pas de l'Escalette, where the most idle will probably choose to dismount from their horses; but it is only a 'mauvais pas' for a few yards; on foot there is no difficulty. The Port de la Picade is not only the frontier of France and Spain but of the provinces of Aragon and Catalonia; and the view from it is, in my opinion, finer than that from the Port de Venasque. Due west are the mountains of the Port d'Oo, which, as seen from this point, are very striking. Then comes, in the SW., the desolate

---

* The glaciers of the Maladetta may well seem the worthiest origin of the mighty Garonne; but if the waters that spring farthest from its mouth are to be considered the true source, the Garonne has its rise in the southeastern extremity of the Val d'Aran in the transition rocks of the Col du Plat de Beret; and within 500 yards of the same spot, just the other side of the col, is the source of the river Noguera, the principal affluent of the Ebro. The extreme length of the Garonne from its source to where it enters the ocean at the Point de Graves is 466 miles; to its confluence with the Dordogne and formation of the Gironde, 385 miles, of which the first 18 miles are in Spain as far as Pont de Roi, where it crosses the chain of the Pyrenees. The elevation of its source in the Plan de Beret is 6,131 feet above the sea; at Viella it is only 2,890 feet; at Toulouse it has fallen to 433, and to 410 at its point of junction with the Canal du Midi, by means of which there is water traffic between the two oceans; at Bordeaux the river bed is only 3 feet 6 inches above the sea. The name Garonne (Garumnus) is compounded of the two Celtic words *Gar*, rapid, and *Avon, Aone, One*, river. Thus we have the river One which joins the river Pique at Luchon, and the Pic de Gar, or rapid peak, at the head of the Luchon valley; and the word Gave, the generic name of rive s in some parts of the Pyrenees, is another form of abbreviation of the same words 'Gar Avon,' Gave.

and sterile Aragonese valley of the Essera, with the Pics Perdiguère, Ramogno, and Aigues Passes, overtopped by the lofty Pic des Posets (11,047 feet); but the most striking view is to the SE. Looking over Catalonia, a huge sea of rugged and lofty mountains, far as the eye can reach, rises one above another, with the singular Pic de Fourcanade conspicuous in the foreground. Having crossed the main chain, keep close under the ridge for the first 40 minutes, which brings you to the **Pas de Monjoyo,** a gap in the chain, opening a view to the E. over the valley of Artigues Tellin. [A path from here leads down to the hermitage of Artigues Tellin in 1 hour 30 minutes. There is no difficulty; descend E. across the pasture, then through the wood, keeping the right bank of the stream.] From this point, 1 hour 20 minutes of easy walking, over an open grassy pasture, from which you gradually bear downward towards the stream, brings you again to the hospice of Luchon.

GEOLOGY OF PORT DE VENASQUE.

From Bagnères de Luchon to the central chain the dominant rock is a friable and glossy schist, sometimes micaceous, but more frequently argillaceous; but at the base of the Superbagnères, immediately behind the lakes, a band of coarse micaceous granite, intermingled with pegmatite (a rock in which large crystals of felspar are imbedded in a matrix of quartz), crops out, and extends eastward up the valley of Burbe towards the Portillon. As far as the Pont Ravi the road traverses granite and gneiss, but a little above the hamlet of Labach it enters upon Silurian strata, and as far as the hospice is continued over a blackish-coloured earth, composed of graphite and foliaceous clay schists. From the hospice to the Port de Venasque, the jagged peaks on either side of the path consist entirely of schistose rocks, those to the W. being clay schists, and those to the E., separating this gorge from that of La Frèche, being principally of talc and mica schists. Passing through the port, you shortly come upon the white limestone rocks of the Peña Blanca, which, extending eastward, are continued in the Pic de Poumero. The principal and central mass of the Maladetta is composed of ordinary close-grained grey granite, but transition rocks compose its outlying buttresses. The granite of the Maladetta, of the variety called passive, is at a glance distinguishable from the coarse-grained eruptive granite of the mountains of Oo, the produce of a later eruption. The granite descends as low as the cavern of the Rencluse; to the east of this the Pic de Rencluse is formed of mica-schist, and to the west the Pic Paderne, of dolomite and saccharoid limestone, similar to that of the Peña Blanca. On passing through the Port de Picade as far as the Col de Monjoyo, the course is over metamorphic schistose rocks; but below this col it again enters upon the dark black earths of the Silurian formation.

The excursion to the Port de Venasque is also full of interest to the plant collector. At the entrance of the forest of the hospice, the *Dianthus barbatus, Impatiens Noli me tangere, Atropa Belladonna,* and several species of *Campanulas* and *Digitalis* are conspicuous among the plants on either side of the road. By the side of the stream, S. of the hospice, may be found the *Asphodelus albus, Papaver cambricum, Myrrhis odorata, Pyrola minor, Hieracium sylvaticum, Viola cornuta, V. biflora, Hypericum nummularium,* and *Geranium pyrenaicum.* In ascending from the hospice to the Port de Venasque, the rocks on your left hand are especially rich in saxifrages. The following varieties may all be found here in the following ascending order: *S. umbrosa, stellaris, Aizöon, aizoides, adscendens, calyciflora, moschata, androsacea, nervosa, oppositifolia muscoides, ajugæfolia, bryoides,* and *grœnlandica*; lastly, on the rocks of Peña Blanca, the *Saxifraga cæsia* and *luteo-purpurea*, and that rare little composite, *Jurinea pyrenaica*.

Section 74.
LUCHON TO VENASQUE.

A. By Port de Venasque; 9 to 10 hours; 33·858 kil. = 21½ miles. On horse or foot; guide not necessary, except as an interpreter on the Spanish side.

	DISTANCE		TIME	HEIGHT		
Port de Venasque	16·444 kil. = 10¼ miles		4 h. 30 min.	2,417 met.	7,930 ft.	(Sec. 73.)
Hospice Espagnol	3·884	2½	1 h. 30 min.	1,707	5,601	
Pont de Ramougno	1·780	1		1,667	5,469	
Pont d'Aguas Passas	1·120	1		1,644	5,394	
Pont de Litayrolles	1·080	0¾		1,631	5,351	
Pont de Campamiento	1·450	1	1 h. 30 min.	1,431	4,695	
Pont de Cubère	4·720	3	1 h.	1,236	4,055	
Chapelle de Santiago	0·580	0¼		1,198	3,931	
Town of Venasque	2·800	1¾	30 min.	1,109	3,639	
	33·858 kil.= 21½ miles		9 hours			

For the route as far as the Port de Venasque, see Sec. 73. A fatiguing and steep zigzag descent down the white limestone rocks of the **Peña Blanca** brings you to the **Hospice de Venasque,** a wretched douane station on the right bank of the Essera. Here all travellers are subjected to an interrogation, and passports and *passe-avants* for horses, if any, are demanded; but in default of these not being *en règle*, remember you are in Spain, and bribery may be used with great effect, and on a not ruinous scale. You will meet with no annoyance, if you answer civilly; and, at a pinch, refreshment, and even a bed, may be here procured, but both the fare and lodging are anything but desirable. Shortly after leaving the hospice, you pass on the right the path leading up to the Port de la Glère; and then, after a very stony and rough ascent, in the course of which you pass several fine cascades, you see a large stone building, the establishment of the **Baths of Venasque,** perched on the mountain at a considerable height on the left bank of the stream. A little below this you cross the Essera by the bridge of the Campamiento. From this the route becomes more easy, and traverses a valley thickly covered with box-trees. It is not, however, till you come to the Pont de Cubère that the valley begins to present a cultivated appearance. Leaving on your right the Pont de Cubère, which, spanning the Essera by a single stone arch, leads up to the Val d'Astos and the Port d'Oo, you pass the little chapel of Santiago, then some saw-mills, the timber for which is floated down the stream from the forests of the Maladetta, and after passing the fort of Venasque, you enter the town by a narrow gateway scarcely larger than the door of a house.

Venasque may be also reached from Luchon, but only on foot.

B. By the Port de la Glère; 8 hours 30 minutes. (Sec. 72.)

C. By the Port Vieux and the Val de Ramougno; 9 hours. (Sec. 69.)

D. By the Portillon d'Oo or the Port d'Oo and the Val d'Astos de Venasque; 16 hours. (Secs. 56 and 57.)

The streets of **Venasque** are dirty and narrow; the two principal posadas are the Maison Brossaou or the Maison Ferras in the Calle Mayor or principal street. You will here be very tolerably lodged, and boarded by the Señora Therese for about 8 or 10 francs per head. The mutton and bread are excellent; indeed, bread-making is one of the few arts in which the Spanish surpass the French. You should also try the vin Rancion, a strong white wine; and especially do not forget to ask for chocolate. There is another posada by the church, of which one Juan is the padrone. The return may be made to Luchon the next day; or, if you have not seen enough of Spain and Spanish, you may proceed the next day to Viella or Las Bordes in Catalonia, and so return on the third day over the Portillon or Bacanère.

Section 75.
VENASQUE TO VIELLA AND LAS BORDES.

On horse or foot; 12 hours. Route A. By the N. side of the Maledetta.

Venasque to	DISTANCE		TIME	HEIGHT	
Port de Picade (ante, Sec. 74)	19·15 kil. = 12	miles	5 h. 30 min.	2,424 met.	7,953 ft.
Œil de Joucou	8	5	2 h.	1,430	4,692
Hermitage of Artigues de Lin	2	1¼	40 min.	1,235	4,052
Las Bordes, or Castel-Leon	5	3	1 h. 10 min.	790	2,326
Viella	10	6¼	2 h.	981	3,219
	44·15 kil.=27½ miles.		11 h. 20 min.		

From the Port de la Picade, descending by the stream that flows E., less than 2 hours brings you to the **Œil de Joucou** (*oculus Jovis*), before mentioned as one of the sources of the Garonne, which here bursts from the rock in four copious springs. From the Œil de Joucou, 30 minutes through the forest along the left bank of the stream brings you to the hermitage of Artigues de Lin (elevation 1,235 metres = 4,052 ft.), a small chapel in which there is nothing much to see. From the hermitage, 1 hour 10 minutes, keeping the pathway along the stream, to **Las Bordes, Castel-Leon.**

From Las Bordes is 10 kilometres = 6 miles, along a fair road, to Viella, which I believe affords better quarters than Las Bordes, where I had the ill luck to put up; it certainly cannot afford worse. [If Viella be the object in view, there is a path diverging at the hermitage of Artigues de Lin, and carried along the right bank of the stream, and through the forest of Baricaoude, direct to Viella. This path is rather difficult to find without a guide, but is 30 minutes shorter than the route through Las Bordes.] However hungry or tired the traveller may be on reaching Las Bordes, do not let him halt there; but if he has no particular reason for visiting Viella, the better plan is to push forward to **Bosost**, 7·5 kilometres = 5 miles, along a good road. Here he will find good food and fair sleeping quarters at the Hôtel de Commerce. At all events, you are not so *entièrement mal* as you will assuredly be at Las Bordes. Of all the bad quarters on the Spanish frontier, Las Bordes is the most detestable. Imposition and high charges are practised equally at all these Spanish inns, but at Bosost I am bound to admit that you get something for your money. After the bed and supper at Las Bordes, the breakfast at Bosost was quite a treat, and I by no means grudged the 5 francs which mine host had the conscience to charge for it. The return to Luchon from Bosost may be made on foot, or on horse, by the portillon (Sec. 77), or, on horse or in carriage, by **Pont du Roi**. N.B.—If returning by Pont du Roi, it is necessary to have brought with you from Luchon a *permit* for your horses, to be shown to the French douaniers at Fos.

Section 76.
Route B.
VENASQUE TO VIELLA BY S. SIDE OF MALADETTA.

	Height met.	feet.	Time
Venasque	1,109	3,639	
Col de Bassivi or Castaneza	2,285	7,497	3 h.
Village of Castaneza	1,465	4,807	4 h. 30 min.
Cheneste	960?	3,117	h. 30 min.
Senet	1,279	4,196	2 h.
Hospice de Viella	1,658	5,440	1 h. 50 min.
Port de Viella	2,456	8,058	2 h.
Viella	981	3,219	2 h. 10 min.

17 hours

Some persons prefer to return from Venasque to Viella on the S. side of the Maladetta, and so make the tour of the mountain, and vary the route. The start should be made very early, or you cannot reckon upon reaching Viella before night, but I should advise no one to attempt this. The valley of Castaneza abounds in many rare plants not to be found on the French side of the Pyrenees, so that the tourist intent on botanising will divide this journey into 2 days.

The best division for this course is as follows :—

1st day to Venasque . (see Sec. 74)
2nd day to village of Castaneza . } 9 hours, walking
3rd day to town of Viella . 10 hours
4th day to Luchon . 7 hours

Horses may be taken from Luchon for the whole course by this route, by which in many parts the pace may be quickened.

Starting as early as possible from Venasque, that you may have leisure to botanise, 1 hour of gentle rise brings you to the village of Cerlé, which is in sight from Venasque. From this to the Col de Castaneza, or Bassivi (height 2,285 metres), is 3 hours, passing a striking conical peak on the right. On arriving at the col, the descent at once commences, and right and left the mountain sides are a perfect garden of rare and beautiful plants. Among these the most striking is the *Vicia argentea* (silver vetch), whose delicately flushed flowers and silver leaves festoon every streamlet. With a very similar flower, though differing in its leaves and thorny spikes, is the *Astragalus aristatus*, which paints the mountain sides; but the greatest prize for the botanist to be found at Castaneza is the *Adonis pyrenaica*, with its dark green leaves and handsome yellow flowers—a very rare plant, and scarcely found elsewhere. The best season for finding the plants of Castaneza in perfection is from the 8th to the 16th of July. The adonis is to be found on the top of the col; but if these specimens should be past flowering, by ascending to the left the mountain sides, which are the outlying spurs of the Maladetta, you will not fail to light upon patches of this beautiful plant, growing in profusion.

On the right, to the S. of the col, there is a little lake, on a hot day affording a refreshing bath; and above this grows a great profusion of the adonis, and also of the moonwort fern, *Botrychium Lunare*; the plants on this side being somewhat later in flowering on account of their northern aspect. The Pic de Bassivi, due S., may be easily reached in 1 hour 30 minutes from the lake. It commands a fine view of the Maladetta range. In the SSE. the most striking mountain is the massive **Pic Turbon**, a mountain probably rich in botanical treasures, but as yet never visited.

A little S. of E. the limestone cone of the **Cotiella** overtops all rivals. This mountain, which is due S. of Plan, might be ascended from that place in a single day. It was first ascended by Count Russell from the town of Venasque; who has given an account of his excursion in the Bulletin of the Société Ramond, to which the reader is referred for a full account.

After passing the col, ascend the rocks to the left; half-way up the mountain, at an elevation of from 6,000 to 7,000 feet, a perfect treasure of plants is to be found in the early part of July; the *Adonis pyrenaica, Vicia argentea, Astragalus aristatus, Ranunculus parnassifolius, Ononis cenisia, Valeriana cordifolia, Teucrium polium, Plantago psyllium, Gnaphalium leontopodium, Alyssum alpestre*, and *Androsace villosa*, being among the rarest. Growing out of some rocks lower down, we found some splendid specimens of the *Lonicera pyrenaica* in full bloom; and yet lower by some 1,000 feet, we came upon a crop of *Arnica montana*, possessing a depth of colour and fragrance only to be produced by the actinism of a Spanish sun. On the **Sommet de Castaneza** (2,870 met.), NNW. of the granges, are beautiful specimens of the *Papaver pyrenaicum* (*Var. rubrum*), also of the *Viola valderia*. This crête may be reached by keeping to the left, and ascending from the Col de Bassivi, or more easily by mounting NNW. direct from the granges to a little brèche. From the top of the col to the village of Castaneza takes about 4 hours' continuous walking, following down the stream. At the bottom of the upper gorge, about an hour from the col, a gloomy valley runs northward into the flanks of the Maladetta. [By this gorge you may pass over to Malibierne, and so to Luchon.] Instead of taking this gorge to the north, after passing the granges of Castaneza (where at a pinch you may pass the night), cross the stream, and, following the road south, you arrive at length at the village of Castaneza. The gorge, as you descend, presents a striking appearance, backed by massive mountains of new red sandstone (grés bigarré), while the rocks by the road side are covered with the silvery tufts of the *Paronychia capitata*, with occasional patches of *Lavandula vera* (common lavender). At Castaneza there is no regular inn or posada, but at a large farm-house, belonging to one of the principal inhabitants, Antonio Frances by name, the English traveller will be entertained well and hospitably. The house of Frances is in the lower village, and near the church.

On leaving Castaneza the next morning by a path striking SE., you presently come upon a mountain completely covered with a species of cistus, growing as high as a laurel bush, with large white flowers. To see this beautiful plant, which we afterwards ascertained to be the *Cistus laurifolius*, growing wild on the mountain side in such profuse blossom, would of itself repay for the visit to Castaneza. Leaving the small town of Vicdaillet below on the right, the path descends to the village of Cheneste, on the banks of the R. Noguera. Here you must cross the river by a bridge, and follow up the stream, keeping on its left bank. From Cheneste to the village of Senet 2 hours. At **Senet** a very tolerable night's lodging may be had at the house of the Señor Roz, in the middle of the village, on the left hand as you ascend. The village of Nethou is, at 20 minutes' distance, perched on the opposite bank of the stream.

[From Senet, instead of proceeding to Viella, you pass in about 6 hours to the Spanish thermal establishment of

Caldas de Bohi, and the next day from Caldas de Bohi to Viella in 6 hours by the Port de Rios; or by a more frequented though less direct route over the Port de Caldas, passing through the village and baths of Artias, in about 8 hours.]

About 1 hour above Senet the wheel-road ceases, but the mule-track is well marked the whole way over the port. The scenery is fine; but the mid-day sun of July is too hot to admit enjoyment in a Spanish valley, and you think only of getting to the hospice of Viella, about 6 hours from Castaneza; the last hour being through beautiful park-like scenery. The hospice of Viella will grudgingly afford the traveller wine, bread, and perhaps an onion, but nothing more. By all means contrive to reach this in time to cross the port, and push on to the town of Viella before nightfall. It occupies 4 hours; a short 2 to the top of the pass, and 2 more to descend. The path is practicable for mules in summer, and sufficiently marked, that there need be no fear of missing it with daylight, even in bad weather. On reaching what appears the top, you must strike to the right for 10 minutes in order to arrive at the port. The view of the different peaks of the Maladetta, as you reach the top, is very fine. The descent to Viella is easy, and at the Hôtel Gillis the beds and fare are excellent.

At Mitg-Aran, half-way between Viella and Bosost on the left bank of the Garonne, the rocks are marked by bullets, which were fired at some French volunteers by a party of Carlists who lay in ambush at the village opposite. Michot, the Luchon guide, was one of the party and escaped. It is not necessary to descend as far as Bosost in returning to Luchon. From Las Bordes a path strikes up to the left under the pine forest of Baricaoude. Winding among the rocks and trees, in 2½ hours this path brings you out at the Spanish douane, near the top of the pass of the Portillon. Certainly no traveller in the Pyrenees with any thought of botany should omit this excursion, which is but little known. I may mention that Haurillon of Luchon is an able and willing guide, knows the locality well, and does not give himself airs.

Section 77.

VIELLA TO BOSOST.—VAL D'ARAN.

Wheel-road.

Distance		Height
		met. ft.
Viella . .		981 3,219
Mitg-Aran .	3·250 kil.	
Pont d'Aubert	1·250	
Las Bordes .	5·500	790 2,326
Bosost . .	6·500	730 2,395

16·500 kil. = 10½ miles; 3 hrs.

A very tolerable road, carried along the right bank of the Garonne as far as Aubert, where it crosses to the left bank of the river at Mitg-Aran (*milieu d'Aran*), so called from its position in the valley. There is a little chapel to the right of the road, and on the opposite side, between the road and the river, a large stone, said to be the remains of a dolmen or Druidical altar. **Bosost** (Hôtel de Commerce, Hôtel de France) is a more civilised town than those in the upper part of the Val d'Aran; decent food may be obtained at the inn, but it is dear. It is completely a Spanish town, dirty as the rest, and for itself not worth visiting. On entering Bosost, horses must be taken to the Spanish douane to have their passports viséd, with a fee of 50 cents. on each animal; without this they will not be permitted to re-enter France. From **Bosost to Luchon** there is a choice of 3 routes.

Route A. By the Portillon; the shortest; on horse or foot; 14·712 kil. = 8¾ miles; time, 4 hours, as follows:—

From Bosost there is a very rough

stony path leading to the SW., which, in 35 minutes, brings you to the chapel of **St. Antoine**. From the rocky knoll on the left of the path just above this, there is a fine view over the valley of Aran. Shortly after the chapel you come to the Spanish douane. From here a tolerably well marked track, winding amongst the shrubs and trees, under which wild strawberries grow in profusion, brings you in an hour to the summit of the pass. The Portillon is one of the easiest passes over the main chain, being only 1,308 metres = 4,287 feet above the sea, and the view from the top is not extensive, shut in as it is to the SSE. by the thickly wooded Pic de Couradilles, 6,511 feet, and to the N. by the Pic de Poujastou, 6,332 feet in height. The summit of the pass is distant from Luchon 10 kil. = 6¼ miles. The descent is much pleasanter than on the Spanish side. In descending from the Portillon, the Pic du Midi de Bigorre is visible due west. The path winds down by the head of the Valley de Burbe, and in an hour brings you to the **Cascade Sidonie**, a delightful douche on a hot summer's day, but rather public, 4·421 kil. = nearly 3 miles from Luchon; and thence to Luchon by the village of St. Mamet, 962 metres = ½ a mile from the building of the Établissement.

The pass of the Portillon is practicable in all weathers; a guide is not necessary. It requires about 3½ to 4 hours.

Route B. From Bosost to Luchon by La Bacanère; on foot.

A far finer view, though attended with more fatigue than the preceding one, will be obtained by returning by the following route. Starting northward from Bosost, 2 miles down the left bank of the Garonne brings you in 40 minutes to the baths and chateau of Lès. Before reaching this, the waters of the Garonne are lost in a deep chasm in the rocks known as the Gouffre de Clèdes. Continue along the road for 30 minutes (2·400 kil.) as far as Pontau, where there is a Spanish douane; and here, quitting the road, and striking up the rocks to the left, 20 minutes will bring you to the village of **Bausen**. From Bausen there is no track marked, but, keeping along the ridge running WSW., a good 2 hours' climb will place you on the summit of the Pic de la Bacanère, 2,195 metres = 7,201 feet, from which there is a magnificent view of the snowy range, as well as of the nearer mountains and valleys. From the Bacanère the return may be made to Luchon in about 5 hours by one of the routes described in Sec. 80.

Route C. By Pont du Roi and St. Béat; good carriage-road all the way; distance 40 kil. = 25 miles.

	Distance	Height	
Bosost		730 met.	2,395 ft.
Lès	3·100 kil.		
Pont du Roi	5·000	593	1,948
Fos	4·500	560	1,837
St. Béat	6·388	522	1,713
Marignac	2·979	521	1,709
Cierp	1·580	490	1,608
Luchon	16·329	629	2,064

39·967 kil. = 24·8 miles

Those who do not like to cross the mountains may yet see some beautiful scenery; but choose a *temps couvert* for this route, or you will be overdone by the heat and dust. Leaving Bosost by the road along the left bank of the stream, in 40 minutes you come to the village and baths of Lès, on the right bank of the stream. In order to reach them, you must cross the bridge, and take the approach up the avenue of lime and sycamore trees; otherwise it is best to continue along the left bank. Murray says quite truly that the baths at Lès afford the best accommodation in the valley; but they are placed rather too near its mouth to be convenient quarters for exploring the upper Val d'Aran. At Pontau, just before reaching Pont du Roi, there is a Spanish douane. **Pont du Roi** is the frontier of the two kingdoms; a new bridge has here been lately constructed over the Garonne to replace the old one, which was carried away by the boiling torrent that eddies between the rocks. On the French side of the bridge there is an auberge, but the supplies are very inadequate, and the cooking is more suggestive of a Spanish than of a French cuisine. If you have not already breakfasted, restrain your hunger till you get to St. Béat, where there is an excellent inn. Before reaching Pont du Roi, on your right, perched on the mountain, you see the picturesque village of Canejan, and you then take leave of the picturesque Spanish costume—the short petticoats of the women, and the long red woollen caps or berrettes of the men, and above all, the fat lazy priests, with their peculiar slouched broad-brimmed hats, constituting about 5 per cent. of the Spanish population. At **Fos**, the first village in France, there is a French douane, and the douaniers are peculiarly suspicious of Spanish cigars. At **St. Béat** (hotel, *chez* Fortan) there is excellent board and lodging, and it is a very pretty spot to stay at for a day or two, if the weather is not too hot. While the horses are baiting here, the church is worth a visit, as well from its architecture as from its picturesque situation. There are also some white marble quarries wrought in the mountain, east of the town known as the Bout de Mont or Pène St. Martin. Other marble quarries are passed in descending to Marignac, on the opposite, i.e. left, bank of the Garonne. At **Cierp** the road enters the main road from Toulouse.

From St. Beat it is worth the while of any botanist to ascend the **Pic du Gar** (1,786 m.). Its limestone rocks have a very rich flora. M. Fourcade, of Luchon, perfectly knows all the localities of the most rare plants. The distances from Luchon up the Valley d'Aran are—

From Luchon to St. Béat . 13 miles
    St. Béat to Bosost . 12
    Bosost to Las Bordes . 4½
    Las Bordes to Viella . 5½

35 mil. = 56 kil.

Section 78.

LUCHON TO ST. BERTRAND DE COMMINGES.

Carriage or horses; good road; tariff price for carriage to St. Bertrand 30 francs; to St. Bertrand and Grotto de Gargas, 35 francs, with a pour-boire.

	Distance	Height met.	ft.
Luchon . .		629	2,064
Barcugnas .	1·340 kil.	620	2,034
Moustajou .	2·240	609	1,998
Antignac .	1·240	611	2,005
Pont d'Antignac	1·240		
Pont de Guran .	5·935		
Pont de Binos .	2·504		
Cierp . .	1·821 10 miles from Luchon	490	1,608
Estenos .	4·480		
Branch road to Siradon . }	2·390		
Junction of road to Bigorre . }	6·060		
Isaourt . .	1·130		
Junction of road to Barousse . }	1·130		
Saint-Bertrand .	2·300	840	2,756

33·730 kil. = 20 miles

Leaving Luchon by the Allée de Barcugnas, to the NE. of the town to Cierp, along the valley of La Pique or Luchon, is 16·320 kil. or rather more than 10 miles. Surmounting the village of Moustajou, an old ruined square tower is seen to the left of the road, similar to that of Castel-Viel in the valley of La Pique, and of Castel-

Blancat in the valley of Larboust. Looking back from this part of the road, you may see the highest peaks of the Maladetta over the main chain, of which the Maupas, Boum, Sacroux, and Sauvegarde are the prominent points. The valley is excessively fertile, being entirely formed of modern alluvium, and probably was anciently a lake basin extending as far as Cier de Luchon. Just before the bridge of Antignac the road is joined by the parallel branch on the other side of the Pique, leading through Juzet and Montauban. At Cierp the limestone rock to the left of the road presents some remarkable contortions. A little below Cierp the Pique flows into the Garonne, and the road continues along the left bank of that river. At Izaourt, immediately on having crossed the stream of the Barousse, which here enters the Garonne, quit the main road, and take the turn to the left, which will bring you to **St. Bertrand** in about 40 minutes.

The church of St. Bertrand, perched on an isolated rock at the entrance of the Val de Barousse, is one of the most richly sculptured edifices in the south of France, and well worthy of a visit when the weather is not fit for mountaineering. St. Bertrand stands on the site of the ancient Lugdunum Convenarum, said to have been founded by Pompey the Great, B.C. 66. The present town and cathedral take their date from A.D. 1167, after the expulsion of the Saracens.

The distance from Luchon is 20 miles, and it is better to sleep at the Hôtel de Comminges, making two days of this expedition. By so doing you will have time to extend your expedition to the grotto of Gargas, a very large stalactite cavern 5½ miles NNW. of St. Bertrand. Overcast and cloudy weather will do for this excursion, if a fine day cannot be spared from the mountains. The site of the church, looking up the valley, is exceedingly picturesque, and it is full of attractions for the architect and archæologist. Among the remains inside the church is a beautiful marble tomb of a bishop of St. Bertrand, of the date of 1352, and on one of the walls, over the remains of a certain prebend, Vital de Ardengost, who died in the year 1334, there is the well-known Rosamond inscription, attesting that he was too much given to the good things of the world:—

Hic jacet in tumbâ rosa mundi, non rosa munda;
Non redolet, sed olet quod redolere solet.

On the walls of the town, over the *Porte de Cabrioles*, there is the wolf of Romulus and an inscription acknowledging the authority of the Roman emperor.

[From the ramparts of St. Bertrand you see in the NW. the wooded mountain on which the grotto of Gargas is situate. If you wish to go there, you must take a carte of admission at St. Bertrand, price 1 franc each person. 2,800 kilometres along the Bigorre road brings you to a bridle-road leading up an avenue on the left to Tibiran (*Tiberiana*), at a distance of 700 metres. From Tibiran the pathway leads up NW. over rock and turf to the grotto, distant from Tibiran 2·300 kilometres, and a short 4 miles from St. Bertrand. You may return by Valcabrère (*Vallis Capraria*), and pay a visit to its church of St. Just, said to date from the close of the 8th century. The church is about 500 metres S. of the town of Valcabrère, and 4,500 kilometres = 3 short miles from the grotto of Gargas.]

St. Bertrand lies only 2 miles off the high road to Toulouse, and 1 mile off the road to Bigorre, so that those who choose may well stop here one night on leaving Luchon, putting up at the Hôtel de Comminges.

Those who are on foot or horseback

may return to Luchon from St. Bertrand by the valleys of Barousse and Oueil (Sec. 85, route D).

Section 79.
MONTAUBAN. — JUZET. — AND TOUR OF THE VALLEY OF LUCHON.
Carriage, horse, or on foot; 3 or 4 hours.

The cascades of Montauban and the garden of the curé form a pretty little afternoon stroll on a hot day. Leaving Luchon to the E. by the Rue de la Pique (that in which are the Casino, Museum, and English church), 1,622 metres = 1 mile, to the church of Montauban: passing upwards through the village, you must knock at a green door you see before you, the door of the garden; the woman who opens it will expect 10 sous for admission. The combination of nature and art in these waterfalls makes it an enjoyable lounge on a hot sultry day. Having reached the top of the garden, pass out through a doorway, and about 250 yards higher you come to another fall. There is a pretty view over the valley of Luchon, and you can continue to ascend the mountain as far as you feel disposed. You will ultimately arrive at the Col de Courets, by which you may reach the summit of the Pic Poujastou (1,930 metres = 6,332 feet). (Sec. 81.)

At **Juzet** there is another cascade similar to that at Montauban. A pleasant drive or ride is what is called the grand tour of the valley of Luchon. Starting by St. Mamet, you traverse Montauban, Juzet, Salles; 1 kilometre beyond which place you strike into the Route Impériale, on the other side of the river Pique, and so return by Barcugnas. The whole distance of the circuit is 13·420 kilometres = 8½ miles, enabling you to visit both the falls at Montauban and Juzet.

Section 80.
PIC DU BACANÈRE (*VACHE NOIRE*).
On foot or horse; 9 hours.

Start not later than 6 in the morning; by so doing you are more sure of the view, and avoid the heat. The range running N. and S., which divides the valleys of Aran and Luchon, and forms a break between the eastern and western Pyrenees, is especially well placed for commanding a fine view from any of its peaks. The Pics Bacanère, Poujaston, Couradilles, and Entecade are all worth ascending, but especially the first and the last.

The best and shortest route is as follows:—

	DISTANCE	HEIGHT	
		met.	ft.
Luchon		629	2,064
Barcugnas	1·340 kil.	620	2,034
Branch road from Juzet	1·092		
Juzet, across the meadows	·980	632	2,073
Sode	1·944	914	2,999
Artigues	3·156	1,241	4,072
Fontaine ferrugineuse	1·100		
Cabane d'Artigues	2·000		
Rochers de Cigalère	2·100	1,888	6,192
Summit of Bacanère	6·090	2,195	7,201

19·772 kil. = 12 miles

After reaching the village of **Juzet**, the path, well marked, mounts by the side of a ravine to the village of Sode, and, passing the church of this village, continues to ascend in zigzags in an easterly direction through a little forest to the village of Artigues; here the track turns sharp to the right, and conducts you to a ferruginous source, known as the *Hount Herido* (cold fountain). From here, keeping along the N. side, and passing close under the **Rochers de Cigalère** on your right, 2 hours will place you on the summit of the Bacanère.

The return is best made by Gouaoux de Luchon as follows:—

	DISTANCE from Luchon	ELEVATION met.	ft.
Summit of Bacanère	19·772 kil.	2,195	7,202
Pales de Burat	2·500	2,150	7,074
Cabane de Gouaoux	1·200		
Gouaoux de Luchon	3·950	939	3,081
Grange de Prat Naou	1·500	846	2,775
Junction with Route Impériale	1·730		
Antignac	1·470	611	2,005
Moustajou	1·240	609	1,998
Barcugnas	2·240	620	2,034
Luchon	1·340	629	2,064

35·942 kil. = 22·32 miles to go and return

From the Bacanère descend northward to a little lake, and here, in some seasons, there is a little cabane where refreshment is sold; but do not reckon upon this, and on no account omit to take provisions from Luchon. From this cabane, you again ascend a little to the NW., and 40 minutes from the Bacanère suffices to reach the summit of the **Pales de Burat**, from which the panorama is even superior to that from the Bacanère, inasmuch as it comprises more of the valley of the Garonne.

From the Pales de Burat descend W., and then S., keeping under the ridge as far as the stream, where there is a cabane, 50 minutes from the top, affording sleeping quarters to those who come here to see the sunrise. Cross the stream, and then take the path, pretty much on a level, through the fir forest, from which you emerge on some rich pasturage that slopes down rapidly to the hamlets of Teiche and Gouaoux.

After traversing this last, follow the path southwards, which passes over some curious-looking rocks of dark Silurian schist, with quartz-crystals imbedded, which, being fractured, have the form of a Maltese cross. The path strikes the main road 1 kilometre below the bridge of Antignac.

The rocks of the Bacanère are well deserving the attention of the geologist. They are chiefly a ferruginous schist (*schiste carburé*). By crossing the easy col 700 metres W. of the Pales de Burat, and descending to the little marshy pool NW. of the summit (which certainly does not merit the title of lake, given it in the map of the État-Major), the fossil collector will find some treasures. The fossiliferous rocks are S. of, and about 200 m. above, the lake. Orthocerates especially are here very abundant.

The Pales de Burat is a grand point for witnessing the sun-rise, but it should be before September, as later the sun does not rise far enough north to be clear of the mountains.

From the Bacanère you may also descend on the east side to Lès in the Val d'Aran (see Sec. 77); or you keep southward along the top of the ridge as far as the Pic de Poujastou, and thence descend to the village of Juzet

Section 81.
PIC DE POUJASTOU.
On foot (or horse to within 30 minutes of the summit); 8 hours.

This is an easy ascent, and, being chiefly under the shade of the forest, it is a favourable one for a hot day. The best route is as follows:—

	DISTANCE		ELEVATION	
Luchon			629 metr.	2,064 feet
Montauban	1·510 kil. = 1 mile	687	2,090
Fountain of Erran	4·920	3		
Mail de Cric, or Col de Simourère	3·020	2		
Summit of Poujaston	2·920	1¾	1,930	6,332

Return through the Forest of Sesartigues.

Fontaine Rouge	2·980	2		
Juzet	4·780	2¾	868	2,846
Montauban	1·920	1¼		
Luchon	1·510	1	629	2,064

23·560 kil. = 14·6 miles whole distance of excursion

Pass right through the village of Montauban, leaving the garden of the curé on your left; or you may pass through this garden, and then, leaving on the left the path to the cascade, keep the track which leads SSE., having the road to St. Mamet below on your right, and in 40 minutes from Montauban you come to the Fountain of Erran (ferré) in the midst of the woods above St. Mamet. From this, changing your course to ENE., 45 minutes brings you to the **Mail de Cric,** also called the Col de Simourère; from this point you look over the valley of Aran as well as that of Luchon. Here, if on horseback, you must descend, and leave your animals in charge of the guide, and 40 minutes more over slippery turf composed chiefly of *festuca eskia* (jonc échinant) will place you on the summit of the Poujastou. You may return the same way, or, if on foot, vary the route by returning on the N. side of the gorge through the forest of Sesartigues and Juzet. From the summit to the Fontaine Rouge takes about 50 minutes; thence an hour, chiefly through forest, brings you down to the Granges de St. Jean, where are some ruins of an ancient village. Here, leaving the cascade on your right, 30 minutes brings you down to **Juzet,** whence to Luchon it is an easy hour.

Section 82.
PIC DE COURADILLES.
On foot or horse; 7 to 8 hours to go and return; total distance, 25·760 kil. = 16 miles.

	DISTANCE		ELEVATION	
Luchon			629 metr.	2,064 feet
Path to Barguères on the road to the Hospice of Venasque	6·102 kil.	1 hr. 15 min.	840	2,756
Cabane de Barguères	2·150	50 min.		
Stream of las Barguères	0·268	5 min.		
Branching off of the path from that to Campsaure	1·090	20 min.		
Clos de Baretgè	1·920	30 min.		
Summit of Couradilles	1·350	30 min.	1,985	6,511

12·880 kil. = 8 miles; 3 hrs. 30 min.

The Pic de Couradilles, another peak on the same chain, S. of the Portillon, offers another charming and easy excursion on a fine day, somewhat shorter than that of the Entecade, but the view is not quite so grand. Starting along the road to the hospice, 1,140 kil., or about 15 minutes after passing the Pont Ravi, and a little before reaching the path that diverges on the right to the Port de la Glère and Cascade des Demoiselles (Sec. 71), a path diverges on the left, and, leaving on the right the Cascade Courège, mounts rapidly to the cheese cabanes of Barguères. At Barguères, after crossing the stream, you come to another cabane, that of Artigon.

From here the direction is nearly due N., leaving to the right the pathway leading to the pasturage of Campsaure.

In descending you may vary the route by following this path, which will bring you out a little below the hospice, passing by the cabane of Roumingaou; from Artigon to the hospice is 1 one hour 15 minutes.

The view from Pic de Couradilles is very fine east and west, overlooking the valleys of Aran and Luchon, and to the south looking full on the mountains of the Port de Venasque, the Port de la Glère, the Crabioules, and the Port d'Oo; but not commanding a view from the Maladetta, the panorama is scarcely equal to that from the Entecade.

Section 83.
PIC DE L'ENTECADE.

On foot or on horseback; 8 hours to go and return; total distance, 22 miles.

	DISTANCE	ELEVATION	
Luchon		629 metr.	2,046 ft.
Hospice	10·422 kil.	1,360	4,462
Divergence of path from that to the Col Monjoyo	3·700		
Cabane de Pouylané	1·820		
Lac des Garces	1·000		
Summit of Entecade	0·900	2,220	7,285

17·842 kil. = 11·07 miles from Luchon

For the route as far as the hospice at the foot of the Port de Venasque see Sec. 71. From the hospice for 50 minutes the route is that to the Port de Picade, in an easterly direction, through the wooded valley of La Frèche. On leaving the trees, the path works up in zigzags to the east, and, 10 minutes after passing a source of excellent water, leaves the path that leads to the Col de Monjoyo. It then shortly crosses the stream, and mounts up its left bank to the cabane of Pouylane, occupied by Spanish shepherds, from which you first get a sight of the summit of the Entecade. Fifteen minutes from this brings you to the little lake des Garces (*des grues*), a little above which you must leave your horses, if not on foot. Ten minutes from the lake places you on the first col, opening a view over the valley of Artigues de Lin; 20 minutes more to a second col, opening a view to the east over the valley of Aran; and thence 10 minutes to the summit of the peak, and the full enjoyment of the panorama. Dotted with villages, the valleys of Aran, Lys, Larboust, Oueil, and Cierp, are spread at your feet; and among the mountains you will recognise many an old friend. Close to you, in the SE., lies the town of Viella, with the port of the same name rising above it; a little more to the south the valley of Artigues Tellin, in which you may well distinguish the Œil de Joucou and its cascade; then, proceeding from left to right, you may remark successively the Pic de Fourcanade, the Pic de Pouméro, the Pic de Nethou, and a large portion of the glaciers of the Maladetta, the Pic des Posets, Perdiguère, Tus de Maupas, and Quairat; more in the distance come the Mont Perdu, the Pic d'Arbizon, and the Pic du Midi de Bigorre; and nearer to you the Monségu, the Monné, and the Montaspé. The Pic de Gar marks due north; then come the pics de Cagire, de Crabère, and the mountains of Ax, beyond which, in the extreme distance, if the air be clear, the Canigou, it is said, may be seen, though it cannot be distant less than 100 miles in a direct line. The return may be made by the same route to the hospice, or more to the north, towards the Pic de Couradilles, traversing the park-like scenery of the plateau de Campsaure, and so reaching the cabane of Artigon (Sec. 82). The Entecade also may be well ascended in crossing by the Col de Monjoyo, for an excursion to Viella or the valley of Artigues de Lin.

Section 84.
PIC D'ANTENAC.

On horse or foot; 7 hours to go and return; total distance, 24·601 kil. = 15·3 miles.

The parallel chain running west of the valley of Luchon presents almost a finer point of view than that of the Bacanère. Of all the easy ascents of the secondary peaks to be made from Luchon, that of the Antenac is most recommended. It should be made as follows:—Leaving Luchon by the Allée des Soupirs to the NE., cross the Pont de Mousquères, and 10 minutes after this quit the main road and turn up a path to the right, which leads up by a zigzag course to the village of **Cazaril**, 4 kil. = 2½ miles from Luchon, and perched at an elevation of 970 metres = 3,182 feet above the sea. Passing upwards through the village of Cazaril, at the top houses turn sharp to the right, by a path at first rough and stony, but which gradually improves, and leads north over the grassy shoulder of the mountain. In ascending you have a beautiful view of the valley and town of Luchon. About 1 hour from Cazaril brings you to some pleasant green knolls, covered with heath and fern, at the entrance to the fir forest. Here, if with horses, you must dismount, as the boughs scarcely permit a horse to pass; altogether the path is a very rough one for horses, but it is just practicable. The trees, principally fir or *sapin* (*Abies picea*), are completely festooned with the long drooping clusters of the rock hair lichen (*Alectoria jubata*). This always grows strongest on the north side of the trees; and, as the direction is north, you may steer by this even without a compass. About 35 minutes brings you through the wood, and you emerge on an open pasture, on which grow a profusion of bilberries. Striking across this by a rough and narrow path, still in the same direction, you come to a spring at the foot of the Tus d'Abacède; here fill your water bottles, for you will find no water on the top. From this spring, 40 minutes of gentle ascent over the turf to the top of the Pic d'Antenac, 1,990 metres = 6,529 feet. The panorama is magnificent, but the mountains have been named from the Entecade (Sec. 83). The descent may be made by either of the following routes:—A. Retrace your steps to the spring at the foot of the Tus d'Abacède, and from thence follow down the path along the stream, through the villages of Sacourvielle and Trebons, passing the tower of Castel Blancat. B. Descend immediately to the W. from the Pic d'Antenac, which brings you to the Col de la Serre, and a little further on to the spring of Cazeneuve; thence, following down the stream, you reach the village of St. Paul, 6·090 kil. = 3¾ miles from the summit, and 4 miles = 6·121 kil. from Luchon. From St. Paul to Luchon, see Sec. 85.

Section 85.
PIC DE MONNÉ AND VALLEY D'OUEIL.
On horse or foot. Guide not necessary.

	Distance	Time	Elevation
Luchon			629 metr. 2,064 ft.
Divergence of road up valley d'Oueil .	4·668 km.	1 h. 0 min.	
Benqué Dessous	1·450	25	1,049 3,442
Benqué Dessus	0·550	15	1,084 3,557
Maylen, hamlet on the right bank of the stream, opposite St. Paul .	1·330	20	
Mayrègne	2·750	30	1,210 3,970
Caubous	1·990	20	1,276 4,186
Cirès	0·420	10	1,306 4,285
Bourg d'Oueil	1·690	20	1,354 4,442
Cabane of Shepherds . . .	1·790	40	
Port de Pierrefitte . . .	1·210	25	1,806 5,925
Summit of Monné . . .	1·390	35	2,147 7,044

19·138 km. = 12 miles; 5 h. 0 min.

The Vallée d'Oueil (*ouaille*, the old patois for sheep) is one of the richest pasturages in the Pyrenees, though its dirty villages have not the air of affluence which one might expect from the fertile fields which surround them. It is also celebrated for its breed of Pyrenean dogs, the average price for a good one ranging from 50 to 200 francs. Start on this excursion at any rate not later than 6 A.M.; but, if you can make up your mind to start by night, there is no difficulty in traversing the valley d'Oueil in the dark, and the view of the sunrise from the Pic de Monné will amply repay you for your broken night's rest. It is still better to take provisions on a horse, and make a night encampment on the summit, which will give an opportunity of exploring the Lac Bordères and Sommet de Lyon. Quitting Luchon by the Allée des Soupirs, follow the road to Oo as far as the second bridge, the Pont de Pons, and 208 metres (5 minutes) above the chapel of St. Aventin, after crossing the stream that comes down from the right, quit the main road, and take that to the right, which leads up the Valley d'Oueil. At Benqué Dessous, overshadowing the churchyard, there is an elm tree of great size—elms being comparatively rare in France. A rough steep road leads up to Upper Benqué, and thence, passing Maylen and across the meadows, you traverse the stream in face of Mayrègne.

[The village of St. Paul is on the left or far bank of the stream. The houses on the right bank form the little hamlet of Maylen. Here you may cross the stream, and pass to Mayrègne along the left bank, through St. Paul; but it is shorter to keep along the right bank. From St. Paul, the Pic d'Antenac, 6,529 feet above the sea, may be reached in 1 hour 40 minutes; distance 6·100 kil. It is a much shorter excursion than that to the summit of the Monné, and the view is, in my opinion, quite equal. From St. Paul the route lies due south over the grassy pastures. You may ride the whole way; and the whole time from Luchon, including the return, need not be more than 6 or 7 hours; this ascent is strongly recommended to ladies (Sec. 84.).]

It is said that at Maylen, frequently during the night, especially when there is thunder about, small phosphoric flames may be seen flickering across the road, and up the mountain side. To see these flames you must cross the bridge, and place yourself opposite to them in the village of St. Paul, as on the spot itself you can never see them. Of course the villagers of St. Paul have a legend to explain this. They say it is the tortured soul of an old miser of Maylen, who scrupled not to enrich himself by despoiling his neighbours and removing their landmarks, and who prays that the boundaries may be replaced on the spots which he indicates by these flames. I have not had an opportunity myself of verifying this phenomenon, but, if it exists, it is probably due to the following cause. Above Maylen there is a deep cavern, in which are accumulated a great quantity of bones and decayed animal matter. A small stream runs from this cavern, and trickles across the road just where these flames are seen; so that these luminous vapours would appear to be the natural effect of the escape of the phosphuretted hydrogen with which these waters must be impregnated. The phosphoric lights, the *feux follets*, that are sometimes seen in burial grounds, are to be accounted for in the same way.

From Maylen St. Paul to Mayrègne 35 minutes. Mayrègne, though the richest village of the valley, has nothing remarkable. [Instead of crossing the stream at Mayrègne, you may continue up the valley along the right bank as far as Cirés, or even to Bourg, the last and principal village of the valley. In thus keeping along the right bank, you come upon large blocks of granite, which must have been transported from the mountains of Oo.] From Marègne to Bourg, 1 hour. At Bourg there is a quasi auberge, but scarcely anything to be had except bread, cheese, and milk. From Bourg to the Port de Pierrefitte, 1 hour 10 minutes. On leaving Bourg, you mount on the left side of the gorge, first skirting the fir wood, and then over the turf and heather; a path all the way.

The Port de Pierrefitte, 5,925 feet above the sea, is the most direct route between Bigorre and Luchon, but practicable as yet only on foot or on horseback. The distance from the top of the col to Luchon is 17·848 kil. = $11\frac{1}{4}$ miles, and to Arreau 13 kil. = 8 miles, making the total distance 31 kil. = $19\frac{1}{4}$ miles. The Col de Pierrefitte is a wild grassy ridge, occupied in the summer months by shepherds and their flocks, and upon this col are generally to be seen some very fine specimens of the Pyrenean dogs; but they are excessively fierce to strangers, and on meeting them, it is well to have their master within hail, as they are not the least afraid of stick or stone.

In crossing to Arreau by this route a guide is not necessary. You at once begin to descend through the fir woods, leaving on your left the Lac de Bordères, from which the stream issues. The track is at first indistinct, but soon merges into a well-marked road. From the top of the col to the village of Bareilles 7 kil. = 1 hour 40 minutes. From Bareilles to Arreau 6 kil. = 1 hour 20 minutes; for the descent to Arreau see Sec. 45. [From the Port de Pierrefitte to the summit of the Monné 40 minutes; a steep ascent along the ridge, practicable for horses, but it is better to dismount. The height of the Monné is 7,044 feet above the sea, and there is no height from which, with so little exertion, so fine a view of mountain and plain, fertile fields and barren glacier, can be obtained. Due west there is a noble view of the Pic du Midi de Bigorre; and beyond this, a little to the north, it is said that, on a favourable day, the Bay of Biscay may be discerned. Arreau lies at your feet, and the Louronaise women in their red capulets may be distinguished as they bustle about the market-place, particularly if it is Thursday, the market-day. In descending from the Monné to Arreau, it is not necessary to return to the Col de Pierrefitte; but, striking westward down the stony slope of the mountain, on reaching the wood, you cannot fail to hit a path which winds through the trees, from which it emerges at the bottom of the gorge.]

For the **return to Luchon,** there are three routes to choose from:—

A. You may return by the way you came, down the **Valley d'Oueil,** easily in 4 hours; making a slight variation by keeping along the left bank of the stream from Mayrègne, and so passing through St. Paul, Sacourvielle, Castle-Blancat, and Trebons. The distance is pretty much the same on either bank of the stream.

B. By the **Valley of Larboust,** 20 kil. = 12½ miles; 4 hours. From the Col de Pierrefitte the track runs nearly due south, skirting the east side of the mountain ridge; and passing under the foot of the Pics de Lyon and Pouylouby, 50 minutes brings you to the **Col de Sarrieste,** height 1,995 metres = 6,545 feet, and distant from the Col Pierrefitte 3·400 kil. = 2¼ miles. From this point there is a beautiful distant view of the cascade of the Lac d'Oo, and of the mountains above it. From the Col de Sarrieste to Jurvielle 4·250 kil. = 2¾ miles. Descending SSE. along the left bank of the stream, you pass the marble quarries of Jurvielle, and reach the village in 45 minutes. From Jurvielle to Garin, passing through the village of Poubeau, 3·014 kil. (35 minutes). From Garin to Luchon, 8·468 kil. = 5¾ miles along the Bigorre road, 1 hour 30 minutes (see Secs. 62, 55).

C. By the crest of the mountains separating the valleys of Oueil and Larboust (only practicable on foot), to the Col de Sarrieste, as in preceding route; thence, instead of descending, keep along the ridge to the east, and, after surmounting the Pic Tirou de la Lit, 1,972 metres = 6,470 feet, and the Rocher de Peyrègne, 1,920 metres = 6,299 feet, you reach the summit of **Bilourtède,** 1,844 metres, and descend upon St. Aventin. You have a glorious view from this ridge of the glaciers of the main chain, but there is some difficulty in finding the track without a guide who knows this part.

D. If you do not care to return to Luchon the same day, you may descend NE. **by the Valley of Barousse,** and, sleeping at St. Bertrand de Comminges (Hôtel de Comminges) (Sec. 57), return to Luchon the next day. The route is as follows:—Passing the col east of the Monné, by the little lake called the Coume de la Laque, or descending, if you please, at once from the summit of the Monné, you pass through the forest of Samaoury, and, following down the stream, 1 hour 30 minutes from the col brings you to the **Chalets St. Néré,** on the left bank of the stream, at the foot of the Pic de Montlas, 1,729 metres = 5,673 feet. At these chalets there are mineral waters, and a very primitive bathing establishment, where you may get a bed, if required. You may here get a tolerable meal, and increase your appetite, if necessary, by a draught from the Source de la Faim, a spring issuing from the granite above the Chalet des Sorcières. Also do not omit to pay a visit to the Roche Damnée, 300 yards in the wood above the Chalet des Chasseurs, one of the most remarkable erratic blocks in the Pyrenees; from the top of this rock you have a charming view of all the valley. Eight miles further down the valley, at Crechet, there is a similar mass of rock; this rock (*grès silicieux*) being found *in situ* on the top of the Monné. From the chalets St. Néré to **Mauléon** reckon 1 hour 15 minutes. On leaving the chalets, continue to descend along the left bank of the stream (*L'Ourse, R.*), and, passing the Cascade Vacqué, the villages of Ferrère and Ourde— where remark the traces of an ancient moraine, and a natural bridge formed by the rock—you reach the picturesque little town of Mauléon (inn, *chez* Grillon). From Mauléon there is a road to the baths of Siradan and Ste. Marie, first mounting the col east to the village of Cazaril. By this route you may return to Luchon in 5 hours' easy walking; the distance is 29.335 kil. = 18¼ miles.

	DISTANCE		ELEVATION	
	km.	mil.	metr.	ft.
Mauléon			581	1,906
Siradan	5·100 =	3¼	483	1,584
Junction of Toulouse road	} 1·145	¾	459	1,506
Luchon	23,090	14¼	629	2,064
	29·335 = 18¼			

From Mauléon to St. Bertrand, 11·150 kilometres = 7 miles, there is a wheel-road. Leaving Mauléon to the S., Troubat is the first village; here there is a grotto about a mile off the road to the E., the grotto of St. Araillé. On the opposite bank of the stream you see the ruined castle of Bramevaque, built towards the end of the eleventh century by Sancho de la Barthe, whose tomb is in the church of St. Bertrand. It once served for a residence to Marguerite de Valois, the wife of Henri IV. Shortly after this you pass to the left bank of the stream by the bridge at Gembrie; and at Crechets, on the left side of the road, opposite the first mill, you see an enormous erratic block (grès silicieux), which can only have come from the summit of the Monné, where that rock is *in situ*. About 1 kil., 10 minutes' beyond Sarp, you come to a cross road, that to the right (E.) leads to Izaourt and Luchon, that to the left to St. Bertrand de Comminges, at a distance of 2·300 kil. = 1½ mile. From St. Bertrand to Luchon 33·730 kil. = 21 miles. For account of St. Bertrand see Sec. 78.

Section 86.

PIC DE NETHOU,

Highest point of the Maladetta.

On foot; 2 days; horses may be taken as far as the cave of the Rencluse; tariff charge for guides, 30 francs each, and provisions must be taken; guide and rope indispensable. Do not omit to take some sort of stewpan for cooking.

	DISTANCE	TIME	ELEVATION	
Luchon			629 metr.	2,064 ft.
Port de Venasque	16·444 kil.	4 h. 30 min.	2,417	7,930
Plan des Étangs	2·650	1 0	1,798	5,899
Rencluse	1·050	0 40	2,083	6,834
		6 h. 10 min. 1st day		

	DISTANCE	TIME	ELEVATION	
Portillon	1·870	2	2,855 metr.	9,366 ft.
Summit of Nethou	3·530	2	3,404	11,168
One hour on summit		1		
Port de Venasque, including halt of 1 hour at Rencluse and 30 minutes on port	9·100	7	2,417	7,930
Luchon	16·444	4	629	2,064

51·098 kil. = 31¾ miles ; 16 h. 2nd day

Of all the mountains of the Pyrenees, the Maladetta is the highest as well as the most massive. It supports the largest glacier fields, the most extensive one — that of the Nethou—being 4,300 metres = 2¾ miles in breadth. Till late years not only had the summit never been reached, but it was not certainly established which one of the jagged peaks bursting through its flattened dome of ice might claim the pre-eminence. It was not till the year 1842 that a French gentleman, M. Franqueville, accompanied by a Russian officer, M. Tchihatcheff, and three guides, Argaro, Pierre Redonnet, *dit* Nate, and Bernard Ursule, accomplished a successful ascent of the Pic de Nethou, and established that to be the highest peak, not only of the Monts Maudits, but also of the Pyrenees, being 11,168 feet above the sea. In later years ascents are not unfrequent, and in fine weather present no very great difficulty to anyone who has been accustomed to the glaciers of the Swiss Alps. Ladies, even, have succeeded in reaching the summit, but it is scarcely fit for them. The excursion is one of considerable fatigue, and entails sleeping out at least one night, at the cave of the Rencluse, on the Maladetta, 6,834 feet above the sea. For the route as far as the Port de Venasque see Sec. 73. [You may come by the Port de Picade, but this lengthens the distance by 1½ mile. From Luchon to the Port de Picade is 18·602 kil. = 11½ miles.] To reach the **Plan des Étangs** from the port, keep down the valley SE. At the Plan des Étangs there is a dirty sort of hut, sometimes used by woodcutters. Here crossing the stream, ascend southwards between two rocky buttresses; on the left side there is a path, which, mounting towards the glacier, turns sharp to the right at the cave of the **Rencluse**. In order to reach the cave, you must cross the stream, which comes down from the Maladetta glacier. 100 yards lower down, these waters disappear under the rock, and reappear in the stream that rises near the Hospice de Venasque, just as the waters of the glacier of Nethou lose themselves in the Trou de Taureau to reappear in the Œeil de Joucou. In the cavern in which the waters disappear are to be found veins of a curious siliceous rock of a green colour. The rock which affords the night's bivouac is on the left bank of the stream, and is overhung by a few stunted fir-trees, at the foot of the Pic Paderne.

On arriving at the cave, the first operation is to cut down some of these to serve for fuel during the night, for, from the contiguity of the glacier, the temperature after sun-down soon sinks below freezing. With a blazing fire and a good supper you may, however, well bid defiance to the cold; and these night bivouacs 'sous les belles étoiles' are certainly not among the least pleasant memories of mountain excursions. In the morning the start had better be made not much later than 4 A.M., as it is a long day's work to reach Luchon. From the Rencluse cross the stream, and, striking south, the course for 2 hours is over rocks mixed with patches of snow. Keeping up the arête that divides the glaciers of Nethou and Maladetta, you presently come to the **Portillon,** a gap in the rocks by which you descend upon the glacier of Nethou. The glacier presents no difficulty; but after about 20 minutes a little rocky islet is reached, and it is here necessary to rope the party together, as a precaution against the crevasses. [This precaution is not superfluous; in the summer of 1861 a guide of our party, named Corrége, disappeared through a crevasse on this glacier and fell 30 feet, but was fortunately rescued. In 1824 the guide Barrau fell through a crevasse on the Maladetta glacier, and his remains were never afterwards seen.] In about 1 hour towards the top of the glacier you come to a depression at the foot of the Pic de Milieu, which, till late years, formed the basin of the Lac Couronné (3,173 m.), which no longer exists, its waters having burst through the ice and found an outlet in August 1857. After this you have to scale the dome, the most steeply inclined portion of the glacier. This is, in fact, part of the actual summit; but in order to reach the highest point, one more difficulty remains to be surmounted, the **Pont de Mahomet.** This is an arête about 60 yards in length, composed of huge fragments of granite piled one upon the other, in a manner to present anything but a secure footing; and below this, on either side, there is a precipice — that to the right hand extending down to the lake and gorge of Malibierne — and the pebbles that your foot dislodges on the left fall down many hundred feet of sheer descent, and are lost in the gaping crevasses of the Nethou glacier. This bit is the most trying to the nerves of the whole expedition; and as you bestride this narrow ledge, crossing it on hands and feet, with the clouds floating around you, you feel verily suspended in mid air. This passage, however, does not occupy many minutes, and your foot is soon securely planted on the summit of the Pic de Nethou. The plateau on which you stand is about 100 feet long by 24 broad, covered with fragments of granite rock. A pyramid of these fragments, piled by the guides, is visible with a glass from the Port de Venasque, and at the foot of this pile there is a box with a book containing the names of all those who have made the ascent, and the date. Here, also, there is a register of thermometrical and barometrical observations. In the course of five ascents to the summit of the Pic de Nethou, the highest temperature I have found here in the shade was on July 17, 1866, at 1 P.M., when the thermometer marked $11\cdot5°$ Cent. There is probably scarcely a night in the year when the thermometer does not descend to the freezing point. On July 19, 1866, Count Russell and Captain Hoskins, with the guide Capdeville, passed the night on the summit. They had not a registering thermometer; but the greatest cold they observed was just above the freezing point, the night being overcast. Registering thermometers have been left both by M. Lézat and myself, but have never long survived the fury of the elements. In the winter of 1857 a registering thermometer left on the Pic de Nethou by M. Lézat marked $-24\cdot2°$ Centigrade $= 11\cdot7°$ below zero, Fahr. while in the plain at Toulouse the extreme cold only reached $-7\cdot5°$ centigrade $= 18°$ above zero, Fahr.

The rock composing the summit of Nethou is the passive or older variety of the granite composing the heart of the chain; this granite is of a whitish colour, and composed of small particles of felspar, quartz, and mica, and altogether distinct from the porphyritic eruptive granite with large oblong felspathic crystals, composing the mountains of Oo, which, from its being commingled with fragments of the transition rocks, is evidently the product of a more recent eruption. The view, if the weather be clear, is of course magnificent, there is no site on the southern side of the chain, except that of the Pic des Posets, at all approaching it. On the flanks of the mountain you see the Lac d'Ereoueil at the foot of the Maladetta peak, the Glacier de Couronné at the foot of the Nethou, but the Lac de Gregonio, at the head of the valley de Malibierne, the largest of the Maladetta lakes, is hid by the interposing Pic of Ereoueil. The number of mountains is almost more than the eye can comprehend; for the first few moments they present a confused chaos of peak beyond peak, but gradually they range themselves into order, and you distinguish a principal chain running east and west, with the long transverse ridges running out into the plain. You may also remark the falling off in the height of the mountains as they recede on either side from this central point. Among the principal, the Mount Perdu, just visible, bears 6° N. of W.; the Pic des Posets, 8° N. of W.; the Vignemale, 12° N. of W.; the Pic du Midi de Bigorre, 38° N. of W.; the Port de Venasque, 11° W. of N.; the Canigou, 17° N. of E.; the Puigmal, 7° N. of E.; the Montarto 18° S. of E., and the Pic Malibierne, 1° W. of S.

From the Portillon, in returning, the best course is to descend by a steep rocky gully on to the **Glacier of the Maladetta,** down which you may descend in a glissade, and thence over the rocks to the Rencluse. The Maladetta glacier is too steeply inclined to admit of its being ascended in the early morning, when the snow is hard, without crampons or the labour of cutting steps. When you reach the Rencluse, you will be quite disposed to appreciate the soup or stew which one of the guides ought to have been preparing during your ascent, and which will invigorate you for the 7 hours' walk that yet remains to reach Luchon. To reach Venasque from the Rencluse takes about 6 hours (see Sec. 74). The return to Luchon may be varied by passing through the Port de la Glère, and ascending the Pic Sacroux (Secs. 72, 69); but this excursion requires 10 hours.

The descent from the Pic de Nethou may also be made on the southern side; passing the crête at the Lac Couronné, and thence descending to the cabanes of Malibierne, which are 4 hours from the summit, 3 hours more to the town of Venasque. This route is even easier than that by the Rencluse, and, except 'ex abundantiâ cautelæ,' does not require a rope, as the snowfields on the S. side scarcely amount to a glacier, and are but little crevassed.

If the weather is fine, one or two nights may be well spent at the Rencluse, to give you an opportunity of examining the geological structure of the mountain, and collecting the botanical specimens to be found at the foot of the glaciers, and especially at the foot of the Pic Paderne. The **Pic of the Maladetta,** 3,312 metres = 10,866 feet, may be ascended by scrambling up the arête running south from the Portillon; but I can especially recommend the **Pic Albe,** 3,280 metres = 10,761 feet, as well worthy an ascent. Ascending from the Rencluse by the stream to the Maladetta glacier, you mount this by the moraine on its eastern side; it is rather rough scrambling, but not very difficult, and on it the *Ranunculus glacialis* and other glacial plants grow in great abundance. Arrived at the top of the moraine (height 2,908 metres), you strike south, across the glacier, which is here not very steeply inclined; but do not omit the precaution to rope, as it was here that the guide Barrau fell into a crevasse and perished. Once across the glacier, you must scale the rocky arête forming the Pic d'Albe, choosing the spot that seems easiest. Exclusive of halts, the pic may be reached in about 4 hours from the Rencluse.

Section 87.
PIC FOURCANADE.

On foot; 2 days; horses may be taken to the Cabane des Aigouailluts. Guide necessary; tariff charge for each guide, 30 francs. Provisions must be taken.

	DISTANCE	TIME	ELEVATION	
Luchon				
Port de Venasque	16·444 kil.	4 h. 30 min.	2,417 metr.	7,930 ft.
Plan des Etangs	2·650	1 0	1,798	5,899
Trou de Toro	1·000	30	2,024	6,649
Cabane des Aigouailluts	600	1 0		
		7 hours, 1st day		
Port Alfred		2 h. 40 min.		
1st fourche		20		
2nd fourche		25		
Pic Fourcanade		15	2,882	9,456
Halt on summit		30		
Base of peak		40		
Col de Poumero		30		
Lac de Poumero		30		
Œil de Joucou		1 55	1,430	4,691
Hospice d'Artigues de Lin	1·650	40	1,235	4,052
(Path becomes a horse track)				
Las Bordes, or Castle-Leon	4·500	1 10		
Spanish douane (on the Portillon)	7·000	1 45		
Portillon	1·250	30	1,308	4,287
Luchon	10·000 km.=6¼ miles	2	629	2,064

13 h. 50 min. 2nd day

This peak, the extreme eastern point of the high ridge of the Maladetta, was first ascended August 1, 1858, by Mr. Alfred Tonnellé, with the guides Redonnet, *dit* Nate, and Antoine Ribis. It is a very wild granite peak not often explored, and consequently you have a good chance of here finding crystals of quartz, tourmaline, and other minerals. Those who do not care to ascend the peak may explore the upper part of this valley, and then, with much less labour, pass over the **Col des Aranais** between the Pics Fourcanade and Poumero.

As far as the Plan des Etangs, the route is the same as that to the Maladetta (Secs. 73 and 86). After reaching the Trou de Toro, continue to ascend the savage gorge to the SE. till you reach the wretched stone **Cabane of the Aigouailluts.** This spot is nearly 7 hours to Luchon, and here you must pass the night, unless you feel strong enough and are sufficiently early to mount the Pic Fourcanade and descend to Las Bordes in the Val d'Aran, which will take about 9 hours more. At the Cabane des Aigouailluts you are fortunately still within the range of a few stunted firs, with which you can make a fire to cheer your night bivouac, and keep off the wolves and bears. You will not be disposed to prolong your slumbers; and starting at daybreak, you soon quit vegetation. Commence along the right bank of the stream coming down from Pic Poumero. In 3 min. cross to the left bank, and mount by a steep stony zigzag path to first plateau. On this remark a curious natural bridge, beneath which the torrent disappears, a phenomenon not unfrequent on these upland granitic plains. Presently cross to right bank, then again to left bank, and in 1 hour 30 min. reach a tiny lake formed by the stream. Here again cross to the right bank, passing a second, and finally a third small lake, from which the ascent is SSE., over frozen snow; and, passing over banks of snow, rocky boulders, and finally a glacier, you come to a gap in the ridge uniting the Pic Fourcanade to the Pic Moulière, called the **Port d'Alfred**. Passing through this, you must descend a little on the other side, in order to attack the Fourcanade from the south, on which side alone it is accessible. A rough scramble over loose moving stones brings you to the first fork. This peak, which, seen from the northern side, is in appearance only two-pronged, is in reality split into four peaks, forming a double fork, more in the fashion of a double tooth. The first fork that you reach is much lower than the principal or northern one; but a little more persevering climbing will place you on the col of the principal fork, from which 20 minutes are sufficient to place you on the highest point, 4 hours 30 min. from the Cabane de Aigouailluts. **The return** may be made by the same road, or, which is preferable, by the Œil de Joucou and the Valley of Aran.

In descending from the peak, steer NE. for a little lake, from which you may reach Viella in 2 hours 30 min.; or else, mounting to the W. by the side of the stream, in 40 min. reach the Col Fourcanade NE., from whence you may either pass WNW. over the Col des Aranais or descend N. to the Lake Poumero, and thence through the forest by the Œil de Joucou, to Artigues Tellin, and so to Las Bordes (Sec. 75). From Las Bordes to Luchon, 5 hours 30 minutes, through Bosost and the Portillon, or by a more direct cut, taking a path to the left, 1 kil. along the Bosost road, which brings you to the Spanish douane station on the Portillon road in about 1 hour 30 minutes. For the route from Bosost to Luchon see Sec. 77.

In this case, if you have brought horses to the cabane of Aigouailluts, you will have given orders to them to go round by the Port de Picade to await you at the hermitage of Artigues de Lin.

Section 88.

LUCHON TO MALIBIERNE AND CASTANEZA.

	TIME
Luchon	
Port de Venasque	4 h. 30 min.
Entrance of Gorge of Malibierne	3 h.
Cabane of Malibierne	2 h. 30 min.
	10 hours.

One of the main attractions of the Pyrenees is the unrivalled and diversified flora extending almost to the mountain top both on the French and Spanish side of the chain.

From Luchon the following expedition is strongly recommended, presenting, as it does, grand, wild, and almost untrodden scenery, with an unrivalled flora, and great facilities for camping out.

Starting from Luchon with guides and provisions, cross the Port de Venasque, and descend by the usual road 1½ hour below the hospice, to the entrance of the gorge of Malibierne, on the left bank of the Essera. Here strike up the gorge, keeping above the right bank of the torrent. The track first mounts among fir-trees, and in an hour emerges on more level ground, and is carried along the same side of the stream to the cabanes of

Malibierne 2 hours from the entrance of the gorge. Here there is excellent camping ground, wood and water in abundance, and perhaps milk. The ascent of the Pic Nethou may be most easily made from here, the distance being rather shorter than that from the Rencluse.

The **Lac Gregonio**, one of the largest of the Pyrenean lakes, at a height of 2,656 metres, may also be visited, taking the same course as that to the Pic Nethou, but turning W. at the foot of the glacier, and mounting the col on the left. From the col to the lake is a very steep descent of 265 metres over rock and snow, not very difficult, but requiring care. From the lake you may pass N., and, leaving the Pic Maladetta on the right, descend from the crête on to the Maladetta glacier, and so reach the Rencluse; but this is a course requiring both a daring and skilful guide. On the moraine towards the bottom of this glacier, which may be easily reached in 2 hours from the Rencluse, the *Ranunculus glacialis*, with white and with red flowers, grows more abundantly than I have ever met it; lower down the *Astrantia minor* is very abundant on the granite rocks.

From the granges of Malibierne, it is easy to pass the col S. to the granges of Castaneza; time about 6 hours; or you may pass eastward to the lakes of Rio Bueno. The route is as follows:—Following the stream upward amid scattered pines, which soon cease, 1 hour from the cabane brings you to a little col, N. of which is a conical mountain that conceals the Nethou. The Pic Posets in the W. is very fine. Passing this col, leave three little lakes on the left; and mounting ESE. over the granite rocks, 1 hour 40 minutes to a second col considerably higher (2,776 m.). From this descend by a steep snow slope, and, leaving a triangular lake on the left, mount a third little col ESE., 50 minutes. From this descend nearly S. over a rough talus, and after passing 3 small lakes on the left, 1 hour 20 minutes will bring you to the **Lakes of Rio Bueno**. The principal lake is the most eastern, which is famous for its trout. They afford capital sport, but of late years great havoc has been committed by nets; the delinquents being Spaniards from the town of Senet, which may be reached from here in 2 hours 30 minutes, descending the col S. of the lake.

The tourist who comes here for sport, must take up his quarters for a night on the spot. There is a cabane by the stream, 15 minutes below the E. end of the lake, or, by descending 1 hour E., above the right bank of the stream; an excellent camp may be formed at the commencement of the forest, with an abundance of fire-wood ready cut to the hand by the Spanish wood-men, who have only the last two years begun to violate with the axe what, when I first visited it, seemed a primæval and virgin forest. 50 minutes below this camp (2 hours below the lake), the path debouches on the Viella road, 1 hour below the hospice, and 1½ hours above Senet. Pass over the Port de Viella, and, sleeping in that place, return the next day to Luchon, over the Portillon.

The Lakes of Rio Bueno, and Malibierne, may also be reached by the Col de Salenque, the lowest depression in the Maladetta range, seen from the Port de Venasque, a little to the east of the Pic de Nethou.

Passing the Port de Venasque as before, descend to the Trou de Toro, and thence 30 minutes along the right bank of the stream, to the Camp des Aigouillats. Here there is no shelter, and a greater abundance of water than wood; but the night can be passed tolerably, if only the weather is propitious. If a storm supervenes, you must console yourself by the reflection that you here witness it in all its grandeur. In the morning follow up the stream S., which issues from the glacier of the Nethou. Keep the right bank for the first 20 minutes; which you should then cross to avoid the very rough stones; and after a little mount, you enter the **Gorge de Barrans**, and in 2 hours attain the col of the same name (2,478 m.).

From here the Col de Salenque is in full view S. To reach it, you must descend a little, and cross the moraine to the foot of the glacier. This is always more or less snow-covered, and not very steep, so that it may generally be ascended without an axe, though occasionally there is difficulty. There are no crevasses. The top of the **Col de Salenque** (2,825 m.) is attained in 2 hours from the Col des Barrans. Here there is a choice of two routes. The surest in thick weather, but not the least laborious, is to follow down the gorge of the **Fèchan**, which opens SE.; and after 4 hours' toil, over almost endless granite boulders, beneath which the stream sometimes completely disappears, a slight track is hit upon, which leads down into the forest, along the left bank of the stream. In this route there is no danger, but considerable fatigue.

The other route, and in fine weather the preferable one, is on a higher level, and more to the S., to the Lakes of Rio Bueno. From the Col de Salenque traverse the rocks to the right, and, without much descending, make for the little brèche which is seen near the shoulder, a little E. of S.; to this brèche 1 hour 10 minutes. From this brèche you must again descend rather to the right, and, making across some rough rocks to the lakes of Salenque, leave these on the right, and cross the stream. Skirting a low crête on the left, in 10 minutes a little col is reached; pass this, and descend S. Then remount another col to the SE., from which you descend into the head of the gorge of Rio Bueno.

The **Pic Malibierne** (3,109 m.), a magnificent observatory for the southern side of the peaks of the Monts Maudits, may be best ascended from the cabanes of Malibierne; from whence the mountain, with the contorted limestone strata that form its upper part, presents a very grand aspect. The ascent is not difficult, occupying about 3 hours 15 minutes. The ascent is for the first hour E.; to the top of the little col leading to Rio Bueno. Thence mount ESE., attacking the final arête from the E. The summit is a crête, running from WNW. to ESE., with a very 'mauvais pas' intervening to guard the western end, which is not the highest. The Pic de Nethou is almost exactly N., the Pic Montarto E., and the Posets a little N. of W.

To descend, the same direction must be at first taken; but when below the summit, you may strike across the arête projected northward from the Pic de Malibierne. This passage is not easy to find; and the descent from the rocks on the other side will not be effected without difficulty; but the rocks are granite, affording good foothold. Once on the snow-beds, there is no further difficulty; you soon reach the region of flowers and pine-trees, and then the cabane, in about 3 hours 30 minutes from the top.

On all sides but the ESE., the Pic Malibierne is precipitous and inaccessible, but, forming an off-shoot from the mountain to the W., are two other smaller peaks, which may be easily reached from the cabane of Malibierne, and deserve the attention of the botanist and geologist.

The westernmost and highest of

these peaks, which I have named Pic Papaver (2,863 m.), especially deserves attention. Captain Barnes found almost on the summit of this peak a fossil trilobite, relief and intaglio, which proved to be the *Ogygia Edwarsii*, giving the date of these rocks as the upper Silurian.

The rocks are of a red ferruginous schist, impregnating strongly with iron all the springs; and on these schists, at from 2,700 to 2,800 metres, the red variety of the *Papaver pyrenaicum* grows in great abundance; a little lower down the *Thalictrum alpinum*, *Viola valderia*, and *Ranunculus parnassifolius*; and, lower still, the *Gnaphalium leontopodium*, *Artemisia mutellina*, *Saussurea alpina*, *Achillea pyrenaica*, *Gentiana glacialis*, *G. nivalis*, *G. Burseri*, *Adonis pyrenaica*, and *Ranunculus amplexicaulis*. All these and a host of other good plants are above the cabane, and on the south side of the stream. In the pastures below the cabane, are the *Nepentha graveolens*, *Swertia perennis*, and the three varieties of aconite. At the entrance of the gorge, close to the Essera valley, the *Ononis arragonensis* is found on the rocks overhanging the stream. These, though but a small portion, may give some idea of the botanical varieties of this gorge. For a better explanation of its geography the reader is referred to the map.

Section 89.

PIC DES POSETS.

On foot; 3 days. Pierre Barrau, Firmin Barrau, and Redonnet (dit Michot) are the only guides who have reached the summit.

This peak, 3,367 metres = 11,047 feet, the second highest of the Pyrenees, is to the west of the Maladetta on the opposite side of the Val d'Essera.

The view from the summit is the finest in the Pyrenees, decidedly superior to that from the Maladetta, though the latter mountain, being the highest in the whole range, as well as more accessible to Luchon, is much better known. There are several ways of reaching the Pic des Posets, that is, the **Cabane de Turmes**, or the **Cabane de Paoules**, two stations answering to that of the Rencluse on the Maladetta, at one of which the night has to be passed, previous to making the actual ascent.

Route A. By the Port de Venasque and down the Valley of the Essera (Secs. 73, 74). At the Pont de Cubère, 8 hours 30 minutes from Luchon, 4 hours from the port, turn sharp to the right over the bridge; a mule path along the stream brings you in 2 hours to the Cabane de Turmes.

Route B. From the Lac d'Oo, passing over the Portillon d'Oo (Sec. 57), 2 hours 30 minutes from the top of the Portillon, 9 hours from the Lac d'Oo, brings you down to the Cabane de Turmes. In descending, you first pass over a steeply inclined snow slope, and then follow the stream down the gorge, leaving a waterfall on your left.

Route C. From the Lac d'Oo, passing over the Port d'Oo, and descending to the Cabane de Paoules, about 8 hours (Sec. 56). The Cabane de Paoules, situate at the foot of the Port de Clarabide (Clara vista), in the summer months is generally tenanted by some Spanish shepherds, from whom milk and bread may be procured. The Cabane de Turmes, though lower down the valley, is generally quite deserted; but at either of these stations the mountaineer must depend on his own resources. The Cabane de Turmes may be reached from the Cabane de Paoules in 1 hour 30 minutes. There is a track keeping along the right bank of the stream.

ASCENT FROM THE CABANE DE TURMES.

The ascent of the Posets is easiest made from the Cabane de Turmes; but it is long, occupying 6 hours. The Cabane de Turmes is a rude stone cabane on the right bank of the stream, but the weather must be very bad indeed before you are driven to take shelter in such a smoke-grimed filthy den; and as there is plenty of wood at hand, most will prefer to light their bivouac fire, and make their resting-place to the lee of one of the boulder stones here scattered about. In the morning start with break of day. The direction is first S. and then SW. A steep little ledge of rock has first to be ascended, and then, passing through some fir-trees, in 45 minutes you come to the foot of the gorge, overgrown with bilberries, above which lie beds of snow. From this point the course lies WSW. up the gorge, and 40 minutes over snow and rocky débris brings you to the top of the **Col de Batticiel**, the first step of the mountain. You have before you a considerable-sized lake, the Lac de Batticiel, and passing to the right of this, 50 minutes brings you up a second col composed of huge granite blocks roughly piled together. Having gained the top of this, the second step of the mountain, a wild rocky plateau, dotted with small tarns, is before you, and on the other side of this rises the arching crest of the Pic des Posets, from here bearing due west. Skirting the rocky arête on the right bank of this plateau, you come to a lake rather larger than the rest; and passing this on your right, 30 minutes over a rocky shoulder brings you to a long gently inclined slope of frozen snow; cross this (50 minutes), and you pass through a gap on to the third step of the mountain, where you find yourself upon the true glacier. Towards the top it is pretty steeply inclined, but there is no difficulty in crossing it, and the only *mauvais pas* is in passing from this glacier on to the arête forming the actual summit, as a considerable *Bergschrund* generally intervenes. Once upon this arête, a steep scramble of 15 minutes up the rocks places you upon the top, which is a mere knife edge of rock running nearly N. and S. The principal mass of the Pic des Posets is of granite; but this last arête, as well as the lower spurs, is composed of a red disintegrated clay schist, the prevailing rock of the Central Pyrenees. Another glacier, flowing down to the W., is bounded by a belt of fir-trees, and beyond these are seen the yellow cornfields above the Spanish village of El Plan. The view from the Pic des Posets comprises all the highest summits of the Pyrenees, the giant forms of the Mont Perdu and Vignemale being pre-eminent in the NW.; beyond these the Pics du Midi d'Ossan, Baletous, Gers, and Gabisos. Then comes the Néouvielle, conspicuous in the NW., with the Pic du Midi de Bigorre seen over it. Full in front a glorious view of abrupt mountains and snowy cols, from the Clarabide to the Perdiguère, and beyond this the menacing peak of the Sauvegarde, with the well-known Ports of Venasque and Picade. To these succeeds the ponderous mass of the Maladetta, with the silver Pic de Nethou marking 10° N. of E. To the south the barren mountains of Aragon rise range upon range, till distance softens their ruggedness into a blue outline. The return to the Cabane de Turmes will take a good 4 hours, from whence you may descend to the town of Venasque in 2 hours; or, again bivouacking, return to Luchon the next day, by one of the routes above described. The shortest ascent of the Pic des Posets is to be made from the Cabane de Paoules; crossing the valley, and continuing up the stream west, you reach the foot of the gorge, down which a glacier flows into the valley; and ascending this south, a sharp pull of 2 hours will bring you to the uppermost glacier on the third step of the mountain, whence the route is the same as that just described. The ascent by this gorge from Paoules was first accomplished by the writer in 1863; it presents no difficulty except the last passage on the rocks, which is the same by either route, and it is considerably shorter than the route from the Cabane de Turmes.

The Pic des Posets has been ascended but about half a dozen times; first, in 1856, by Mr. Halkett; again in the same year by Mr. Behrens, and in 1861 by the writer. Upon the jagged storm-beaten rocks of the summit there is no trace of vegetation, not even the humblest lichen, and on my last visit I could find no trace of the record left of previous ascents. Not only the cards, but the stone cairn raised over them, had been swept away by the wind, or, as the guide graphically expressed it, "le mauvais temps les a mangé."

From the summit of the Posets the shortest descent to Venasque is by the gorge of Éristé; time about 6 hours.

The Pic des Posets is a favourite haunt of the izards, the Pyrenean chamois, which are always to be found on some part of this mountain. In addition to these, ptarmigan (*perdrix blanches*) are pretty frequent on the rocks just below the snow line, and in one of the lower lakes there are said to be large trout. An outing of two or three days, or even a week, might be well spent on the Pic des Posets and the neighbouring mountains of the Port d'Oo. Those of the party intent upon sport would be pretty sure of finding game, and the rest would find ample subjects to interest them in the magnificent scenery and unrivalled flora. A tent might even be taken, or at all events sleeping bags; and enjoyable would be the reunion for the evening meal, and the story of each day's adventures. The head-quarters might be made either at the Cabane de Turmes, or the Cabane de Paoules, whence the Spanish town of Venasque is within easy reach, to fall back upon for fresh supplies of provision, or in case of bad weather. In any case a retreat is always open to Luchon over the Port de la Glère, or the Port de Venasque, as both these passes are practicable in all weathers during the summer months—the first only on foot, but the latter even on horseback.

Section 90.
CALDAS DE BOHI AND PIC DE MONTARTO.

East of the Maladetta is an extensive granite region, sprinkled with innumerable lakes, some of considerable size, and at great elevations. Many of these lakes abound in fine trout; and on the mountains ptarmigan and izards are pretty plentiful.

Caldas de Bohi is the best head-quarters from which to explore this district, and the most direct route to it is as follows:—First day to Viella; second day, 1½ hour from Viella, up the Valley d'Aran, as far as the village of Artias. Here turn up the gorge S., first along the left, in 50 minutes crossing to the right bank of the stream. 1 hour 30 minutes from Artias the gorge bifurcates, the right branch mounting SW. to the **Port de Rieux**, which leads to the hospice of Viella. Leaving this, take the branch to the left, which mounts steeply by a rough stony path, in 2 hours 40 min. from Artias, reaching a little shallow lake. Leaving this on the right, again mount by a track ESE., 30 min., to another long narrow little lake, and thence 50 min. winding among the boulders to the **Port de Caldas** (height about 2,440 m.), 4½ hours from Artias. From the port descend 10 min. SSE. to the large circular lake of Les Religieuses, in which, I believe, there are no fish. Coast above the western shore of the lake, and then descend SW. There is a rough track, which there is no fear of missing, as stone markers have been set up by the pilgrims, who twice a year, viz. on August 15 and September 8, cross by this port in great numbers, to pay their devotions to Nôtre Dame of Caldas de Bohi. Two hours from the port brings you to another much lower lake, the **Lac de los Caballeros,** in which there is very fair fishing. Skirt this along its eastern shore, crossing the stream which issues at the S. end of the lake. From this lake to the hospice of Caldas 50 min.; 3 hours from the port. The establishment of Caldas is half secular half religious. There are hot sulphur baths of considerable efficacy, and a shrine of the Virgin, of great sanctity. The virtues of the two combined are infallible.

Spanish pilgrims and invalids from Senet and the lower valley frequent this spot during the summer months, but it is not lively for a lengthened stay.

The principal hot source is in the wood, above the left bank of the stream, 10 minutes S. of the building. There is no town or village, but the whole establishment, which is a large quadrangular building, is under the charge of a 'padre administrador,' Señor Martin Musoles, who speaks French, and is very obliging and affable to strangers. The beds here are very clean, and the fare quite sufficient. The bill, which is made out by the 'padre,' is very moderate; and, if camping out, a good supply of wine, bread, and meat, may be procured by sending down to Caldas. At the lake Tramesane, 1 hour above the Lac de los Caballeros, from which it bears NE. and SSE. from the Lac des Religieuses, there is excellent fishing. N. of this lake a very rough mule-track mounts to the **Port de Salardu.** a little to the E. of the Port de Caldas, leading down to the village of Salardu, in the Val d'Aran, 3·3 kil. above Artias.

From Caldas de Bohi a path over the col ENE. leads to Senet, in 5 hours.

From Caldas the **Pic de Montarto** may be conveniently ascended, in an easy day of about 9 hours, including the return to Caldas, or the descent on the W. to the hospice of Viella. A good walker may reach Viella. The best route is as follows: Take the path to the Port de Caldas, and 30 min. above the Lac de los Caballeros cross the stream coming down from the port, and mount the gorge from the WNW.; 1 hour 20 min. up this brings you to a little tarn at the foot of a large snow slope. Ascend this SSW., and, on reaching the top, which amounts to glacier, make for the SE. ridge of the mountain, and ascend whichever peak you consider the highest. I am inclined to think it is the centre one of this ridge, but there are at least five peaks on this mountain, so nearly equal that, without exact measurement, it is impossible to assign the pre-eminence. They all range between 2,900 and 3,000 metres.

If intending to cross into the Viella valley, at the top of the glacier make for a little brèche on the right of the southern extremity of the ridge, running N. and S. From this, 10 min. down a little rocky cheminée to the western snow-slope, which is steeply inclined in its upper part, and may require a few steps: then make NW. across the snow to the rocks, and thence, continuing the same direction, descend to the little stream coming down from the N. Follow this stream, keeping the right bank, 15 min. to a small lake, and 30 min. farther to a cabane and the Lac **Becibere**. This lake also is full of fish, and as there is wood about the lake, and a cabane, or, if preferred, rocks, affording shelter, what more can the sportsman desire? And I should by all means advise him to encamp here for a night or two. The lake, though small, is very picturesque, at a height of about 2,120 metres. From the lake to the hospice of Viella 1 hour 40 min. Leaving on the left the cascade, just below the W. end of the lake, a rough track descends over the stones, keeping close under the rocks, rather away from the stream for the first 20 min., when it again approaches the stream, and winds down among the forest, emerging on the Viella road, just above the confluence of the path to Rio Bueno.

The rocks of the Montarto being all granite, the flora is not very rich, but there are good *Carices*, and the *Saxifraga geranioides* abundant, which is not found in the Central Pyrenees; also *Saxifraga nivalis*.

North of the Pic de Montarto, at a distance of 1,500 metres, is one of the largest lakes in the Pyrenees, full 2 kil. in length, with an island in the centre. It seems to be quite unknown, and I have given it the name of the **Lac de l'Isle**. In many respects it reminded me of Wastwater, but it is more desolate and savage. In it I believe there are no fish, but it is well worth a visit. It may be reached in 3 hours 15 min. from the Lac Becibère. Pass over a brèche at the western extremity of the northern arête of Montarto. The gap to be taken is not the nearest to the peak, but the second. To the top of this col 2 hours, thence 1 hour 15 min. over snow and rocks to the lake. The north border is precipitous, but along the south border is a faint track, though in August we here saw neither sheep, shepherd, nor cabane. 1 kil. W. of this lake is another, equally desolate, separated by a low arête.

From the Lac de l'Isle you may descend upon the path of the Port de Caldas, and so to Artias in 3 hours.

Section 91.

LES MINES DU CAP DE GUERRI AND PIC CRABÈRE.

This course, independently of the great interest it has for anyone investigating the mineral deposits of the Pyrenees, will yield to very few for its intrinsic beauty, and the grand panorama it commands. I strongly recommend it to anyone who has a couple of spare days.

The first night sleep at Fos, an easy afternoon drive, ride, or walk. (Sec. 78.) From Fos continue the road as far as Pontau, the station of the Spanish douane, 3 kil. beyond Pont du Roi. There cross the Garonne, and mount the valley E. along the left bank of the stream of the Toran, 2 hours, as far as the village of **St. Jean de Toran**, the confluence of two torrents. Take that to the right, and, following it up S. and then E., 2½ hours brings you to the summit of the Cap de Guerri, SSE. of the little lake of the same name, by the side of which is the cabane of the chief miner and the first attack on the lode, that of Gabriela. Observe, as you mount, the profuse quantity of yellow gentians which cover the mountain sides. These are gathered in great quantities, and used, I am sorry to say, as a substitute for hops, for making beer. 3 kils. E. of the cabane, at about the same level, is the third and richest attack, that of Estrella. The lode here seems very rich, extending from W. to E. These mines have been conceded by the Spanish government to M. Carvalho, the French consul for Persia at Bordeaux. In richness of the mineral, and facility for working it, they are inferior to none in the Pyrenees, but they have two drawbacks—1st, being at an elevation of 2,000 metres, the mines cannot be worked for above five months in the year; 2ndly, the difficulty of constructing a road for transporting the metal as far as Pontau. At present, however, there is a horsepath, and the mines may be easily reached in 6 hours from Fos.

WNW. of the eastern attack, that of Estrella, at a distance of 600 metres, is the **Étang de Liat**, a considerable piece of water. N. of this is the **Tuc de Canejan** (2,654 m.), and a little more to the W. the menacing black precipices of the **Pic de Crabère** (2,650 m.). E. of the Tuc de Canejan is the **Portillon del Albi**, and a little farther on, ENE. of the lake, the **Port de la Orqueta** (2,545 metres); and still more to the E. the **Pic de Mauberne**, the highest of the group (2,880 m.). From both these ports, del Albi and de la Orqueta, the descent may be made to the village of **Seintein** (height 760 m.), in Ariège, in 2½ hours. The Port de la Orqueta is the easiest, being a mule-path. On the N. side of these ports, just above their point of junction, are the lead mines of Bentaillou (1,885 m.), the most extensive and profitable working in the Pyrenees. From Seintein there is a wheel-road, 12 kil., to Castillon; thence 13 kil. more to St. Girons. Hotels, chez Ferrière, de France. From St. Girons to Aulus (Hôtel Souquet), 33 kil.; there is a diligence service daily.

From St. Girons to Tarascon (Hôtel Ginestet), 51 kilometres wheel-road. At Massat (26 kilometres) there are some bone caverns in the calcareous mountain on the right bank of the river, N. of the town.

The pedestrian will find this route a very interesting one, by which to enter Ariège. From Fos to Seintein is 11 hours' walk, exclusive of stoppages; for further details see Sec. 92.

If the ascent of the Pic Crabère is contemplated, the izard-hunter, or botanist (for there is here a fine field for both), will do well to have brought provisions, and make arrangements for sleeping either at the cabane of the mine of the Cap de Guerri or at one of the shepherds' cabanes W. of the Étang de Liat.

If the tourist wishes to return to Fos the same day, and is on foot, he may agreeably vary the route from the Étang de Liat. After observing the singular chasm at the E. end of the lake, where the waters disappear, as those of the Trou de Toro, skirt the southern shore, passing two other pieces of water nearly on the same level. After passing these, the precipitous rocks which must be descended offer some difficulty. The best passage is found by keeping rather to the N., as near as possible to the roots of the Pic Crabère. Below this, 10 minutes across a rough stony talus brings you to the forest; through which there is a path, high above the right bank of the stream, that in 1½ hours emerges at the village of St. Jean de Toran, at the junction of the path by which you ascended. The gorge through this forest is very wild; and bears are here killed in considerable quantities.

Section 92.

ARIÈGE AND EASTERN PYRENEES.

For those whose time is limited, the Eastern Pyrenees offer small attractions in comparison with the mountains around Luchon. A few days at Biarritz is the usual termination of the Pyrenean season ; but as it may be an object with some to go eastward, a few of the by-routes over the mountains eastward from Luchon are here given.

LUCHON TO CASTILLON.

Route A, by **St. Béat**, 57 kil. = 35½ miles, as follows:—

	DISTANCE	
From Luchon to St. Béat	21 kil.=13 miles,	carriage-road
From St. Béat to St. Lary	24 15	a mountain-path
From St. Lary to Castillon	12 7½	carriage-road
	57 kil.=35½ miles	

For the route to St. Béat (Sec. 77). On leaving St. Béat, passing under the mountain called Bout du Mont, and following up the left bank of the stream, you pass the village of Boutx, from whence the path mounts through a forest of beeches to the Col de Mendé, 1,331 metres = 4,347 feet, between the Pic Cagire, 1,912 metres = 6,272 feet, on the north, and the Tus d'Estang, 1,814 metres = 5,952 feet, on the south. From this col, descending east, 1 hour 30 minutes brings you to the village of Couledoux and 2 hours more (11 kil.) to the village of St. Lary, surrounded by beech forests.

Route B, by **Melles** and **Seintein,** as follows:—

	DISTANCE	
From Luchon to Fos	27 kil.=16¾ miles,	carriage-road
From Fos to Seintein	7 hours,	mountain-path
From Seintein to Castillon	12 kil.=7½ miles,	carriage-road

For the route from Luchon to Fos, see Sec. 77.

On leaving Fos, after passing under the old tower of Pomorin, and crossing the stream of the Serial, turn to the left, and mount, 10 minutes to the village of Melles. Thence continue to ascend the gorge of Maudan, first E., and then, inclining SE., keep on the right bank of the stream. One hour from Melles, the Pas de Cho (2,117 m.) is seen opening to the S., leading over into the valley of Canejan. Leave this on the left, and a steep pull up the escarped buttress of the Pic Crabère will place you on the **Col d'Aoueran**. On the E. side of this col is the Lac d'Araing (1,880 m.); to which you must descend; and then follow down the stream that issues from the lake, the direction being first N.; 30 minutes from the lake the chapel of the izard is passed on an eminence on the left. Pilgrimages are annually made to this chapel on August 5. Below this chapel the direction of the path gradually changes to the E., passing through forest as far as the village of Seintein, which is 2 hours from the lake. From Seintein there is a good road down the valley of Biros to **Castillon.** At Castillon, a night's lodging may be had at a pinch. From Castillon to St. Girons (hotels, chez Ferrière, de France), 13 kil. = 8·07 miles, carriage road. Castillon to St. Gaudens (Hôtel de France) by Aspet, 40 kil. = 25 miles, carriage-road. St. Gaudens to Toulouse, 88 kil. = 54¾ miles.

Section 93.

UPPER VAL D'ARAN.

Viella to Esterri; mule-path, 8 hours.

The Valley d'Aran is very thickly populated, especially in its upper part. It contains 39 villages, all Spanish, though more naturally by their position placed under France.

The issues of the Upper Val d'Aran

to the east are by three principal ports: the Col d'Espot, the Port de Paillas or Bounaigo (good water), so called from a spring on the top, and the Col de Peyreblanca (white stone), so called from a rock placed as a mark on the top. Of these three the **Port de Paillas** is the one most frequented. From Viella to the Port de Paillas 4 hours, and from the top of the Port to Esterri the descent occupies 3 hours. The track is well marked, and there is a hospice on the eastern side of the port. The Port de Paillas separates the basin of the Garonne from that of the Ebro.

Section 94.
VIELLA TO CONFLENS.
ROUTE A.
By the Col de Peyreblanca and Port de Salau, 12 hours; mule-path.

	DISTANCE		HEIGHT	
	kil.	miles	met.	feet
Viella . .			981	3,219
Belren . .	0·800			
Escunan .	0·850			
Cazaril . ·.	0·600			
Artias . .	3·500			
Gesa . .	1·700			
Salardu .	1·600			
Source of the Garonne	7·500		1,872	6,142
Col de Peyreblanca			1,889	6,198
Hermitage of Montgarri	8·000		1,737	5,699
Port de Salau	15·000(?)		2,052	6,733
Conflens .	7·000		898	2,946

46·550 kil. = 29 miles

From Viella to Salardu 2 hours. Leaving the village of Tredos and the route to Port de Paillas on the right, a zigzag track leads over rocks to the col of the Plat de Beret, or Peyreblanca. On the south side of this the waters of the Garonne have their rise in two little pools of water, at an elevation of 1,872 metres = 6,142 feet, **les Yeux de la Garonne** (see Sec. 52). On the NW. side of this col, at a distance of 600 metres from the source of the Garonne, issues the spring that gives rise to the river Noguera at an elevation of 1,912 metres = 6,273 feet. It may be observed that the word 'Noguera' is an anagram of the word 'Garoune,' the Catalan way of pronouncing Garonne. The Aranais describe the different characters of the two streams in the following Catalan distich :—

> Garonna per Aran, braman ;.
> Noguera per Lous, tout doux.

'The Garonne roars down by the valley of Aran; the Noguera, all gentle, by that of Lous.' The path continues, along the stream of the Noguera, northward for some time, but making an angle to the east at the **Hospice of Montgarri.** At the Hospice of Montgarri there is bread and wine, and at a pinch the night may be passed here; it does not do to start late for the Port de Salau, as it is not easy to find. From the Hospice de Montgarri it is 3 hours of descent along the stream, which soon loses its original character of gentle, and, about 4 kil. = 2½ miles above the hamlet of Lous, again turns to the north, and in 2 hours more reaches the Port de Salau. This port offers a very easy passage; and 2½ hours from the summit, in a NE. direction, will bring you to the village of **Conflens** (chez Bardou). 1 hour before reaching Conflens is the hamlet of Salau, with an indifferent auberge. From Conflens (898 m.), where there is a douane station, 28 kilometres to Saint-Girons.

ROUTE B.
By the Port d'Aula, 12 hours; mule-path.

About 2 hours below the Hospice de Montgarri, shortly after passing the hamlet of Mongossou, and about 1 hour higher up than the track to the Port de Salau, there is a gorge leading N. to the **Port d'Aula**, height 2,237 metres = 7,340 feet, between the Pic d'Areou to the left and the Tuc de

Berbegué to the right. From the Port d'Aula there is a fine view of the Pic de Montvallier, a little to the NW., height 2,840 metres = 9,318 feet. A little below the col you come to the Prat Mataou (Pré du Massacre), the scene of a bloody combat between the Catalans and Arriégeois. Below this the path traverses another wild plateau, and the Lac d'Aréou, surmounts the Col de Pauze, and then, turning to the E., descends by the valley of Salat to Conflens.

From Conflens to St. Girons 28 kil. = 17½ miles, passing through the village of Seix 10 kil. = 6¼ miles below Conflens.

For another mountain route into Ariège, see Sec. 91.

Section 95.
MONTVALLIER.
2,840 metres = 9,318 feet.

A mountain, of which the form and height much resemble that of the Pic du Midi de Bigorre. Its advanced position to the N. renders it conspicuous from nearly all the plain of the Haute-Garonne, and the number of points from which it can be seen sufficiently evince what a panorama it must command. Though a mountain of only the second rank, it is impossible to persuade the peasants of Ariège that it is not the highest point of the chain; just as the native of the Eastern Pyrenees believes in the pre-eminence of the Canigou.

The mountain is best ascended from the large village of Seix, on the road from St. Girons to Conflens, 10 kilometres below the last.

From Seix take the road to Conflens for 3 kilometres, and then turn SW. up the gorge of Estours. 1 hour to the hamlet of Estours; and thence 1 hour by a good path along the right bank of the stream to the granges of Artigues. After passing these, mount W. across the pasturages to the Col de Cruzous (2,316 m.), an opening on the N. of the peak, which must be attacked from the W., leaving on the right the lake of Cruzous, and on the left the peak of Montvaillerat (2,652 m.). 1 hour 20 minutes from the col to the summit; 6 hours from Seix.

Section 96.
ESTERRI TO VICDESSOS.
By the Port de Tabascain.

After passing the mountains east of Viella by the Port de Paillas, or the Col de Peyreblanque to Esterri, in the valley of the Noguera, you may cross again into France by the Port d'Aulus (2,237 metres = 7,330 feet) to Aulus; or by the Port de Tabascain (2,319 metres = 7,509 feet) to Vicdessos. Of these two passes the last is the most recommended, from the glorious view it affords of the snow-draped Montcalm. The journey, however, is rather too long to be undertaken in one day; and it is advisable to break it, and submit to make your quarters for one night at Esterri or Tabascain, in the valley of Cardos.

From Vicdessos (Hôtel de la Renaissance) there is a path over the mountains to Aulus (Hôtel Souquet) in 5 hours.

	DISTANCE	
	kil.	miles
Vicdessos to Tarascon (hotel, Gabach-Ginestet)	15	9½
Tarascon to Foix (hotels Lacoste, Rousse)	16	10
Foix to Toulouse (hotels Souville, de l'Europe)	82	51

113 kil. = 70¼ miles.

ESTERRI TO AULUS.

From Esterri you may also pass into France more to the W. by a mule-path, leading over the **Port d'Ustou** and descending upon St. Lizier d'Ustou (auberge, chez Galli), a village in which the education of bears is the principal occupation of the inhabitants.

From St. Lizier d'Ustou, so named from the forests that were burnt to give place to cultivation, there is a mountain-path E. over the Col de Latrape (1,122 m.) leading in 2½ hours to Aulus.

From St. Lizier to St. Girons, wheel-road, 28 kilom.

South of Aulus (776 m.), 33 kilometres from St. Girons, there is a cascade, described by Count Russell as being the finest in the Pyrenees, and of which he says:—'The roar may be heard at a distance of 8 kilometres.' To reach the cascade, start from Aulus SSE., and ascend a rough stony path along the left bank of the torrent of Arse; 1 hour 20 minutes to the foot of the cascade, composed of three separate falls, which at a little distance appear united in one, of near 200 metres in height. To ascend from the foot to the top of the fall requires 40 minutes. Cross the stream, and take the narrow path which winds up the right bank. The plateau on the top is a desolate amphitheatre, strewn with huge granite boulders, and cradling small lakes. S. of this is the **Port Guillou** (2,342 m.), from which you may descend in 3 hours to the Spanish village of Tabascan.

Section 97.

ASCENT OF THE MONTCALM AND THE PIC D'ESTATS.

Vicdessos (Hôtel de la Renaissance), a village situate at a height of 695 metres = 2,280 feet, on the left bank of the Vicdessos, is the centre of a mining district, where those who are curious in such matters may study the working of the forges Catalans. The mines of Rancié, to the south of the village, are the most productive.

The Montcalm, 3,080 metres = 10,105 feet, may be ascended most readily from Vicdessos. From the village of Auzat, 10 minutes from Vicdessos, the Montcalm is in view at the head of the valley. Here leave on the right the gorge leading up to the **Port de Saleix** (1,801 m.); and so to Aulus in 5 hours. Two hours from Vicdessos, following the path along the stream, brings you to the Pont de Marc, where the valley divides into two branches. The eastern branch leads up the savage gorge of Artigues; and the western branch, which is the one to be taken, conducts across some fields, which are still cultivated, to the Granges d'Amperrot. Here M. de Chausenque, who ascended the mountain in 1829, made his quarters for the night; but the traveller with a sleeping bag would probably select some spot higher up the mountain. From the Granges d'Amperrot you cross the torrent, and immediately on reaching the right bank the actual ascent commences. Leave on the right the gorge leading to the Port de Tabascan, and in 1 hour pass on your left a cascade. A mount of 1 hour 30 minutes, SSW., brings you to the cabanes of Pijeol (1,704 m.), and 40 minutes further to the plateau on the top, and the cabane de Subra. Thus far horses may arrive, but no farther. Here the Montcalm rises grandly in the SSW. more than 1,000 metres above you. Having crossed the torrent, attack the peak directly, leaving on the left the gorge and rocks of Rioufred, by which also the ascent may be made, but it is longer. There is no real precipice; and a climb of 2½ hours from the cabane of Subra over snow and schistose arêtes will land you on the summit; good 6 hours from Vicdessos.

From the summit, a huge flattened dome (3,079 m.), there is a splendid view, especially of the Maladetta in the WSW. In the SW., at a distance of only 500 metres as the crow flies, is the sister Pic d'Estats (3,150 m.), which may be reached with the greatest ease in 30 minutes, by following the crête which unites it to the Montcalm. South of these two peaks, and between them, two little lakes nestle glittering in the sun, exempted alone by their southern aspect from a covering of éternal ice, for their elevation is probably not less than 2,700 metres. The precipices of these two lofty peaks are not formidable, and on neither of them are there regular glaciers. Indeed, east of the Maladetta, there are no glaciers in the Pyrenees; and the mountaineer need fear nothing, but a long course and fatigue. From the Pic d'Estats the return to Vicdessos may be made the same way in 5 hours. The descent might probably be also made to the Port de Tabascan.

From Vicdessos to Foix 29 kilogrammes. Diligences daily.

Section 98.
TOULOUSE TO AX.

The railway now being open between Toulouse and Foix, this is the most convenient and easy method of entering Ariége, and the Eastern Pyrenees. Trains run in 3 hours 15 min.; price of places 9 fr. 30 c., 6 fr. 95 c., and 5 fr. 10 c. There are three daily diligences between Foix and Ax during the season; from Toulouse to Ax the distance is as follows:—

Toulouse	DISTANCE	
	kil.	miles
Foix	82	51
Tarascon	16	10
Ussat	3	2
Les Cabannes . . .	7	4
Ax	16	10

124 km.=77 miles

At **Foix** (Hôtel Lacoste, good and reasonable) the most striking object is the ancient chateau of Gaston Phœbus, now converted into the county gaol. Of the three towers, the low square one on the north is the most ancient. The round tower perched at the other end of the rock is the most lofty and striking. Its height is 42 metres. It was built or at any rate repaired by Gaston Phœbus, in 1361.

Diligences daily from Ax to Foix, (42 kil.), leaving at 1 p.m., arrive 6.15, fare in banquette, 5 fr. 25 c.

At Tarascon, 16 kil. from Foix, the road branches, that to the SW. leading to Vicdessos. (15 kil.)

Ussat (hotels, Cassagne, de la Renaissance) is separated from the road, being on the right bank of the Ariége. From Les Cabannes the **Pic St. Barthélemy**, or Pic de Tabe, though only 2,349 metres = 7,707 feet, presenting an admirable panorama, may be conveniently ascended; 6 hours to ascend, 4 hours to return. The ruined castle of Lordat, the largest in the Eastern Pyrenees, is also worth a visit. **Ax** is a little town and thermal establishment, 710 metres = 2,329 feet above the sea (hotels, Boyé, Sicre; the latter is the best: room and board 6 francs per day). Ax is one of the most remarkable thermal sites in the Pyrenees; and the surrounding scenery is extremely beautiful, presenting charming combinations of water, rocks, hills, and mountain crests, some of which are snow-clad. The waters are the hottest in the Pyrenees, and the valley is literally a vast boiling cauldron. The last official medical report states there are eighty-four warm sulphurous springs, and you have only to bore a hole in the ground to obtain hot water. The junction between the limestone-slate and the great granite chain, which rises immediately to the east of Ax in huge mountain masses, forming the geological axis of the Pyrenees, occurs precisely in the centre of the town. The hottest of the springs attains a temperature of 76° centigrade = 168·4° Fahrenheit.

Section 99.

AX TO THE VALLEY OF ANDORRE.—Route A.

Ax to Andorre by the Port de Saldeu, 15 or 16 hours' journey. Better to make 2 days, and sleep at Canillo, 9 hours 30 min.; carriage-road as far as Merens, 8 kil. = 5 miles; thence mountain-path practicable for horses.

Ten minutes from Ax, the road, which ascends the right bank of the Ariége, crosses the stream by a stone bridge of a single arch, and, after again crossing, in 1 hour 30 minutes brings you to the village of Merens, height 1,085 metres = 3,560 feet. Here there is a passable auberge, which you will not find at Hospitalet. After this the valley is less confined, but more and more sterile; and continuing to ascend by a rough road in a southwest direction, two hours more brings you to Hospitalet, the last French village, 1,411 metres = 4,629 feet above the sea level, and 17 kil. = 10½ miles from Ax. On leaving Hospitalet, cross to the left bank of the stream; and ascending the gorge, 15 minutes brings you to a douane station, close to the bridge of Cerda. Here the path divides into two branches; that to your left, crossing the Ariége, leads to the Col de Puymorin and Puycerda (see *post* route B). That to the right continues to ascend (SSW.) along the left bank; and winding its way upwards by a stony and steep track along the flanks of the mountains, with not a tree or shrub to break their desolation, 2 hours 30 minutes brings you to the Rochers d'Avignolles or Pourtailles, where the Ariége takes its rise.

Observe that all this part of the Pyrenees is very destitute of forests; but the pasturage reaches as high as 2,700 metres; and there are a great quantity of rare and good plants. Above Hospitalet, in ascending the Port de Saldeu, notice among others the *Ranunculus aconitifolius, R. lacerus,* and *Gentiana pyrenaica*. None of them found in the Central Pyrenees; also some rare Senecios.

At the source of the Ariége two gorges open; that to the left leads, by a longer though somewhat easier route, into the valley of Andorre, over the Port de Framiquel; that to the right, which is the continuation of the track you have been following, is the one to be taken leading to the **Port de Saldeu,** 2,500 metres = 8,202 feet. After clearing the ridge, and crossing a narrow plateau, the course lies west, and, following the stream of the Embalire, you descend into the upper valley of Andorre. The narrow gorge to your left, with its dark forests, is the issue on the Spanish side of the Port de Framiquel. Facing you rises the snowy summit of Mount Rialp. Two hours from the top of the port brings you to the wretched hamlet of Saldeu, where the stream bends to the south. Below Saldeu, leaving on the right the gorge that mounts to the Port de Fontargente (2,252 m.), by which you may pass to Les Cabannes, follow the course of the stream along its right bank. In 1 hour more you reach the village of Canillo, where you may procure a lodging for the night. On leaving Canillo, cross the stream of the Embalire, and passing by the chapel of Merichel, and the villages of Encamp and Las Escaldas, 3 hours from Canillo brings you to Andorre, the capital of the Republic. A little before reaching Andorre, the road crosses the stream of the Ordino, which comes down from the gorges of the Port de Rat, and the Port de Siguier, by either of which you may pass into the head of the valley of Vicdessos.

The fashionable 'quartier' of Andorre is the *plaza* with the fountain. Ask for the house of Don Guillem, on the W. side of the square, where you will be hospitably received, though not very luxuriously entertained. In this square is the church, but the building most deserving a visit is *La Casa de la Valle* ; where the council - general have met from time immemorial, and which also serves for the state prison and hall of justice. Over the portal of this building, which might be taken for a porte-cocher leading into a stable, are the arms of Andorre, surmounted by the inscription *Domus consilii, sedes justitiæ*. Underneath are some Latin verses. In the interior, are the council chamber, a chapel, the archive-room, and the kitchen, with a huge fire-place in the centre, and a vaulted chimney opening over-head, after the Spanish fashion.

From Andorre to Urgel, a town of 3,200 inhabitants, there is a road, or rather mule-path, for there are no wheel-roads in Andorre, along the river Embalire ; time 6 hours.

From Andorre follow the right bank of the Embalire ; 40 minutes to Santa Coloma (Santa Columba), one of the six parishes, where there is a Gothic church. 30 minutes below this cross to the left bank of the Embalire ; and 2 hours from Andorre, reach San Julia de Loria (Lauredia), the grand shopping emporium of the valley, and where, under cover of its neutral position, the smuggling trade is carried on to a large extent. Here the temperature is less rigorous, and the vegetation of a more southern climate begins to show, such as hemp and tobacco ; and the hill sides are planted with vines. 20 minutes from San Julia, the cascade of Auvina is passed on the left ; and an hour below this the road enters Spain, the frontier being marked by a post of *carabineros*, by the side of a forge and saw-mill. 2 hours more, always on the left bank of the river, to the **Seu d'Urgel**, situate on the right bank of the river Ségre, just above the confluence of the Embalire.

The see of Urgel confers on its bishop the further title of prince of Andorre. The cathedral, having been re-constructed, was consecrated by Bishop Sizibut, A.D. 819.

Before taking leave of Andorre, a short history of its constitution may not be without interest.

In the eyes of many the Valley of Andorre is a land of sentiment and romance. Let them travel in it, and they will change their opinion.

In the first place, it is not a republic, but rather an aristocratic federation of the valley, retaining much of the spirit of feudalism. It is about 40 kilometres long by 36 in breadth. The federation consists of six parishes : Canillo, Encamp, Ordino, Massana, Andorra, and San-Julian, with a population in all of about 6,000, that of Andorra, the capital, being 850.

Andorra, with all the surrounding country, was emancipated from the Moors by Charlemagne, A.D. 778, who granted rights over this valley, and the tenths of its revenue, to the bishop of Urgel, reserving half of the tenths from the town of Andorre to an individual, for especial service rendered in the war. This half-tenth is to this day appropriated to Don Guillem Plandolit, the representative of that family, and the portion is called ' le droit Carlovingien.'

The rights over Andorre, conceded to the bishop by Charlemagne, were confirmed to him by his son, Louis-le-Debonaire, who also granted certain privileges to the inhabitants of the valley, secured by a charter, which is the foundation of their independence. This charter is carefully preserved in the archives of the valley. It is in Latin, signed by Louis, and some other counts and bishops, and dated 805.

Subsequently disputes ensued between the counts and bishops of Urgel, concerning their rights over the valley, and Bishop Bernard, in 1094, being worsted in arms, called in the aid of Raymond Roger, Count of Foix.

It was the old story of the horse calling in the man to help him against the stag. The bishops found the counts of Foix more exacting in their claims, and more formidable to refuse, than the counts of Urgel.

At last, A.D. 1278, by the mediation of the bishop of Valence and others, an arrangement was made, and an agreement signed, known in Andorre as the *Acte de Paréage*. By this deed it was stipulated, that Bernard Roger, Count of Foix, and his successors, should exercise the rights of lordship over the valleys of Andorre, *par indivis*, with the bishops of Urgel; that the counts of Foix should receive tribute (questia, *quête*) from the valleys, jointly with the bishop; and that they should each delegate an officer named a 'viguier,' with the exercise of criminal and civil justice. This decree was signed by the bishop, Pierre d'Urgis; by Roger Bernard, Count of Foix; by Pedro, King of Arragon, and approved and sealed by the Pope Martin IV.

The house of Foix having become united to that of Bearn, A.D. 1266, in the person of this Roger Bernard, who married Constance de Moncade, the heiress of Bearn, his successors added to their other titles that of *Princes souverains par indivis de la Vallée d'Andorre*; until both these counties, as well as the kingdom of Navarre, in the person of Henry IV., passed to the House of Bourbon, and merged in the crown of France, A.D. 1589.

The privileges of the Andorrans as an independent and neutral state have been recognised on several occasions by the sovereigns both of Spain and France. The kings of France, successors of Henry, as *Princes souverains par indivis de la vallée*, have always respected the rights of the inhabitants, and conformed to the ancient usages established by the counts of Foix and bishops of Urgel; while the government of Andorre has always respectfully acquiesced in the nomination of the French 'viguier.'

In 1793 the amicable relations of Andorre with France were for a few years interrupted; but in 1806, by a decree of Napoleon, of March 27, the inhabitants of the valleys of Andorre were reinstated on their former footing, at their own urgent request. The decree is thus worded:—

'Vu la demande des habitants des vallées d'Andorre, tendante à être rétablis dans leurs anciens rapports d'administration, de police, et de commerce avec la France, etc.:

'Article 1. Il sera nommé par nous, sur la présentation du ministre de l'intérieur, un viguier, pris dans le département de l'Ariège, et qui usera de tous les priviléges que les conventions ou l'usage lui avaient attribués.

'Art. 2. Le receveur-général du même département recevra la redevance annuelle de 960 francs.

'Art. 3. La faculté est accordée aux Andorrans d'extraire annuellement la quantité de grains, et le nombre de bestiaux dont l'arrêt du conseil de 1767 leur avait garanti l'extraction.

'Art. 4. Trois deputés des Andorrans nous prêteront serment, chaque année, entre les mains du préfet de l'Ariège, que nous autorisons à cet effet par le présent décret.

'Art. 5. Nos ministres de l'intérieur, des finances, et des relations extérieures sont chargés, etc.

The authorities composing the government of Andorre are the council-general, composed of 24 members, elected by the six parishes, half of whom retire annually in succession; and a syndic, who is the president of the council. The council decides by the majority of votes; and the syndic is charged with the execution of their decrees. The syndic also annually renders an account of the receipts and expenditure of the state. There are neither taxes nor customs in Andorre; and the revenue consists solely of the proceeds from the letting of the pastures to the Spaniards of Urgel, and from the sale of the superfluous wood. The expenditure is naturally also restricted. From the viguier, downwards, there is no paid public officer. The annual contributions of 960 francs to France and 450 francs to the bishop of Urgel have the first claim on the public purse. Then comes the expense of maintaining the *casa de la vallée*, or house of parliament, and the porter's wages. This is about all. The army costs nothing. All must serve if called upon; and every head of a family is bound to keep a musket, and a certain quantity of powder and ball. Every parish supplies a captain and two officers, and the two viguiers are the generals-in-chief.

The administration of justice, civil and criminal, emanates from the French government and the bishop of Urgel. Each of these appoints a chief magistrate, or viguier, with this difference: the viguier of the préfet is a Frenchman, and appointed for life; that of the bishop, a native of Andorre, and capable of being dismissed after three years. When a crime is committed, the accused is arrested and information given to the viguier d'Urgel, who, being Andorran is, in the country. The viguier present investigates the case, and communicates it to his colleague in France. If the charge is grave, the viguiers summon a general council, which appoints two of its members to sit as a criminal court with the two viguiers. If the charge is lighter, the viguier alone decides. In no case is there a jury. There are no written laws, or fixed penalties; but the viguier hears the case, decides, and sentences, according to his conscience. There is no appeal in criminal cases, and the sentence is carried out within the 24 hours. The capital punishment is death by the garrotte. It is inflicted in bad cases of murder.

Civil cases are decided by two *bailes*, who are named one by each viguier, out of a list of six presented by the syndic. In civil cases there is an appeal from the decision of the 'baile,' to a judge of appeal, who is appointed alternately by the préfet of Ariège and the bishop of Urgel. The judge of appeal has no fixed emolument, but custom gives to him, and to him only (for all the other offices are gratuitous), 15 per cent. on the value of the object in litigation, which he takes good care to levy before putting the successful party in possession. Appeals, however, are not very frequent, for more often than not the judge, being able to decide without going to Andorre, will not undertake the journey; and the litigants cannot be at the trouble and expense of going to seek him.

The manners of the people are patriarchal. The right of primogeniture is strictly observed, and all the younger members of the family are hangers-on and dependants. In the case of daughters alone, the eldest inherits, and the man who marries her takes the name of the family. Hence the property is handed down with little change from generation to generation. There are none absolutely indigent. The poorest always have a chimney corner and a meal in one of the more affluent houses.

Education is at a low ebb in Andorre; a large proportion of the people can neither read nor write. The chief of the family seldom quits his property. He is clothed and fed exactly the same as his servants, and eats at the same board. He indulges in no personal luxuries, and, if wealthy, employs his superfluous income in improving his property and increasing his flocks.

The Andorrans are strict protectionists. Those proprietors who have grain more than sufficient for their own households are forbidden by law to export it, or sell to anyone but a native of the valleys, while another salutary law provides against forestallers and regraters.

Breeding of cattle, rearing of mules bought at the fairs of Tarascon and Toulouse, letting of sheep pastures, and the working of iron foundries are the sole sources of public as well as of private revenue.

For six months of the year cut off from intercourse with France by snow-choked ports, in customs, language, and geographical position, the inhabitants of Andorre are Spaniards. They owe the independence of their territory during ten centuries partly to the double patronage under which they have placed themselves, partly to the natural barriers by which they are protected; but they owe it still more to the insignificance and poverty of their territory, which the policy of their council has never sought to extend, and so arouse the covetousness of their powerful neighbours.

With their desires bounded as their horizon, the Andorrans, to use the words of Southey, are 'plain men, who have neither manufactories to corrupt, ale-houses to brutalise, nor newspapers to mislead them.' If deficient in that ambition which is the mainspring of all progress, they have not ill chosen for their own happiness, and afford a striking example to nations more advanced, that content is natural wealth, luxury artificial poverty.

Section 100.
URGEL TO PUYCERDA.
40 kil. = 25 m.; time 8 hours.

From Urgel to Puycerda there is a good mule-track, ascending the river Sègre, and traversing the fertile valley of Cerdagne. Cerdagne, like Andorre, became independent after the invasion of the Moors, but did not remain so. In 1196 it became a province of the kingdom of Aragon, and was subsequently absorbed in the Spanish monarchy. By the treaty of the Pyrenees, in 1659 a small portion in the NE. was ceded to France. The Cerdagne Française, however, exists in defiance of all natural boundaries, and the emperor would no doubt be glad to exchange it for the Spanish Val d'Aran.

At Belver, an old feudal-looking town on the left bank of the Sègre, 5 hours from Urgel, there is a fine view, and a posada.

Section 101.
PUYCERDA TO AX.
38 kil. = 23·6 miles, mule-path; 10 hours.

The fact of its being the ancient capital of Cerdagne does not excuse the dirtiness of Puycerda. It is a wretched half-Moorish half-Spanish town, with the streets much resembling those of Fontarabie; and the traveller who has been compelled to sleep here will be ready for an early start. Puycerda is perched on a little monticule, at a height of 1,242 metres. Descending from this by the NW., after winding through a labyrinth of low earthen walls which surround the town, you strike the road, which, after passing (1 hour) the Tour de Carol, built by Charlemagne in honour of his victory over the Moors, ascend above the left bank of the stream as far as Porta, where it turns more to the N. Before reaching Porta, notice a very wild gorge on the left, by which there is a rather difficult pass to Andorre. 50 min. from Porta to the village of Portet, 4½ hours from Puycerda, in the gorge of Fontvive. To reach the village of Portet, where there is an auberge, cross the stream to the left bank. From Portet mount NW. 1 hour 10 min., to the **Col de Puymorins** (1,931 m.), between the Pic de Fonfrède, 2,554 metres = 8,380 feet, to the S. and the Pic Sabarthe, 2,549 metres = 8,363 feet, to the N., forming the limits of the departments of Ariège and the Pyrénées Orientales. One hundred yards from the path, on the left, is a douane station. Descend from the col NW. and afterwards N. to Hospitalet, which is reached in 1 hour 10 min. Thence to Ax, 17 kil. along a good road.

Section 102.
ANDORRE TO FOIX,
By the Port de Siguier; 15 hours.

The return from Andorre may be varied by the above route, which is also the most direct. Leaving the town of Andorre by the N., for 15 min. continue along the right bank of the Embalire; but on reaching the stream of the Ordino, turn up this gorge NNW. There is a good mule-path, first along the left, and afterwards along the right, bank of the stream. On reaching **Massana** (*Massiana*), one of the six parishes, 1 hour from the entrance to the gorge, the stream bifurcates. Take the branch to the right, 40 min. more, NE. along the right bank of the Rialp, to **Ordino** (*Hordinavi*), another of the parishes, where a good breakfast may be had, but where the right of getting what they can out of strangers seems to be the construction put on the 'droit Carlovingien.' From Ordino mount N. along the stream; 1 hour to the hamlet of Llors, which you leave on the left; and 50 minutes farther to Sarrat, the last houses. Here the gorge splits into two branches. The western branch, to the left, leads up to the Port Neuf and the Port Vieux, or d'Arbeille, both leading down to Vicdessos. To reach the Port de Siguier, take the branch to the right. The path mounts NNE. among rocks and stunted pines, first along the right, and afterwards along the left, bank of the torrent. After mounting 30 minutes, the trees disappear. Continue along the left bank of the stream, direction N., till 1 hour further the stream bends rather to the W. By the side of the stream the *Ranunculus aconitifolius* is abundant. Do not quit the watercourse till you find yourself in a desolate and rather striking amphitheatre at the foot of the Port de Siguier. From here there is a zigzag track, mounting N. to the port in 40 minutes. The height of this port is 2,594 metres. 10 minutes' easy descent brings you to a collection of little semi-frozen tarns, and by these, surrounded by patches of snow, and not more than 100 metres below the port, is a wretched shepherd's cabane, tenanted only for two months in the year, and probably the highest in the Pyrenees. The Port de Siguier is a good 6 hours from Andorre, and it is 5 hours of rough walking down to the village of Siguier.

In descending from the port, 1 hour below the cabane is a lake, the Lac de Peyregrand, which you must pass on the right, skirting it by a very stony track. Below this, a steep descent, and then the gorge bends to the W. After passing a limpid pool formed by the torrent, 3 hours below the port, when, as you fancy, almost on a level with the lower valley, you suddenly find yourself on the upper brink of a precipice, which you have to outflank, first mounting a little to the right, and then descending by a bad stony zigzag above the right bank of the torrent. Below this there is no difficulty: on reaching the main valley, follow the path N. At Siguier there is a passable auberge, on the right-hand side; thence 1 hour 30 minutes to Tarascon; carriage-road (9 kil.).

Those who like a climb may ascend the **Pic de Signier**, or Mont Rialp (2,903 m.), W. of the Port de Siguier; but this will probably entail sleeping in the cabane below the port on the French side. From this cabane it takes about 5 hours to reach the summit and return. From the cabane cross the ridge a little W. of the port, and then descend a little, making for the southern arête of the mountain, which must be crossed, and the summit finally attacked from the SW. There is no difficulty, and a grand view, especially of the Pic d'Estat and the Montcalm. There are some good plants on the mountain, *Anemone sulfurea*, a yellow tulip, *Tulipa celsiana*, and *Loiseleuria procumbens*.

Section 103.
AX TO MONTLOUIS AND CABANASSE.

In the whole range of the Pyrenees, there is no spot offering such charming head-quarters to the botanist as Cabanasse, in the Eastern Pyrenees. The mountains will not compete in grandeur with those of the Central Pyrenees; but any of them may be ascended without too much difficulty; they abound in rare and beautiful plants; and at Cabanasse, which is close to the celebrated Vallée d'Eynes, there is a capital little country inn, inferior to that of Gavarnie only in the scale of charges, which are about half; the accommodation and food is rather better; and the elevation exceeds that of Gavarnie by 200 metres. The master of the inn, M. Vaillant, when a boy, in 1825, accompanied Bentham in his botanical researches in this part of the chain.

For approaching Cabanasse **the route by Lac Lanoux** is strongly recommended; 2 days, on foot.

First day, to Lac Lanoux, 8 hours.

At Merens, 1 hour 30 minutes from Ax, a guide, or porter, may be taken. From here continue 1 hour along the road, and after passing a cascade on the left, just before reaching the zigzags, quit the wheel-road, and mount E. by a path on the left bank of the stream, in the gorge of Bésines. *Luzula nivea* abundant. 1 hour 50 minutes, always on the same bank of the stream, to a little triangular lake (height 2,069 m.). [Here for the first time I saw the *Gentiana pyrenaica*, the most beautiful of its tribe, which in the Eastern Pyrenees is as common as the *G. verna* elsewhere, on the upland pastures from 1,800 to 2,200 metres. It does not exist in the Central Pyrenees, though stated by Lapeyrouse to have been found in the Cirque de Gavarnie. The Port de Siguier is the most western station where I have observed it.] From the little lake, leaving the Pic d'Auriol on the left, and the Pic Pedrous (2,831 m.) on the right, 1 hour to the uppermost cabanes of Bésineilles (height 2,200 m.). From these cabanes, the direction is ESE.; and an easy ascent of an hour will place you on the Col de Bésines (about 2,350 m). In mounting the col, remark the profusion of *Saxifraga geraniodes*, and *S. pentadactylis*, both rare plants. From the col you have before you the **Étang de Lanoux**, nearly 3 kilometres long (height 2,154 metres), perhaps the largest sheet of water in the Pyrenees; and on the other side of this in the SE. rises the sharp pyramid of the Pic Carlitte. All round the shores of the lake the country is most savage and desolate, not a tree in sight.

Do not descend at once upon the lake, where the ground is boggy, but traverse the rocks obliquely, making for its southern extremity. On these rocks, in the beginning of July, we found some good plants: specially, *Gagea minima*, and the same yellow tulip which I saw on the Pic de Siguier. 1 hour, 30 minutes from the col to a shepherd's cabane at the SW. extremity of the lake, on the left bank of the stream. This cabane is eulogised by M. Russell rather more than it merits; but you may pass the night here very well, of course having brought your own provision. I believe there are some excellent fish in the lake. By following down the stream of the Fontvive, which flows by the cabane, in 2 hours you reach Portet.

Second day, to Cabanasse.

Start early, 6 A. M. at the latest, that you may have time to ascend the Pic Carlitte, and to botanise. Cross the stream of the Fontvive, and mount due E. and then SE. over rocks and beds of snow for 1 hour 20 minutes, as far as a small lake, which is frozen 9 months of the year. From here another hour of steep but not difficult climbing will place you on the Col Carlitte (2,600 m.), due S. of the pic. From here, even, there is an extended view over a very savage country; but it is quite worth while, being so near the summit, to mount the **Pic Carlitte** (2,921 m.): first, because it is the monarch of the Eastern Pyrenees.; secondly, because there are some very good plants to be found between the col and the summit: *Luzula lutea, Papaver pyrenaica, Androsace argentea* and *A. pyrenaica, Primula latifolia* and *pedicularis rostrata*; all these, except the poppy, in flower at the end of June. The rocks of the Carlitte are schist, but all the country to the E. is a sterile waste of granite sprinkled with mountain tarns. From the Lac Lanoux the Pic Carlitte appears menacing; but there is no difficulty in the ascent. From the col mount N., descending from the arête, where necessary, rather upon the eastern side; a short hour to the top. [Two kilometres SSW. of the Pic Carlitte is another col, the Col Rouge, so called from its red schistose rocks; from it you may descend on the E. into the gorge of Fontvive, and so to Portet.]

From the summit of the Pic Carlitte, descend towards the col, and then E. over frozen and crevassed beds of snow, at the bottom of which you traverse a very wild bit of country with numerous tarns, the Étangs de Carlitte, on the sandy borders of which grows the *Subularia aquatica*.

Two hours from the col, you reach the last of these tarns, the Lac Noir, still at an elevation of 2,000 metres, though apparently now in the region of the plain. On these upland pastures the *Gentiana pyrenaica* grows in profusion; but here it is scarcely in flower till July. From the Lac Noir continue E., striking the head of **La Têt** river, which you reach in 30 minutes. Descend the gorge, finding your way through the forest as best you may, and keeping on the right bank of the stream. The path through this forest is not very easy to find, but it contains such a rich harvest of plants that the botanist will not regret that it is not more trodden. After 2 hours through the forest, leave the stream below to the left, and ascending and descending, sometimes among green and open glades, sometimes among fir-trees, keep a general direction SE. You emerge on the main road just outside the gates of Montlouis, 6 hours from the col. It is not necessary to enter the town of Montlouis, where the inns are bad, and the gates of the town are locked every night. Take the first turning of the road to the right, which will bring you to Cabanasse, 1·5 kil. S. of Montlouis. The height of Cabanasse is 1,550 metres, that of Montlouis is 1,603 metres (371 m. above Barèges) being the highest inhabited town in France. The house of Vaillant, which cannot be too much praised, is on the left as you enter the village of Cabanasse.

Section 104.
AX TO CABANASSE BY QUERIGUT.

50 kil., mule-track as far as Querigut (25 kil.), thence to Cabanasse (25 kil.), wheel-road.

Those who do not care to sleep in a shepherd's cabane should take this road, and sleep the first night at either Querigut or Formiguères. All that country is very rich in botany, and deserves exploring. Leave Ax by the E. and follow the left or southern bank of the river Ode; 2 kil. to a forge catalan on the borders of the stream. The neighbouring heights are all pierced with holes, resembling ancient excavations. From the forge, above which is the village of Ascous, continue along the same bank, remounting the main valley, which gradually bends to the NE. and becomes a narrow defile. The **Col de Paillers** (1,972 m.), at the head of this, is surmounted without difficulty. The col is a vast grassy plateau dominated on the S. by the serrated crags of **Llaurenti**; one of the most rich botanical regions in the Pyrenees. To the E. the Sonne torrent descends into the valley of the Aude. Follow this, descending by a zigzag path, and keeping always on the north side of the gorge. In 1 hour 30 minutes you come to the village of Mijanes, and 1 kil. beyond this to the village of Rouze (973 m.). Perched on the opposite bank of the stream appears the old castle of Usson. On leaving Rouze, turn S. at a right angle, and traversing the Sonne R. mount a valley running S., parallel to that of the Aude, but more to the W. After passing the village of Pla, 2 kil. from Rouze, 2 kil. farther you reach Querigut, the 'chef lieu de canton.' At one of these villages, the botanist should sleep a night, for the sake of exploring the chain of Llaurenti, said to be the richest in plants of the whole Pyrenees.

From Querigut the road mounts SSE. in long sweeps on the east side of the gorge towards the chain separating Ariége from the Pyrénées Orientales. On attaining the col (1,600 m.), the fertile valley of **Capsir** is below you on the S., and, descending into this, the road leaves perched on the left the village of Puyvalador (1,458 m.) 7 kil. from Querigut. A little farther it traverses the streams of the Fontrabiouse and the Galba descending from the W., and at 5 kil. from Querigut reaches **Formiguères** (1,480m.) a little town of 804 inhabitants, the ancient capital of the valley of Capsir. The church is remarkable for its antiquity. Tradition assigns the year 873 as the date of its foundation.

At Formiguères, as in all the villages of Capsir, it is always cold, even at midsummer, on account of their elevation.

From Formiguères, follow the main road into Spain, direction S. This traverses the pine forest of Matte, and 5 kil. from Formiguères traverses the **Aude**, here but a tiny stream. Thence it ascends over a high tableland to the col forming the water shed between the Aude and the Têt rivers (1,720 m.). Beyond this the road passes a little lake on the right, and descends by a desolate valley to the village of Llagona (1,688 m.); and thence 2 kilometres to Montlouis on the other side of the Têt river, 12 kil. from Formiguères.

From Formiguères to Olette, there is a wheel-road, 25 kilometres.

From Montlouis to Villefranche 29 kil.; thence to Vernet 5 kil.

Section 105.

VALLÉE D'EYNES AND CAMBREDASE (2,750 m.).

10 hours on foot.

Combining the flora of the granitic, the schistose, and the limestone rocks, there is scarcely a good plant of the Pyrenees that may not be found in the Valley d'Eynes. It is impossible, of course, to get all in flower at the same period, but the end of June seems as good a time as any for visiting it.

Leaving Cabanasse by the road to Spain, at the Col de la Perche (*Gentiana pyrenaica* abundant, and *Ranunculus Angustifolius*), strike SSW.; and leaving the village of Eynes below you on the right, make for the stream and ascend the rocks on the left (the N. side of the gorge), mounting through the wood. The rocks at first are granite, passing into schist, and on the summit into limestone. There is no difficulty in passing along the arête, proceeding from cairn to cairn, but on the N. side the rocks are too precipitous to admit of descending. Among the plants found on this arête are the *Alyssum diffusum, Iberis garrexiana, Papaver pyrenaicum, Artemisia mutellina, A. spicata* and the *Senecio leucophyllus*, a very beautiful as well as rare plant. Continue along the arête SE. for nearly 2 hours, and then descend from the Pic d'Eynes (2,786 m.) upon the **Col de Nuria**. On the Spanish side, 2 hours below the col, is the hermitage of Nuria, where refreshment and shelter for the night may be procured. Returning from the Col de Nuria, along the left bank of the stream, you find the *Anemone sulfurea*, and *Primula latifolia*, and much lower down the *Adonis pyrenaica*. For the lower part of the valley, the end of June is the best season, but on the high crête of Cambredase, where there is always some snow, the plants are not much developed till the middle or end of July.

The **Pic de Puigmal** (2,909 m.) may also be ascended from Cabanasse; but to go and return in a single day is a very long course; it would probably be better to descend to the hermitage of Nuria, and there pass the night. To reach the Pic Puigmal, take the southern arête of the Vallée d'Eynes, SE.; descending upon the Col de Llo (2,558 m.), 5 hours. From this col the hermitage of Nuria is in sight. From the Col de Llo continue to steer S. 1 hour 30 minutes, over the easy and round-backed mountains to the Pic de Ségre (2,795 m.), and thence 40 minutes to the summit of the Puigmal; seven good hours from Cabanasse.

The Puigmal is the highest mountain in the Pyrénées Orientales, and presents a view, striking from its desolation, over the Carlitte country. Like all the other mountains in this part of the chain, it is of easy access; a compass is the only guide required.

Section 106.

PERPIGNAN TO PUYCERDA.

100 kil. = 62 miles; carriage-road; diligences daily.

	DISTANCE		HEIGHT	
	kil.	miles	met.	ft.
Perpignan			30	98
Prades	42		320	1,050
Villefranche	7		392	1,286
(branch road to Vernet)				
Olette	9		665	2,182
Montlouis	20		1,513	4,964
Col de la Perche	2		1,621	5,318
Sallagossa	8			
Bourg-Madame	10		1,140	3,740
Puycerda	2		1,242	4,085

100 kil.=62 miles

For the route from Paris to Perpignan, see Sec. 1.

On leaving Perpignan (hotels, du Midi, du Nord, and de l'Europe) by the gate of Notre Dame, to the west, the road is carried through a fertile country along the right bank of the river La Têt. At **Prades**, the chief town of the arrondissement, the Hôtel Januari is recommended as both clean and good. On leaving the fortified town of Villefranche, the road crosses to the left bank of the river. In the hedge-rows, by the road-side, the pomegranate, *Pomum granatum* and jessamine, *Jasminum fruticans*, are seen growing wild. The *Jasminum officinale*, though common, seems hardly indigenous. At Olette (hotels, du Midi, de la Fontaine) there are some sulphurous waters, and a thermal establishment, but on a very mild scale. From Olette you take leave of the olive-trees and vineyards of the lower valley of La Têt, and the road gradually ascends to the strongly-fortified town of Montlouis on the right bank of the river. There is a hotel in the town (Jambon); but the traveller is on every account advised to give the preference to that of Vaillant, in the suburb of La Cabanasse, situate 1 kil to the SE. of the town, below the citadel. The road passes through this suburb before ascending to the Col de la Perche. From the summit of this col Mount Canigou is seen due east. From the col the road descends in a SW. direction, by a series of gentle windings, to the small town of Sallagossa, on the left bank of the river Sègre. On leaving Sallagossa, cross this river, and, continuing in the same direction, come to Bourg-Madame (Hôtel Jambon), the last town in France. From Bourg-Madame you cross the Regur river, and ascend by a stony road, whose badness at once reminds you that you have crossed the frontier, to the Spanish town of Puycerda.

Section 107.
PERPIGNAN TO VERNET.

54 kil.=33½ miles; carriage-road; daily public conveyances, price 8 francs.

As far as Villefranche the road is the same as that to Puycerda (Sec. 106). On leaving Villefranche for Vernet, distant 5 kil. = 3 miles, turn up the valley to the south. Vernet is a pretty retired little thermal establishment, 620 metres = 2,034 feet above the sea, placed to the NW. of one of the spurs of the Canigou (hotels, Thermes des Commandants, the principal establishment, where the accommodation and food is very good, but rather dear; Thermes Mercaders, and various lodging-houses). The waters are sulphurous, very similar to those of Luchon.

The ascent of the Canigou, which is only 2,785 metres = 9,144 feet, can be conveniently made from Le Vernet. The ascent may be made on divers sides. The most direct and best route is as follows:—On leaving Vernet, ascend along the right bank of the stream; 20 minutes brings you to the village of Casteill, and thence 30 minutes, by a zigzag path up the rocks to the left, to the ruined abbey of St. Martin du Canigou (*Asplenium fontanum* and *Cistus laurifolius*). From hence there is a fine view, looking back upon Vernet. Beyond these ruins continue to ascend in a SE. direction, following a path leading to the summit of an arête which separates two gorges. About 2 kil. (25 minutes) from St. Martin, this path suddenly comes to an end, and the inclination becomes steeper. You then have a stiffish climb over rocks and trunks of trees, making for the top of a ridge which interposes to hide the summit of the Canigou (*Luzula nivea, Lilium pyrenaicum, Gentiana Burseri, &c.*). It takes 2 hours from St. Martin to attain this first plateau, from whence you have a view of the double-peaked summit of the Canigou, with the naked rocks here and there intermingled with patches of snow. On this plateau there is a spring of excellent water, the last you will meet. From here the path winds for another hour round the W. shoulder of the mountain almost on a level, till finally it mounts among some dwarfed fir-trees to the plateau of Cadi. Here, on leaving the last stunted firs, the yellow flowers of the beautiful *Anemone sulphurea* are very abundant. Then 1 hour over a chaos of huge granite boulders, SW. of the summit, descending to the granges of Cadi. Among these boulders on July 2 there is still a quantity of snow; but do not fail to remark the beautiful composite plant, *Senecio leucophyllus*, which here grows in profusion, with its beautiful silver leaves, and golden flowers just bursting, though not in full flower till the end of July. From the granges of Cadi, 1 hour 15 minutes to the top. Make NNE. over the snow beds to a sort of cheminée, where steps have been cut in the rock. [Horses may be brought as far as the foot of this cheminée (30 minutes from the top) by following the gorge S. from Casteill, and the path over the Col du Cheval-Mort, to the granges of Cadi (5 hours from Vernet).]

On the top of the Canigou, there is a rough cabane which will serve perfectly for shelter in case of a sudden storm.

From its imposing form and isolated position, the Canigou was once considered the highest of the chain, though it is only a mountain of the third rank, and there are more than a hundred peaks exceeding it. There is a magnificent panorama from the summit, including all the eastern part of the chain, and a large horizon of sea from NE. to SE.; but I question whether the giants of the central chain can be distinguished from it, even on the clearest day. It is quite worth while to pass the night at the granges of Cadi, or better still in the cabane on the summit, for the sake of seeing the sun rise out of the Mediterranean. The buttresses of the Canigou are all of granite, but the actual summit seems to consist of gneiss. On the summit I observed no rare plants; *Erysimum alpinum, Hutchinsia alpina, Draba hirta*, and some saxifrages.

The descent may be made on the north side, but in places it is rough, and almost difficult, on account of the deep ravines between the buttresses of the Canigou. Follow the arête projected N. from the peak, keeping rather below it on the west. Here is a profusion of flowers: *Senecio leucophyllus, Gentiana verna*, and *G. acaulis*; a little lower, *Anemone sulfurea, Genista cinerea*, and *Rhododendron*; and below this the fir forests. 2 hours to a cabane and cows; from this descend into another ravine on the N., much wooded; for 20 minutes keep among the trees above the left bank of the stream, and then cross the ravine to the right bank, where the path is excellent. The path first winds among a luxuriant belt of the *Cistus laurifolius*; a very handsome shrub, with large white flowers, and leaves somewhat resembling those of an apple-tree. Lower down the path is again bad; and descending the stony ground to reach the bottom, you come upon several good plants, especially the *Lavandula latifolia*.

The descent to Vernet by this route occupies near 5 hours. You may also descend on the other side to Prats de Mollo, on the south of the mountain. You must first descend a little in a SW. direction to the Pla-Guilhem, from which to Prats de Mollo is 4 hours (Sec. 108). By those accustomed to mountains, the ascent of the Canigou may very well be made without a guide; but those who require one will do well to take one from the village of Casteill rather than from Vernet. Michel Nou, of Casteill, is spoken of as an excellent guide, and well acquainted with the geology and botany of the mountain.

Section 108.
VERNET TO PRATS DE MOLLO BY THE PLA-GUILHEM.

Mule-path; 9 hours; 14 hours if the ascent of the Canigou is made in crossing.

Vernet to Casteill, 2 kil., see *route* 107. On leaving Casteill, ascend the desolate valley leading due south; and a little beyond the Col du Cheval Mort (2 hours 30 minutes from Vernet), leaving to the right the path to the Canigou, the path ascends the rocks to the summit of the ridge that separates the valleys of the Têt and the Tech rivers, and unites the Canigou to the main chain, by the mountain of Costabona. This is the plateau of Pla-Guilhem, 1,850 metres = 6,070 feet in height, and to cross takes a good hour, still ascending. On reaching the eastern side of this plateau, you find an arête projecting between two gorges; that to your left is the one to be taken; it requires 2 hours to descend and reach the stream of the Moline, which flows along the bottom. This you must cross near some houses; and then climbing for 15 minutes on the opposite side of the valley, to pass above some precipitous rocks, you keep above the left bank of the stream for some distance, and descending between some singularly-formed rocks, at the junction of the Moline and the Tech, 3 hours 30 minutes from the Col du Pla-Guilhem, reach Prats de Mollo, the *chef lieu* of the canton, where there is a tolerable auberge. Prats de Mollo is situate on the left bank of the Tech, 798 metres = 2,618 feet above the sea. From Prats de Mollo, in 1 hour 30 minutes, ascending the mule-path along the left bank of the Tech, you may reach the baths of La Preste (water mild sulphurous), distance 6 kil. = 3¾ miles.

Section 109.
PERPIGNAN TO AMÉLIE LES BAINS.

38 kil. = 23¾ miles; carriage road; diligences daily.

Leaving Perpignan by the Porte St. Martin, the direction is south, the first stage being Le Boulou, on the left bank of the Tech, 22 kil. = 13¾ miles.

At Bolou the carriage-road to Figuères in Spain branches off to the left by the Fort de Bellegarde. From Boulou to the bridge of Ceret, 8 kil. = 5 miles. Here the road crosses the river Tech by a single arch, 45 metres = 148 feet in span, with holes pierced through the buttresses to lighten the masonry. The bridge is very ancient, but of uncertain date; and like all old bridges, the roadway is very narrow.* From the bridge of Ceret the road continues to mount by a gentle ascent along the left bank of the Tech to Amélie les Bains, a thermal establishment of sulphurous waters, placed about 200 metres = 656 feet above the sea. From Amélie les Bains to Arles, 4 kil.=2½ miles; from Arles to Prats de Mollo, 19 kil.=12 miles. From Prats de Mollo to La Preste, 6 kil. = 3¾ miles.

---

* The bridge over the Taff river at Pont-y Prydd, completed in 1755 by William Edwards, with a span of 140 feet, is constructed on exactly the same plan; and mentioned by Smiles (Engineers, vol. i. p. 272) as being peculiar in its construction. On each side of the arch the haunches above the pier are perforated with three tunnels, to relieve the pressure upon the piers.

The total distance from Perpignan to La Preste is 67 kil. = 41¾ miles. There is a carriage-road for 6 kil. beyond Arles, afterwards only a mule-path.

[Canigou may be ascended from Arles by way of Corsavi, but the route is much longer and more difficult than that from Vernet. From Prats de Mollo to Vernet, 9 hours (Sec. 108).]

Section 110.

The best months for seeing the Pyrenees are August and September, as at that period of the year one may most depend upon fine weather. True it is that in the Pyrenean valleys the heat is greater, and the air less bracing, than in Switzerland; but among the mountains the climate is truly delightful.* The rate of living is not expensive. Forty pounds will suffice for a tour of 6 or 7 weeks, including the journey there and back again; and in this time the principal scenes of the Pyrenees may be visited. Guides are not generally necessary to an experienced mountaineer, except on 2 or 3 of the highest mountains; but in lieu of these a map and compass is of course indispensable. The best general map of the whole chain of the Pyrenees is that published in 1861 by M. Lézat; but the one drawn by myself and published with this volume, will, I hope, be found to give more clearly, as well as more accurately, the details of that part of the range which, as comprising the grandest scenery, is most usually visited. The extent of country comprised in it is in length 90 miles = 146 kil. Above all, in making an excursion across the mountains, never omit to have a small flask of brandy, and some bread or biscuit in the pocket. The cabanes are few and far between, and it is a great satisfaction to have something to fall back upon.

Throughout the chain, and especially on the Spanish side, there is a great deficiency of hotel accommodation on the mountains, so that a sleeping bag is almost an indispensable part of his kit to anyone who would see and thoroughly enjoy the grander parts of the Pyrenees. There is generally no stint of wood for fuel, at about 2,000 metres; and more may be seen of the mountains in four or five days camping out than in three weeks of hotel life, with an occasional excursion. Besides the bag, a tin saucepan with lid, frying-pan, and a few spoons ought to be taken. Fresh meat may be provided for two days' consumption; but a good supply of fat bacon stowed in tin boxes is the most useful form of animal food. It always contributes to the meal, whether eaten as rashers, or used for frying fish, or making soup. This, with bread and wine, tea, coffee, chocolate, sugar, salt, and pepper, is all that is absolutely necessary, though other little extras will of course be added. An extra shirt, two pairs of socks, towel, pair of espadrilles, and perhaps a light overcoat; is all that should be taken in the way of clothing. All the eatables should, as far as possible, be packed in tin boxes, as otherwise the contents of the '*bisac*' are often turned out in a most deplorable plight, especially after a wet night. Each man engaged as porter ought to carry 15 kilogrammes. For

---

* From statistical returns, the salubrity of the air of the Pyrenees appears to be quite equal to that of the southern provinces of France; but strangers, on first coming to the Pyrenees during the hot season, are very liable to be attacked by a mild form of diarrhœa, probably owing to some peculiarity in the water. It is not, however, in general more than an inconvenience of one or two days' duration; and I have always found the following simple prescription most effective in putting a stop to it:—

Wine-glass of camphor julep, or 4 drops of spirit of camphor.
Teaspoonful of sal volatile.
10 to 15 grains of soda.
8 to 10 drops of laudanum.

Four doses of this may be taken ready made up in a case bottle.

CONCLUDING REMARKS.

the porters 5 francs per day, with food, is ample pay. The chief guide should have something more, but 6 francs, with a remuneration by way of 'bon-main,' on his return, ought to be sufficient.

Those who are luxurious should carry with them a tin pot of composition for dressing boots. The following recipe will render the leather soft and pliable and almost waterproof:—

- 1 pint of twice boiled linseed oil.
- 2 oz. of beeswax.
- 1 oz. of Burgundy pitch.
- 2 oz. of spirits of turpentine.
- 6 oz. of best rendered tallow.

Melt by degrees in an earthen pipkin over a slow fire, and, when thoroughly mixed, pour into tin cans.

With all this, and much more, the kit will never be found complete, unless the traveller has brought with him the counsel of Seneca. 'Necessarium est parvo adsuescere. Multæ difficultates locorum, multæ temporum, etiam locupletibus occurrent. Magna pars libertatis est bene moratus venter et contumeliæ patiens.' (Senec. Epist. Lucil.)

The principal charm of the Pyrenees consists in the unrivalled scenery; but in the way of sport there are also some attractions, though the wild animals of the Pyrenees are fast disappearing. The lynx, once not uncommon in the woods, is pretty nearly, if not quite, extinct. Bears are sometimes found in the woods, and killed, but more frequently during the winter than in summer. At any rate, of all deceptions practised upon the inexperienced traveller in the Pyrenees, the '*chasse à l'ours*' is the greatest and most unsatisfactory. Wolves are more numerous, but not often seen in summer, though in hard winters they have been known to descend to the very streets of Luchon. Both the wolves and the bears, however, unless molested, are perfectly harmless to man. The sportsman who does not mind roughing it, and a bivouac at the foot of the snows, may be pretty sure of finding izards (*Antelope rupicapra*), and perhaps a bouquetin (*Capra ibex*), on the wild mountains between the Vignemâle and the Maladetta. As far as my experience goes, the Pyrenean izards are more numerous and less shy than their Swiss cousins, the chamois. During the months of August and September, izards are constantly brought into the town of Luchon, where they are sold to the hotel keepers for from fifteen to thirty francs. There are four or five different kinds of eagle in the Pyrenees; but the king of them all is the *Gypaetus barbatus*, closely akin to, if not identical with, the Lämmergeier of the Alps. One or more of these is generally seen sailing over the higher peaks; and just below the snow line, on the less frequented mountains, ptarmigans (*perdrix blanches* or *lagopedes*) are tolerably numerous. In the woods a favourite chase is that of the *coq de bruyère*, or caipercailzie, but these are few and far between. You may also find the *gélinotte*, a smaller bird, more like a species of pheasant, and excellent eating; and the little bustard, *outarde* (*Otis tetrax*). In the more open valleys you sometimes come upon a solitary stork; and occasionally one of these birds that has been captured alive may be seen exposed for sale in the Allée d'Étigny at Luchon. In several of the lakes there are very large trout, but not easily captured with an artificial fly, though the smaller fry, which abound in every stream, may be readily taken on favourable days.

In conclusion, I would repeat that these pages are not written for that class of persons who love *to see* with their ears. The surpassing grandeur of these scenes defies the strongest powers of pen, pencil, or even photograph; and to form any adequate conception of their beauties, you must go yourself. Learn all you can beforehand of the places you are about to visit, and then trust to yourself and your own emotions, and your soul shall have a feast. Unlike the works of art, which, in losing their novelty, have lost their greatest charm, these magnificent scenes are ever revisited with increasing pleasure. The traveller, on his first acquaintance with the picturesque scenery of the Pyrenees, can scarcely distinguish what most excites his admiration—the deep solitude of the mountains, and the sense of independence and self-reliance inspired by them; the beauty and contrast of forms; or the vigour and freshness of the air, which braces up all his energies, and almost renders him insensible of fatigue. But perhaps the most exquisite enjoyment is reserved for him to whom the scene recalls the happy memories of bygone years. Nature, though ever changing her aspect, always and everywhere speaks to her admirer with a familiar voice. To him, the tiny flowers that carpet the soil; the old moss and fern that cover the rocks; the refreshing cadence of the torrents that gush down from the mountains; the harmonious accordance of the tints reflected by the waters, the verdure, and the sky—all alike recall sensations which he has already felt.

Section III.
CLIMATE.

The climate of the Pyrenees can only be described generally according to the zones of elevation; for though there is a general tendency to a warmer and drier temperature as we proceed eastward along the chain, each of the many watering places has a climate of its own, according to its height above the sea, its contiguity to the high mountains, its exposure to the sun, and the direction and depth of the valley in which it nestles. Considered, however, with relation to the elevation above the sea, the climate of the Pyrenees, as of all mountain chains, is most interesting to study. For man here finds an epitome of all the zones of temperature and vegetation which would present themselves to the traveller starting from these mountains and journeying to the poles. How delightful, on a warm summer morning, to start from some flowery valley, and traversing the seasons in the inverse order of their development, to arrive by mid-day at icebound regions and polar snows. According to De Candolle, 200 metres (656 feet) of elevation operates on the vegetation in the same manner as a degree of latitude farther to the north.

The climate of the Pyrenees becomes gradually warmer as we advance from the Atlantic to the basin of the Mediterranean; the mean annual temperature of the eastern being at least 6 degrees Fahrenheit higher than that of the western portion of the chain. In consequence, the olive, which is not seen in the Basses Pyrénées, will grow on the slopes of Canigou to the height of 1,300 feet. On the same mountain the vine will reach 1,800 feet, the chestnut, 3,000; millet and buckwheat, 5,400; the fir-tree is found at 6,400; the birch, at 6,600; the rhododendron, at 8,350; and the juniper is found on the very top, at 2,787 metres, or 9,144 feet.*

---

* In order not to complicate the table on the following page, the simple latitudes have been given, without regard to the isothermal lines.

It must be remembered that the *mean* only of the heat and cold is given—for the year, for the coldest month, and for the hottest month. The extreme temperatures would take in a much larger range of variations. A self-acting alcohol thermometer left by M. Lézat on the summit of the Pic de Nethou—11,168 ft.—during the winter of 1857 and 1858, gave the extreme cold at this height—24·2 centigrade=11 degrees below zero Fahrenheit; while in the plain at Toulouse the extreme cold only reached—7·5 centigrade=18 degrees above zero Fahrenheit.

Climate and Vegetation of the altitudinal zones of the French side of the Pyrenees, compared with the corresponding latitudes; and the mean temperature of each zone for the year, the hottest month, and coldest month.

Altitudinal Zone of Pyrenees	Corresponding Latitude	Mean Temperature Fahrenheit			Range of Vegetation
		of the Year	hottest Month	coldest Month	
From foot of Pyrenees to 1,200 feet	43° to 45°	59 to 55·4	75·2	41	Olives, vines, maize and all kinds of cruciferous and umbelliferous plants.
1,200 to 3,000 feet	45° to 50°	53·6 to 50	70·5	36	Forests of oak, chestnut, beech; abundant crops of cereals. This zone is the limit to the growth of maize and the vine.
3,000 to 6,000 feet	50° to 60°	48 to 41	65	26·6	Firs of every variety appear in the forests, mixed with the other trees. Potatoes, millet, barley, and oats, walnut-trees, and apple and pear-trees; but the vine will no longer ripen.
6,000 to 8,950 feet	60° to 70°	39 to 32	59	8·6	This zone may be divided into two divisions: 1st. From 6,000 to 7,000 feet=lat. 61°, the forests of fir still abound, and barley and oats can be grown; but the trees gradually become stunted, and dwindle away, and their place is taken by rich open pastures, composed chiefly of leguminous, rosaceous, and cyperaceous plants. Above this point the class of plants known as *annuals* is not found; the summer and autumn being too short to ripen the grain and perpetuate the stock, if ever a peculiarly favourable and less cold spring has allowed any chance seed to germinate. The *perennial* plants, on the other hand, which are already fully developed, are enabled to put forth their flowers on the first fine days of spring; and an occasional season sufficiently favourable to ripen the seed is sufficient to preserve the species, as the same roots reproduce year after year. 2nd. From 7,000 feet to the line of permanent snow, 8,950 feet= latitude 62° to 70°. In this zone a few stunted fir and birch-trees represent the whole forest growth; and the open pastures generally become more and more scanty in leguminous plants, till at last nothing but tufts of juniper, rhododendron, and broom are seen, succeeded, as the ground becomes more and more barren, by little bright clusters of such flowers as the gentian, *Potentilla nivalis*, *Ranunculus glacialis*, and the different varieties of saxifrage.
8,950 to 11,168 feet	70° to 80°	30	42	5	The vegetation gradually reduced to lichens and mosses, and a few cryptogamous plants. Under the glaciers sometimes are found a peculiar kind of fungi, *Uredo nivalis*.

Thus we see that the decrease of temperature, as we ascend the mountains, is just as regular as the decrease from the lower to the higher latitudes. But just as in the arctic regions a considerably higher temperature may be attained during the summer months, because the days are excessively long (so much so that in the highest of these latitudes the sun does not set for months together), and this protracted continuance of the sun above the horizon not only thoroughly warms the earth's surface, but diffuses a considerable store of caloric in the atmosphere; in like manner, on high mountain tops a high degree of temperature is often reached, because the air, being more rarefied than that of the plains, gives a freer passage to the rays of the sun, which at the same time rests upon them for a longer period. Not only is it a fact that the rays of the sun earliest gild, and latest leave, the highest summits but it is also to be remarked that foggy days on the culminating peaks are much less frequent than on the intermediate uplands in the chain. The zone most liable to fogs in the Pyrenees is that lying between 4,000 and 7,200 feet of elevation. How often have I found myself emerge on one of the higher elevations in a bright sunshine and scorching heat, while a thick pall of heavy and chill mist hung over all the valleys; and curious it is on such occasions to look down on the undulating fleecy sea of cloud, with its jutting islets of black rock. On one occasion I remember our party had ascended the Brèche de Roland, and from thence had a fine view of the Vignemâle and other mountain tops, while the Emperor Napoleon, who had chosen that day for his excursion to Gavarnie, was not able even to get a sight of the waterfall, by reason of the mist that filled the Cirque.

From the greater tenuity of the air, and consequent brightness of the rays on the loftier summits, result several phenomena. The disposition to nose-bleeding, the scarifying of the unprotected portions of the skin, and the tendency of the eyes to inflame, are results with which mountain travellers are only too well acquainted. A more pleasing illustration of the chemical power of the light in this rarefied atmosphere is exhibited in the sensitiveness of photographic papers and the excessively vivid colouring of the small flowers, such as those of the saxifrage, and gentian, and other phanerogamous plants which grow just below the zone of eternal snow.* The more powerful direct action of the solar rays in the higher regions of the atmosphere has a sensible effect upon the climate of the mountains. While in the valleys the mean temperature of the soil is greater than that of the air, on mountains the reverse is the case. On the summits of the high mountains the hottest moment of the day is half an hour or three quarters of an hour after midday; while in the plains of the south of France the greatest heat does not arrive till two hours after the moment of the sun attaining its highest point.

* About one-third of the heating rays are absorbed in traversing a cloudless sky; while the chemical rays of light lose about two-thirds of their force in being transmitted perpendicularly through our atmosphere; that is, when the sun is in the zenith. We may therefore conceive how greatly their force is increased on the tops of the higher mountains. At an elevation of 18,000 feet, for instance, where the barometer only stands at half its height, the chemical rays will be transmitted with only a loss of one-third of their power; i.e. they will have twice the effect they have at the sea level. At the summit of the Pic de Néthou, where the barometer stands at 20 inches = 510 millimetres, instead of two-thirds, the chemical rays will only have lost half their force. We know that without light colour cannot be developed in plants, and it is to the same cause, namely, the increased power of the solar rays, that Mr. Darwin refers the fact that tropical shells and birds, and shells that inhabit shallow waters, are more brightly coloured than those that are brought from further north, or from greater depths. (Origin of Species, p. 132.)

Section 112.
LENGTH OF DAY.

The latitude of the Pyrenees, 42° 30', being 9 degrees south of that of London (51° 30'), there is a considerable difference in the length of day at the two places during the winter and summer months. At the spring and autumn equinoxes—the 21st of March, and the 21st of September—the length of day is very nearly the same; but at the summer solstice—the 21st of June—the sun rises 38 minutes later, and sets 39 minutes earlier in the latitude of the Pyrenees than in that of London, making the day at the Pyrenees 1 hour 17 minutes shorter; while, on the other hand, at the winter solstice—the 21st of December—the sun rises 37 minutes earlier, and sets 37 minutes later in the latitude of the Pyrenees, making the shortest day at Luchon 1 hour 14 minutes longer than that of London.

The following table gives the hours of sunrise and sunset; the height of the sun above the horizon at mid-day; and the length of a shadow projected by a rod of 1 metre = 39⅓ inches at mid-day for 42° 30', the latitude of the Pyrenees, and the longitude of Luz, which is exactly that of London.

Day.	Month	Sun rises	Sun sets	Meridian height of Sun above the horizon	Length of Mid-day Shadow of 1 metre	In latitude of London, 51° 30'	
						Sun rises	Sun sets
		h. m.	h. m.		Metres	h. m.	h. m.
1	January	7 33	4 35	26° 36'	2·137	8 8	3 59
11	,,	7 32	4 45	29° 12'	2·032		
21	,,	7 27	4 57	32° 43'	1·927		
1	February	7 17	5 11	34° 39'	1·812		
11	,,	7 5	5 25	36° 44'	1·707		
21	,,	6 50	5 38	39° 49'	1·602		
1	March	6 38	5 47	41° 54'	1·518		
11	,,	6 24	5 59	44° 38'	1·418		
21	Spring Equinox	6 3	6 11	47° 31'	1·309	6 2	6 13
1	April	5 43	6 24	50° 12'	1·193		
11	,,	5 26	6 36	52° 26'	1·099		
21	,,	5 10	6 47	55° 28'	0·984		
1	May	4 57	6 57	57° 45'	0·879		
11	,,	4 42	7 9	60° 2'	0·674		
21	,,	4 32	7 18	63° 0'	0·669		
1	June	4 24	7 38	65° 45'	0·554		
11	,,	4 20	7 36	68° 21'	0·449		
21	Summer Solstice	4 22	7 39	70° 58'	0·343	3 44	8 18
1	July	4 24	7 40	68° 4'	0·448		
11	,,	4 31	7 36	65° 51'	0·551		
21	,,	4 40	7 28	63° 38'	0·654		
1	August	4 51	7 19	60° 19'	0·767		
11	,,	5 2	7 6	57° 66'	0·870		
21	,,	5 13	6 52	55° 13'	0·973		
1	September	5 24	6 35	52° 35'	1·087		
11	,,	5 35	6 17	50° 20'	1·190		
21	Autumn Equinox	5 47	5 59	47° 29'	1·293	5 46	5 59
1	October	5 58	5 41	45° 16'	1·396		
11	,,	6 9	5 24	42° 13'	1·490		
21	,,	6 21	5 7	40° 10'	1·602		
1	November	6 35	4 52	37° 32'	1·716		
11	,,	6 48	4 39	34° 39'	1·819		
21	,,	7 1	4 31	32° 26'	1·922		
1	December	7 12	4 27	29° 33'	2·025		
11	,,	7 22	4 25	27° 20'	2·128		
21	Winter Solstice	7 29	4 28	24° 2'	2·245	8 6	3 51

The hours given in the above table only apply, however, to places situate at the foot of the chain, the duration of the day being a little longer on the culminating points of the mountains; while in the valleys, on the other hand, overshadowed by surrounding mountains, though the daylight continues, the direct rays of the sun illumine the earth for a much shorter period.

The sun rises earlier and sets later on the different summits of the Pyrenees than it would do at their base, assuming that to be the sea level, in the following proportion:—

	Altitude		At the Equinox		At the Summer Solstice		At the Winter Solstice	
	Metres	Feet	Min.	Sec.	Min.	Sec.	Min.	Sec.
For	1,000 =	3,281	5	31	6	44	4	18
,,	2,000 =	6,562	7	50	9	32	5	68
,,	3,000 =	9,843	9	32	11	40	6	24
,,	3,404 =	11,168	10	18	12	32	7	84

So that on the summit of the Pic du Nethou (3,404 metres) the equinoctial day is 20 minutes 36 seconds longer than it is at its base.

By the preceding table we see also that the shadows of all objects, that of ourselves, for example, is a little more than a third of their height at mid-day on the summer solstice; more than twice their height at the winter solstice; and a little more than their height in the spring and autumn:—the shadows exactly coincide with the height on the 19th of April and 24th of August. From these observations we shall see that even at mid-day, and in the summer solstice, the central chain, like a huge wall, casts a shadow towards its northern base exceeding a third of its height, supposing that base to be on the sea level; while in the winter months that shadow is lengthened to twice its height. That of the Maladetta, for example, would extend 1,174 metres, or more than three-quarters of a mile, on the 21st of June, and 7,631 metres, or not much short of 5 miles, on the 21st of December. Hence we shall easily be able to conceive how there are many points of the north side of this chain which for many months never see the sun, by reason of its not rising sufficiently high to overtop the dominating peaks to the south that keep them in shadow. This is the reason of the long continuance of the snows and greater extent of glacier on the French side, many of these beds of ice and snow being reached by the sun only during a few weeks. Thus there are certain lakes in the high mountains south of Luchon, as that of the Port d'Oo, the Portillon, and the Port Viel, where the ice is never entirely melted; because, independently of the cold engendered by their lofty situation, these lakes are dominated by peaks and crags which overtop them by 2,000 feet, while on account of their very contiguous position to the foot of these crags, the distance not exceeding 1,600 feet, they can only receive the sun's rays when the shadow projected by these crags is less than 1,600 feet; so that by a rough calculation it may be shown that the sun only reaches them for the first time about the 6th of May, and shines on them for the last time about the 7th of August; for these three months, moreover, it is only during an exceedingly short portion of the day that the sun can touch them. Some of them, indeed, the lake of the Port Viel for example, abut so close on the precipitous wall of rock that the southern half of those icy waters have never for a single moment felt the warmth of the sun's ray.

Section 113.
BAROMETER.

The weight of the column of the atmosphere equals 32 feet of water, or 30 inches of mercury. The atmosphere extends in height probably about 50 miles, diminishing in density in geometrical progression. For example, at the height of 18,000 feet, the atmosphere is diminished to half its volume, at the height of 36,000 feet to one quarter, at the height of 54,000 feet to one-eighth, and so on.

The mean height of the baroemter at the level of the sea, in lat. 45° at the temperature of melting ice, is 29·922 English inches = 76 French centimetres, or 760 millimetres (30 inches is = 762 millimetres).

1 inch = 25·4 millimetres ∴ 1 millimetre is equal to about 1·25th of an inch, or rather less. The barometer sinks 1 millimetre for 11 metres = 36 feet of elevation.
,, 1 centimetre = ·39371 English inch for 360 feet.
,, 1 decimetre = 3·937 English inches for 3,600 feet.

The two tables subjoined will show the relation between the weight of the atmosphere, the fall of the barometer, the point of water boiling, and the mean temperature. The first table gives those proportions in French metres and grammes, centigrade degrees of the thermometer, and millimetres of the barometer; the second table gives them in English weights and measures.

The weight of the atmospheric column is 1,032 French grammes on every square centimetre, or 14·73 pounds on every square inch.
∴ Assuming the surface of a man's body to present 17,500 square centimetres = 1¼ square metres, the pressure on it at the level of the sea will be 18,060 kilogrammes.

The decrease of heat in the atmosphere, and diminution of pressure in ascending, is roughly as follows:

	French Measures.	
The mean annual temperature	falls 1 degree Cent.	each 180 metres of elevation.
The mercurial column	falls 1 millimetre	each 11 metres of elevation.
The boiling point of water	falls 1 degree Cent.	each 290 metres of elevation.
The pressure on the human body = a surface of 1¼ square metres	diminishes 13,600 grammes	each 11 metres of elevation or each millimetre of fall in the barometer.

ENGLISH MEASURES.		
The mean annual temperature	falls 1 degree Fahr.	each 328 feet of elevation.
The mercurial column	falls 1-10th of an inch	every 90 feet of elevation.
The boiling point of water	falls 1 degree Fahr.	each 528 feet of elevation.
The pressure on the human body	diminishes 29½ lbs.	every 35 feet of elevation.

By aid, therefore, of the barometer, mercurial or aneroid, sympiesometer, or boiling-point thermometer, we are enabled to estimate the height with considerable accuracy.

As many travellers carry an aneroid barometer, or some other hypsometric instrument, in their mountain excursions, I subjoin one or two tables, by means of which heights may be calculated on the spot with much facility and accuracy.

Table I. gives the equivalent in metres of a vertical column of the atmosphere to a millimetre of the barometer, at different elevations and temperatures. It is based on the formula of Laplace, with 18,336 as the barometric coefficient, and 0·004 as the coefficient for the expansion of air in its ordinary state of humidity; only below 0° cent., where the force of vapour can be but small, I have substituted 0·0037 as the coefficient of expansion.

In favourable conditions, that is, 2 hours after sunrise, or at sunset, from these tables the heights may be deduced with wonderful accuracy; but at other hours the horary equation will interfere, producing a slight excess in the height in the day time, and a defect at night. The excess is greatest in the summer months, and from noon to 2 P.M. The defect is greatest in the winter months, and from 2 to 4 A.M.

It often happens that the traveller wishes to ascertain the height of some particular point above a known station; for instance, the absolute elevation at which a certain plant grows, or of the first snow, or the vertical distance of an erratic block above the plain.

To resolve such questions use the table as follows:

Example.—Suppose the barometer at the lower station to read 723·1; and at the upper station 664·5, the temperature of the air being respectively 12° and 8° centigrade. The difference of the barometer, supposed to be reduced to the same temperature, is 58·6 millimetres.

Then Table I. gives for 72 centimetres at 11° centigrade . . . 11·59
for 66 centimetres at 8° centigrade . . . 12·50
2 / 24·09

Half sum, or mean . . 12·045

And 58·6 × 12·045 = 705·8 metres.

TABLE I.—Height in Metres of a Column of the Atmosphere equivalent to 1 Millimetre of the Barometer, at different Elevations and Temperatures.

Temp. of the Air Cent.	Barometer, in Centimetres									
	76	75	74	73	72	71	70	69	68	67
	Metres	Metres	Metres	Metres	Metres	Metres	Metres	Metres	Metres	Metres
0°	10·52	10·66	10·80	10·94	11·10	11·26	11·42	11·59	11·75	11·93
2	10·60	10·74	10·89	11·03	11·19	11·35	11·51	11·68	11·85	12·03
4	10·69	10·83	10·97	11·12	11·28	11·44	11·60	11·77	11·94	12·13
6	10·77	10·91	11·06	11·20	11·37	11·53	11·69	11·86	12·04	12·22
8	10·85	11·	11·15	11·29	11·46	11·62	11·78	11·96	12·13	12·32
10	10·94	11·08	11·23	11·38	11·55	11·71	11·87	12·05	12·22	12·41
12	11·02	11·17	11·32	11·47	11·63	11·80	11·97	12·14	12·32	12·51
14	11·11	11·25	11·41	11·55	11·72	11·89	12·06	12·23	12·41	12·60
16	11·19	11·34	11·49	11·64	11·81	11·98	12·15	12·33	12·51	12·70
18	11·27	11·43	11·58	11·73	11·90	12·07	12·24	12·42	12·60	12·79
20	11·36	11·51	11·67	11·82	11·99	12·16	12·33	12·51	12·69	12·89
22	11·44	11·60	11·75	11·90	12·08	12·25	12·42	12·61	12·79	12·99
24	11·53	11·68	11·84	11·99	12·17	12·34	12·51	12·71	12·88	13·08
26	11·61	11·77	11·93	12·08	12·26	12·43	12·61	12·79	12·98	13·18
28	11·70	11·85	12·01	12·17	12·35	12·52	12·70	12·88	13·07	13·27
30	11·78	11·94	12·10	12·25	12·43	12·61	12·99	12·98	13·16	13·37
32	11·86	12·02	12·18	12·34	12·52	12·70	12·88	13·07	13·26	13·46
34	11·95	12·11	12·27	12·43	12·61	12·79	12·97	13·16	13·35	13·56
36	12·03	12·19	12·36	12·52	12·70	12·88	12·06	13·25	13·45	13·65
38	12·12	12·28	12·44	12·60	12·79	12·97	13·15	13·35	13·54	13·75

Temp. of the Air Cent.	Barometer, in Centimetres									
	66	65	64	63	62	61	60	59	58	57
	Metres	Metres	Metres	Metres	Metres	Metres	Metres	Metres	Metres	Metres
0°	12·11	12·30	12·49	12·69	12·89	13·10	13·32	13·55	13·78	14·02
2	12·21	12·40	12·59	12·79	13·	13·21	13·43	13·66	13·89	14·14
4	12·31	12·50	12·69	12·89	13·10	13·31	13·54	13·77	14·	14·25
6	12·40	12·60	12·79	13·00	13·20	13·42	13·64	13·88	14·11	14·36
8	12·50	12·69	12·89	13·10	13·31	13·52	13·75	13·98	14·22	14·47
10	12·60	12·79	12·99	13·20	13·41	13·63	13·86	14·09	14·34	14·59
12	12·69	12·89	13·09	13·30	13·51	13·73	13·96	14·20	14·45	14·70
14	12·79	12·99	13·19	13·40	13·62	13·84	14·07	14·31	14·56	14·81
16	12·89	13·09	13·29	13·50	13·72	13·94	14·18	14·42	14·67	14·92
18	12·98	13·19	13·39	13·61	13·82	14·05	14·28	14·53	14·78	15·04
20	13·08	13·28	13·49	13·71	13·93	14·15	14·39	14·63	14·89	15·15
22	13·18	13·38	13·59	13·81	14·03	14·26	14·50	14·74	15·	15·26
24	13·27	13·48	13·69	13·91	14·13	14·36	14·60	14·85	15·11	15·37
26	13·37	13·58	13·79	14·01	14·24	14·47	14·71	14·96	15·22	15·48
28	13·47	13·68	13·89	14·11	14·34	14·57	14·82	15·07	15·33	15·60
30	13·57	13·78	13·99	14·22	14·44	4·68	14·92	15·18	15·44	15·71
32	13·66	13·87	14·09	14·32	14·55	14·78	15·03	15·28	15·55	15·82
34	13·76	13·97	14·19	14·44	14·65	14·89	15·14	15·39	15·66	15·93
36	13·86	14·07	14·29	14·52	14·75	14·99	15·24	15·50	15·77	16·05

Temp. of the Air Cent.	Barometer, in Centimetres									
	56	55	54	53	52	51	50	49	48	47
	Metres	Metres	Metres	Metres	Metres	Metres	Metres	Metres	Metres	Metres
−10°	13·69	13·93	14·18	14·46	14·76	15·04	15·33	15·64	15·91	16·30
−8	13·80	14·04	14·29	14·57	14·87	15·16	15·45	15·76	16·08	16·47
−6	13·90	14·14	14·40	14·68	14·98	15·27	15·57	15·88	16·20	16·59
−4	14·	14·25	14·51	14·79	15·09	15·39	15·69	16·	16·32	16·72
−2	14·10	14·35	14·62	14·90	15·20	15·50	15·81	16·12	16·45	16·80
0	14·21	14·46	14·73	15·02	15·31	15·61	15·92	16·24	16·57	16·93
2	14·32	14·58	14·85	15·14	15·43	15·73	16·05	16·37	16·70	17·07
4	14·43	14·69	14·97	15·26	15·56	15·86	16·17	16·50	16·84	17·20
6	14·54	14·81	15·09	15·38	15·68	15·98	16·30	16·63	16·97	17·34
8	14·65	14·92	15·21	15·50	15·80	16·10	16·43	16·76	17·10	17·47
10	14·77	15·04	15·32	15·62	15·92	16·23	16·56	16·89	17·23	17·61
12	14·89	15·15	15·44	15·74	16·04	16·35	16·68	17·02	17·36	17·74
14	15·00	15·27	15·56	15·86	16·17	16·48	16·81	17·15	17·40	17·88
16	15·12	15·38	15·68	15·98	16·29	16·60	16·94	17·28	17·63	18·01
18	15·23	15·50	15·70	16·10	16·41	16·73	17·07	17·41	17·76	18·15
20	15·35	15·62	15·91	16·22	16·53	16·85	17·19	17·54	17·90	18·28
22	15·46	15·73	16·03	16·34	16·65	16·98	17·31	17·67	18·03	18·42
24	15·58	15·85	16·15	16·46	16·78	17·10	17·44	17·80	18·16	18·56
26	15·69	15·96	16·27	16·58	16·90	17·23	17·57	17·93	18·30	18·69
28	15·71	16·08	16·39	16·70	17·02	17·35	17·70	18·06	18·33	18·73
30	15·92	16·20	16·50	16·82	17·15	17·48	17·83	18·19	18·56	18·96

TABLE II.—Similar to the preceding, but adapted to Feet, Inches, and the Fahrenheit Thermometer. The example given for Table I. will serve to explain the use of this table.

Barometer Reading in Inches	HEIGHT OF A COLUMN OF AIR CORRESPONDING TO ONE-TENTH OF AN INCH IN THE BAROMETER								
	Temperature of the Air, Fahrenheit, being								
	40°	45°	50°	55°	60°	65°	70°	75°	80°	85°
	Feet	Feet	Feet	Feet	Feet	Feet	Feet	Feet	Feet	Feet
18·5	144·6	146·1	147·7	149·3	150·9	152·5	154·0	155·7	157·2	158·8
19	140·8	142·3	143·8	145·4	146·9	148·4	150·0	151·5	153·1	154·6
19·5	137·1	138·6	140·1	141·6	143·1	144·6	146·1	147·6	149·1	150·6
20	133·7	135·2	136·6	138·1	139·6	141·0	142·5	143·9	145·4	146·9
20·5	130·5	131·9	133·3	134·7	136·1	137·6	139·0	140·4	141·8	143·3
21	127·3	128·7	130·1	131·5	132·9	134·3	135·7	137·0	138·4	139·8
21·5	124·3	125·7	127·0	128·4	129·7	131·1	132·4	133·8	135·1	136·5
22	121·5	122·9	124·2	125·5	126·8	128·2	129·5	130·8	132·2	133·5
22·5	118·8	120·1	121·4	122·7	124·0	125·3	126·6	127·9	129·2	130·5
23	116·2	117·5	118·8	120·0	121·3	122·6	123·8	125·1	126·4	127·7
23·5	113·7	115·0	116·2	117·5	118·7	120·0	121·2	122·5	123·7	124·9
24	111·3	112·6	113·8	115·0	116·2	117·4	118·6	119·9	121·2	122·3
24·5	109·1	110·3	111·5	112·6	113·8	115·0	116·2	117·3	118·6	119·8
25	106·9	108·1	109·3	110·4	111·6	112·8	113·9	115·1	116·3	117·4
25·5	104·8	105·9	107·1	108·2	109·3	110·5	111·6	112·8	113·9	115·1
26	102·7	104·0	105·0	106·1	107·2	108·4	109·5	110·6	111·7	112·8
26·5	100·9	100·1	103·1	104·2	105·3	106·4	107·5	108·6	109·7	110·8
27	99·0	98·2	101·2	102·3	103·3	104·4	105·5	106·6	107·6	108·7
27·5	97·2	96·5	99·3	100·3	101·4	102·5	103·5	104·6	105·6	106·7
28	95·4	94·8	97·5	98·6	99·6	100·7	101·7	102·8	103·8	104·8
28·5	93·8	93·1	95·8	96·9	97·9	98·9	99·9	100·9	101·9	103·0
29	92·1	91·6	94·1	95·1	96·2	97·2	98·2	99·2	100·2	101·2
29·5	90·0	90·0	92·6	93·6	94·5	95·5	96·5	97·5	98·5	99·5
30	89·1	89·1	91·0	92·0	92·9	93·9	94·9	95·9	96·8	97·8
30·5	87·6	87·6	89·5	90·4	91·4	92·3	93·3	94·2	95·2	96·1

TABLE III.—Table for correcting Mercurial Barometer with brass scale, in Millimetres, and reducing it to 0° Centigrade.—Subtract the number opposite the ascertained Temperature, under the number of Centimetres marked by the Barometer. For a Temperature below 0° the Correction must be added.

Barometer, in Centimetres
Correction in Millimetres

Temperature Centigrade	47	48	49	50	51	52	53	54	55	56	57	58	59	60	61	62
2°	·15	·15	·16	·16	·16	·17	·17	·17	·18	·18	·18	·19	·19	·19	·20	·20
4	·30	·31	·32	·32	·33	·34	·34	·35	·36	·36	·37	·37	·38	·39	·39	·40
6	·46	·47	·47	·48	·49	·50	·51	·52	·53	·54	·55	·56	·57	·58	·59	·60
8	·61	·62	·63	·65	·66	·67	·68	·70	·71	·72	·74	·75	·76	·78	·79	·80
10	·76	·78	·79	·81	·82	·84	·86	·87	·89	·90	·92	·93	·95	·97	·99	1·00
12	·92	·94	·95	·97	·98	1·00	1·02	1·04	1·06	1·08	1·10	1·12	1·14	1·16	1·18	1·20
14	1·06	1·09	1·11	1·13	1·15	1·18	1·20	1·22	1·24	1·26	1·28	1·30	1·33	1·36	1·38	1·40
16	1·21	1·24	1·26	1·29	1·32	1·34	1·37	1·40	1·42	1·45	1·48	1·50	1·53	1·56	1·58	1·60
18	1·37	1·41	1·43	1·46	1·48	1·51	1·54	1·57	1·60	1·63	1·66	1·69	1·72	1·75	1·78	1·80
20	1·52	1·55	1·58	1·61	1·65	1·68	1·71	1·74	1·78	1·81	1·84	1·87	1·90	1·94	1·97	2·00
22	1·67	1·70	1·74	1·77	1·81	1·85	1·88	1·91	1·95	1·99	2·02	2·06	2·09	2·13	2·17	2·20
24	1·82	1·86	1·90	1·94	1·98	2·02	2·05	2·08	2·12	2·16	2·20	2·24	2·29	2·33	2·37	2·40
26	1·98	2·02	2·05	2·09	2·13	2·18	2·22	2·26	2·31	2·35	2·39	2·43	2·47	2·52	2·56	2·60
28	2·13	2·17	2·21	2·26	2·31	2·35	2·40	2·44	2·49	2·53	2·58	2·62	2·67	2·72	2·76	2·80
30	2·28	2·32	2·37	2·42	2·47	2·52	2·57	2·62	2·66	2·71	2·76	2·81	2·86	2·91	2·95	3·00
32	2·43	2·48	2·53	2·58	2·63	2·69	2·74	2·79	2·84	2·89	2·94	3·00	3·05	3·10	3·15	3·20
34	2·58	2·63	2·69	2·74	2·80	2·86	2·91	2·97	3·02	3·07	3·13	3·18	3·24	3·29	3·34	3·40
36	2·73	2·79	2·84	2·90	2·96	3·02	3·08	3·14	3·19	3·25	3·31	3·37	3·43	3·49	3·54	3·60
38	2·88	2·94	3·00	3·07	3·13	3·19	3·25	3·32	3·37	3·43	3·50	3·56	3·62	3·68	3·74	3·80
40	3·03	3·09	3·16	3·21	3·29	3·36	3·42	3·49	3·55	3·62	3·68	3·74	3·81	3·87	3·94	4·00

Barometer, in Centimetres
Correction in Millimetres

Temperature Centigrade	63	64	65	66	67	68	69	70	71	72	73	74	75	76	77	78
2°	·20	·21	·21	·21	·22	·22	·22	·23	·23	·23	·24	·24	·24	·25	·25	·25
4	·41	·41	·42	·43	·43	·44	·45	·45	·46	·46	·47	·48	·48	·49	·50	·50
6	·61	·62	·63	·64	·65	·66	·67	·68	·69	·70	·71	·72	·73	·74	·75	·76
8	·81	·83	·84	·85	·87	·88	·89	·90	·92	·93	·94	·96	·97	·98	·99	1·01
10	1·02	1·03	1·05	1·07	1·08	1·10	1·11	1·13	1·15	1·16	1·18	1·19	1·21	1·23	1·24	1·26
12	1·22	1·24	1·26	1·28	1·30	1·32	1·34	1·36	1·38	1·40	1·41	1·43	1·45	1·47	1·49	1·51
14	1·42	1·45	1·47	1·49	1·51	1·54	1·56	1·58	1·60	1·63	1·65	1·67	1·69	1·72	1·74	1·76
16	1·63	1·65	1·68	1·70	1·73	1·76	1·78	1·81	1·83	1·86	1·89	1·91	1·94	1·96	1·99	2·01
18	1·83	1·86	1·89	1·92	1·95	1·98	2·00	2·03	2·06	2·09	2·12	2·15	2·18	2·21	2·24	2·27
20	2·03	2·07	2·10	2·13	2·16	2·20	2·23	2·26	2·29	2·32	2·36	2·39	2·42	2·45	2·49	2·52
22	2·24	2·27	2·31	2·34	2·38	2·41	2·45	2·49	2·52	2·56	2·59	2·63	2·66	2·70	2·73	2·77
24	2·44	2·48	2·52	2·56	2·60	2·63	2·67	2·71	2·75	2·79	2·83	2·87	2·91	2·94	2·98	3·02
26	2·64	2·69	2·73	2·77	2·81	2·85	2·90	2·94	2·98	3·02	3·06	3·11	3·15	3·19	3·23	3·27
28	2·85	2·89	2·94	2·98	3·03	3·07	3·12	3·16	3·21	3·25	3·30	3·34	3·39	3·43	3·48	3·52
30	3·05	3·10	3·15	3·20	3·24	3·29	3·34	3·39	3·44	3·49	3·54	3·58	3·63	3·68	3·73	3·78
32	3·25	3·31	3·36	3·41	3·46	3·51	3·56	3·62	3·67	3·72	3·77	3·82	3·87	3·93	3·98	4·03
34	3·46	3·51	3·57	3·62	3·68	3·73	3·79	3·84	3·90	3·95	4·01	4·06	4·12	4·17	4·23	4·28
36	3·67	3·72	3·78	3·84	3·89	3·95	4·01	4·07	4·13	4·19	4·25	4·30	4·36	4·42	4·48	4·53
38	3·87	3·92	3·99	4·05	4·11	4·17	4·23	4·29	4·35	4·42	4·43	4·54	4·60	4·66	4·73	4·78
40	4·07	4·13	4·20	4·26	4·33	4·39	4·46	4·52	4·58	4·64	4·71	4·78	4·84	4·91	4·98	5·04

TABLE IV.—Table for correcting Mercurial Barometers with brass scales, in Inches, and reducing them to 32° Fahrenheit.—Subtract the number opposite the ascertained Temperature, under the number of Inches marked by the Barometer.

Temperature		Inches of Barometer																
Fahrenheit	Centigrade	14	15	16	17	18	19	20	21	22	23	24	25	26	27	28	29	30
		355·6	381·0	406·4	431·8	457·2	482·6	507·8	533·4	558·8	584·2	609·6	635·0	660·4	685·8	711·2	736·6	762·0
								Millimetres										
29°	—		—	—	—	—	—	—	—	—	—	·001	·001	·001	·001	·001	·001	·001
30	—		—	—	—	—	—	—	—	—	—	·003	·003	·004	·004	·004	·004	·004
31	—		—	—	—	—	—	—	—	—	—	·005	·006	·006	·006	·006	·007	·007
32	0·00		·001	·002	·002	·002	·002	·002	·003	·003	·003	·008	·008	·008	·008	·009	·009	·009
33	0·55		·002	·003	·004	·004	·004	·004	·005	·005	·005	·010	·010	·011	·011	·011	·012	·012
34	1·11		·004	·005	·005	·006	·006	·006	·007	·007	·007	·012	·012	·013	·013	·014	·014	·015
35	1·66		·006	·007	·007	·007	·008	·008	·009	·009	·009	·014	·015	·015	·016	·016	·017	·018
36	2·22		·007	·008	·008	·009	·009	·010	·010	·011	·011	·016	·017		·018	·019	·020	·020
37	2·77		·008	·009	·011	·010	·011	·011	·012	·013	·013	·018	·019	·020	·021	·021	·022	·023
38	3·33		·010	·011	·012	·012	·012	·013	·014	·015	·015	·020	·022	·022	·023	·024	·025	·026
39	3·88		·011	·012	·013	·014	·014	·015	·016	·017	·017	·023	·024	·024	·025	·026	·027	·028
40	4·44		·013	·014	·015	·016	·016	·017	·018	·019	·019	·025	·026	·027	·028	·029	·030	·031
41	5·00		·014	·015	·016	·017	·018	·020	·020	·021	·022	·027	·028	·029	·030	·031	·033	·034
42	5·55		·015	·017	·018	·019	·020	·021	·022	·023	·024	·029	·030	·031	·033	·034	·035	·036
43	6·11		·017	·018	·019	·020	·022	·023	·024	·025	·026	·031	·032	·034	·035	·036	·038	·039
44	6·66		·018	·019	·021	·022	·023	·024	·025	·027	·028	·033	·035	·036	·037	·039	·040	·041
45	7·22		·020	·021	·023	·024	·025	·026	·028	·029	·030	·035	·037	·038	·040	·041	·043	·044
46	7·77		·022	·023	·024	·026	·027	·028	·030	·031	·032	·038	·039	·041	·042	·044	·045	·047
47	8·33		·023	·024	·026	·027	·028	·029	·031	·032	0·34	·036	·041	·043	·045	·046	·048	·050
48	8·88		·025	·026	·027	·029	·031	·031	·032	·035	·036	·038	·040	·042	·044	·045	·047	·049	·051	·052
49	9·44		·026	·027	·029	·031	·032	·032'	·034	·035	·036	·038	·040	·042	·044	·046	·048	·050	·051	·053	·055

Temperature		Inches of Barometer																
Fahrenheit	Centigrade	14	15	16	17	18	19	20	21	22	23	24	25	26	27	28	29	30
		355·6	381·0	406·4	431·8	457·2	482·6	507·8	533·4	558·8	584·2	609·6	635·0	660·4	685·8	711·2	736·6	762·0
								Millimetres										
50°	10·00		·029	·030	·032	·034	·036	·038	·040	·042	·044	·046	·048	·050	·052	·054	·056	·058
51	10·55		·030	·032	·033	·035	·037	·039	·042	·044	·046	·048	·050	·052	·054	·056	·058	·060
52	11·11		·032	·033	·035	·037	·039	·041	·044	·046	·048	·050	·053	·055	·057	·059	·061	·063
53	11·66		·033	·034	·036	·039	·041	·043	·045	·048	·050	·053	·055	·057	·059	·061	·064	·066
54	12·22		·034	·036	·038	·040	·043	·045	·047	·050	·052	·055	·057	·059	·062	·064	·066	·068
55	12·77		·036	·037	·039	·042	·044	·046	·048	·051	·054	·056	·059	·061	·064	·066	·069	·071
56	13·33		·037	·039	·041	·043	·046	·048	·050	·053	·056	·059	·061	·064	·066	·069	·071	·074
57	13·88		·039	·041	·043	·046	·048	·050	·052	·055	·058	·061	·064	·066	·069	·071	·074	·076
58	14·44		·041	·043	·045	·048	·050	·052	·055	·057	·060	·063	·065	·068	·071	·074	·076	·079
59	15·00		·043	·045	·048	·050	·052	·055	·056	·059	·062	·065	·068	·070	·073	·076	·079	·082
60	15·55		·045	·048	·050	·052	·054	·056	·058	·061	·064	·067	·070	·071	·075	·078	·081	·084	·085
61	16·11		·047	·049	·051	·053	·055	·058	·060	·062	·065	·069	·072	·075	·078	·081	·084	·087
62	16·66		·048	·050	·052	·054	·057	·060	·062	·064	·067	·071	·074	·077	·080	·083	·086	·089	·090
63	17·22		·049	·051	·054	·056	·059	·062	·064	·066	·070	·073	·076	·079	·082	·086	·089	·092	·093
64	17·77		·051	·052	·055	·057	·060	·063	·065	·068	·072	·075	·078	·081	·085	·088	·091	·095
65	18·33		·052	·054	·056	·059	·062	·064	·067	·070	·074	·077	·080	·084	·087	·090	·094	·097	·098
66	18·88		·053	·055	·057	·061	·064	·066	·069	·073	·076	·080	·083	·086	·089	·093	·096	·100	·101
67	19·44		·054	·056	·059	·063	·065	·068	·072	·075	·078	·082	·085	·088	·092	·095	·099	·102	·103
68	20·00		·056	·058	·061	·065	·068	·071	·074	·077	·080	·084	·087	·090	·094	·098	·101	·105	·106
69	20·55		·057	·060	·063	·066	·070	·072	·075	·078	·082	·086	·089	·093	·096	·100	·104	·108	·109
70	21·11		·058	·061	·064	·068	·071	·075	·077	·080	·084	·088	·091	·095	·099	·102	·106	·110	·111
71	21·66		·060	·063	·066	·070	·072	·077	·080	·082	·086	·090	·093	·097	·101	·105	·109	·113	·114
72	22·22		·061	·064	·067	·071	·074	·079	·082	·086	·090	·093	·097	·101	·105	·109	·113	·117

BAROMETER.

Temperature		Inches of Barometer															
Fahrenheit	Centigrade	14	15	16	17	18	19	20	21	22	23	24	25	26	27	28	29	30
		355·6	381·0	406·4	431·8	457·2	482·6	507·8	533·4	558·8	584·2	609·6	635·0	660·4	685·8	711·2	736·6	762·0
							Millimetres											
73°	22·77		·063	·065	·069	·073	·076	·081	·084	·088	·092	·095	·099	·103	·107	·111	·115	·119
74	23·33		·064	·067	·070	·074	·077	·082	·086	·090	·094	·097	·102	·106	·110	·114	·118	·122
75	23·88		·065	·068	·072	·076	·079	·084	·087	·092	·096	·100	·104	·108	·112	·116	·120	·125
76	24·44		·067	·070	·073	·078	·081	·086	·089	·094	·098	·102	·106	·110	·114	·119	·123	·127
77	25·00		·068	·071	·075	·079	·083	·088	·091	·096	·100	·104	·108	·112	·117	·121	·126	·130
78	25·55		·070	·073	·076	·081	·084	·090	·093	·098	·102	·106	·110	·115	·119	·124	·128	·133
79	26·11		·071	·074	·078	·082	·086	·091	·095	·100	·104	·108	·113	·117	·122	·126	·131	·135
80	26·66		·072	·075	·079	·084	·088	·093	·097	·102	·106	·110	·115	·119	·124	·129	·133	·138
81	27·22		·074	·077	·081	·086	·089	·095	·099	·103	·108	·112	·117	·122	·126	·131	·136	·141
82	27·77		·075	·078	·082	·087	·091	·097	·100	·105	·110	·114	·119	·124	·129	·134	·138	·143
83	28·33		·077	·080	·084	·089	·092	·099	·102	·107	·112	·117	·121	·126	·131	·136	·141	·146
84	28·88		·078	·081	·085	·090	·093	·100	·104	·109	·114	·119	·124	·129	·134	·139	·144	·149
85	29·44		·080	·083	·087	·092	·095	·102	·106	·111	·116	·121	·126	·131	·136	·141	·146	·151
86	30·00		·081	·084	·088	·094	·097	·104	·108	·113	·118	·123	·128	·133	·138	·144	·149	·154
87	30·55		·082	·086	·090	·095	·098	·106	·110	·115	·120	·125	·130	·136	·141	·146	·151	·157
88	31·11		·084	·087	·091	·097	·100	·107	·112	·117	·122	·127	·133	·138	·143	·149	·154	·159
89	31·66		·085	·089	·093	·098	·104	·109	·114	·119	·124	·129	·135	·140	·146	·151	·156	·162
90	32·22		·087	·090	·094	·100	—	·111	·116	·121	·126	·131	·137	·142	·148	·153	·159	·164

Table V.

Hypsometrical Tables* for calculating altitudes from the observed temperatures of boiling water and of the air.

Boiling point of water in degrees centigrade.	Corresponding height of Barometer reduced to zero, in millimetres.	Equivalent in millimetres of Barometer of 1-10th of a degree c n:.	Approximate height in metres above the sea, temperature of the air being zero, centigrade	Difference between each 5 tenths of a degree	Value of each tenth of a degree in metres of altitude, at zero	Approximate height in metres above the sea; temperature of the air being 10° centigrade	Value of each tenth of a degree in metres of altitude at 10 cent.
87°	468·17	1·81	3858·1		30·85	4012·4	32 09
87·5	477·33	1 84	3703·8	154·3	30·75	3851·9	31 98
88	486·64	1·87	3550·	153·8	30·65	3692·	31·88
88·5	496·09	1·90	3396·7	153·3	30·55	3532·6	31 77
89	505·70	1·93	3243·9	152·8	30·45	3373·6	31 67
89·5	515 47	1·97	3091·6	152·3	30 35	3215·2	31 56
90	525 39	2·00	2939 8	151 8	30 25	3057·4	31·46
90·5	535·47	2·03	2788·5	151·3	30·16	2900·0	31 36
91	545 71	2·06	2637 6	150·9	30.07	2743·1	31·27
91·5	556 12	2·09	2487·2	150·4	29·98	2586 7	31·18
92	566·69	2.13	2337·3	149·9	29·89	2430·8	31·09
92·5	577 43	2·16	2187·9	149·4	29·80	2275·4	30 99
93	588·33	2 19	2039·0	148 9	29 71	2120·5	30 90
93·5	599 41	2 23	1890·5	148 5	29·62	1966·1	30 80
94	610 66	2·26	1742·4	148 1	29 53	1812·1	30 71
94·5	622·09	2·30	1594·8	147·6	29·44	1658·6	30 62
95	633·69	2·33	1447·6	147·2	29 36	1505·5	30 53
95·5	645·48	2·37	1300 8	146·8	29·27	1352·8	30·44
96	657 44	2·41	1154·4	146 4	29·18	1200 5	30·35
96·5	669 59	2·44	1008·4	146·0	29·09	1048·7	30 25
97	681·93	2·48	862·9	145 5	29·00	897·4	30·16
97 5	694·46	2·52	717 9	145·0	28·91	746 6	30·07
98	707.17	2·56	573·4	144·5	28·82	596·3	29·97
98·5	720 08	2 60	429 4	144 0	28·73	446·5	29 88
99	733·19	2 64	285 9	143 5	28·64	297·1	29·79
99·5	746 50	2·68	142·8	143·1	28·56	148·3	29 70
100	760·	2·72		142·8			

As this table only gives the altitude of a column of air at 0° and 10° centigrade, if taken at any other temperature a correction must be made, by adding to the height obtained from the fourth column the coefficient for expansion, viz. $\frac{1}{273}$, or 0·004 for every degree centigrade.

This is most readily done, by taking the double sum of the temperatures at the upper and lower station, multiplying them into the height obtained and dividing the produce by 1,000, which is done by striking off the three last units. The remainder is to be added to the height.

* These Tables have been computed from Regnault's Tables, revised by Moritz, as given in the Smithsonian Series. It will be seen that they vary somewhat from those arranged by Colonel Sykes; but I think they will be found more accurate in theory and to give a nearer approximate result.

HYPSOMETRY. 169

EXAMPLE.

		Metres
Boiling point on Pic des Posets 89°·73 cent.	89°·5 = 3091·6 0°·23 = −69·8	3021·8
Boiling point at Lac d'Oo 95°·67	95°·5 = 1300·8 0°·17 = −49·8	1251·0
Temperature at upper station 10·7		1770·8
Temperature of lower station 17·3		

28·0 × 2 = 56 × 1771 = 99·176 gives 99·2 Correction for temperature

1870·0
1497· Height of Lac d'Oo.
3367·0 Height of Pic des Posets

The preceding Table, adapted to the English measurement of feet, inches, and the Fahrenheit thermometer.

TABLE I.—Showing the Barometer pressure and elevation, corresponding to any observed temperature of boiling water between 214° and 180° Fahr.

Boiling point of pure water in Degrees Fahrenheit	Approximate height above the level of the sea	Value of each degree in Feet of Altitude	Proportional part for 1-10th of a degree	Corresponding height of barometer at 32° Fahrenheit	Difference in 10ths of an Inch	Value of each 1-10th of an inch at 32° Fahrenheit in Feet	Value of each 1-10th of an inch at 50° Fahrenheit in Feet
214°	—	—	—	31·132	—	—	—
213	—	—	—	30·522	6·10	—	—
212	0	—	52	29·922	6·00	87·5	91·3
211	520	520	52	29·331	5·91	89	93·3
210	1,042	522	52	28·751	5·80	91	94·6
209	1,566	524	52	28·180	5·71	93	96·7
208	2,092	526	53	27·618	5·62	95	98·9
207	2,620	528	53	27·066	5·52	97	101·2
206	3,150	530	53	26·523	5·43	99	102·6
205	3,682	532	53	25·990	5·33	101	105·0
204	4,216	534	53	25·465	5·25	103	107·2
203	4,752	536	54	24·949	5·16	105	108·3
202	5,290	538	54	24·442	5·07	107	111·7
201	5,830	540	54	23·943	4·99	109	114·0
200	6,372	542	54	23·453	4·90	112	116·5
199	6,916	544	54	22·971	4·82	114	118·9
198	7,462	546	55	22·498	4·72	117	121·4
197	8,011	549	55	22·033	4·65	119	124·0
196	8,562	551	55	21·570	4·57	122	126·5
195	9,115	553	55	21·126	4·50	125	129·6
194	9,670	555	56	20·685	4·41	128	132·7
193	10,227	557	56	20·251	4·34	130	134·8
192	10,786	559	56	19·825	4·26	132	137·2
191	11,347	561	56	19·407	4·18	135	140·
190	11,910	563	56	18·996	4·11	138	143·1
189	12,476	566	57	18·592	4·04	142	147·1
188	13,044	568	57	18·195	3·97	145	150·4
187	13,615	571	57	17·800	3·89	148	153·5
186	14,199	574	57	17·423	3·83	152	157·6
185	14,776	577	58	17·048	3·75	156	161·8
184	15,356	580	58	16·680	3·68	159	164·9
183	15,939	583	58	16·318	3·62	163	169·0
182	16,526	587	59	15·962	3·56	167	173·2
181	17,116	590	59	15·613	3·49	171	177·3
180	17,709	593	59	15·270	3·43	175	181·5

TABLE II.—Table of Multipliers to correct the Approximate Height for the Temperature of the air.

Temperature of the Air	Multiplier	Temperature of the Air	Multiplier	Temperature of the Air	Multiplier
32°	1·000	52°	1·042	72°	1·083
33	1·002	53	1·044	73	1·085
34	1·004	54	1·046	74	1·087
35	1·006	55	1·048	75	1·089
36	1·008	56	1·050	76	1·091
37	1·010	57	1·052	77	1·094
38	1·012	58	1·054	78	1·096
39	1·015	59	1·056	79	1·098
40	1·017	60	1·058	80	1·100
41	1·019	61	1·060	81	1·102
42	1·021	62	1·062	82	1·104
43	1·023	63	1·064	83	1·106
44	1·025	64	1·066	84	1·108
45	1·027	65	1·069	85	1·110
46	1·029	66	1·071	86	1·112
47	1·031	67	1·073	87	1·114
48	1·033	68	1·075	88	1·116
49	1·035	69	1·077	89	1·118
50	1·037	70	1·079	90	1·121
51	1·039	71	1·081	91	1·123

To use the Tables I. *and* II.

From Table I. take the approximate height due to the boiling point at the upper and also at the lower station. Multiply the difference between them by the multiplier found in Table II., corresponding to the mean of the temperature of the air in the shade at the two stations.

EXAMPLE.

```
                                                                Feet
Boiling point on Pic des Posets   . 193°·51 Fah.  193° = 10,227
                                                  ·51 =   −284
                                                         ──────
                                                          9,943
Boiling point at Lac d'Oo         . 204°·20 Fah.  204° =  4,216
                                                  ·20 =   −107
                                                         ──────
                                                          4,109
                                                         ──────
                                                          5,834

Temperature at top .     .    .    51° Fah.
Temperature at Lac d'Oo  .    .    63
                                  ─────
                                  2)114

     Mean        .    .    .    57 gives multiplier 1·052
                           Feet
     5834 × 1·052 = 6,137
Height of Lac d'Oo   4,911
                    ──────
                    11,048  height of Pic des Posets above the sea.
```

HYPSOMETRY. 171

TABLE III.—For correction of account on exposure of stem.

Difference between Temperature of Column of Mercury and of the Bulb	Corresponding Multiplier
70°	0·006
80	0·007
90	0·008
100	0·009
110	0·009
120	0·010
130	0·011
140	0·012
150	0·013
160	0·014
170	0·015
180	0·015
190	0·016

To use Table III. *with the Zeometer.*

In the event of a considerable portion of the mercury in the stem being outside of the vessel containing the boiling water, a correction must be added to the reading, to increase it to the degree it would have attained if the entire instrument had been submitted to boiling heat. To find this correction, multiply the number of degrees along which the exposed column of mercury extends, by the multiplier in Table III. corresponding to the approximate difference between the average temperature of the tube and that of its bulb.

It will be sufficiently near to the truth, if we estimate the temperature of the tube to be a few degrees higher than that of the air, for an error of 10 degrees cannot make a difference of more than 20 feet in the calculated altitude when the zeometer is employed.

EXAMPLE.

Reading of the thermometer	194°
First graduation on the exposed stem	175°
Length of exposed column	19°

Reading of thermometer . 194°
Temperature of air . . 60°

134° difference gives Multiplier 0·011

19° × 0·011 = 0·209

Corrected reading, 193·7 + 0·209 = 193·9

In order, however, to calculate the height with real exactness, recourse must be had to tables. For my own part, I prefer to use Bailey's formula, which requires logarithmic tables. But admirable tables for calculating heights, which dispense with the use of logarithms, are given in the Meteorological Tables of the Smithsonian Institution, edited by M. Guyot. Those prepared by Delcros are adapted to the metrical and centigrade scales; those by Guyot to the English barometer and the Fahrenheit scale. From my own experience, I am inclined to prefer those of Delcros, as giving the most expeditious and true results. In this very complete collection there are many other most valuable tables to assist meteorological and physical calculations; but the book is too ponderous a volume to be carried about.

Of instruments, of course, the most accurate is the mercurial barometer; but to those who object to carrying one of these, which even in its most portable form is an inconvenient and fragile instrument, I can strongly recommend the little zeometer, or portable boiling point apparatus, constructed by Casella, of Hatton Garden. A careful observer will, I think, find the results given by the zeometer as accurate and unvarying as those given by the barometer, to which it is vastly superior in compactness and portability. The only drawback to this instrument is the time required for each observation, and the difficulty sometimes experienced in keeping it alight. The mountain sympiesometer and aneroid barometer are both handy and useful instruments for determining intermediate heights. For measuring great intervals of vertical altitude, they can hardly be depended on; but from the facility with which they are read, they are great auxiliaries in taking the intervening points; and if verified from time to time with the mercurial barometer, or boiling point, I do not think they will be found to involve any serious error. The great difficulty in all these calculations is ascertaining the mean temperature of the intervening column of the air, and the consequent correction.

Mr. Glaisher, in his balloon ascent, sufficiently ascertained, that the decrease of temperature is not uniform; but that in contradiction to La Place's hypothesis, it becomes less and less rapid in ascending from the surface of the earth. Even supposing any regularity in this decrease, the inequality of the earth's surface among mountains must necessarily interfere with it; but in ascending a mountain side, the decrease may be fairly reckoned at $1°$ Cent. for every 180 metres of ascent, or $1°$ Fahr. for every 328 feet.

In exact calculations, the wet and dry bulb thermometers should be used, and allowance made for the vapour tension of the atmosphere.

All aëriform bodies, whether vapours or gases, expand $\frac{1}{491}$ part; or 0·00206 for every increase of temperature $= 1°$ Fahr. The expansion is $\frac{1}{273}$ part or ·00367 for every increase $= 1°$ Centigrade. The density of aqueous vapour being only 0·622, or $\frac{5}{8}$ that of dry air at the same temperature and pressure, the air becomes lighter and more buoyant in proportion to the amount of moisture its contains. In order, therefore, to deduce accurately the height from an observation with the barometer, the humidity of the air should also be taken into consideration, by comparison of the wet and dry bulb thermometers. Most useful psychrometrical tables are given by M. Guyot, in the Smithsonian series.

The diminution of vapour, however, in the upper strata is much more rapid than that of the air. Assuming the air to be throughout saturated with moisture, and the reduction of temperature for ascent to be on an average 3° Fahr. for each 1,000 feet, Colonel Strachey, from observations by himself and Dr. Hooker, makes the following computation:—

Proportion of Vapour at various Altitudes.

Height in feet	Calculated — The temperature of air at the sea level being				Observed		Force of Vapour deduced from Glaisher's Balloon Observations	
	80°	60°	40°	20°	By Dr. Hooker, in Sikim—Therm. at sea level being 70° to 90°	By Mr. Welsh, in a Balloon—Therm. at surface being 50° to 70°	Inches	Millimetres
0	1·00	1·00	1·00	1·00	1·00	1·00	·39	9·91
2,000	·82	·81	·79	·77	·82	·88	·30	7·62
4,000	·67	·65	·62	·58	·68	·77	·22	5·59
6,000	·54	·52	·48	·44	·62	·58	·16	4·06
8,000	·44	·41	·36	·34	·52	·45	·11	2·79
10,000	·35	·32	·28	·26	·42	·35	·08	2·03
12,000	·28	·25	·21	·19	·35	·30	·05	1·27
14,000	·22	·19	·16	·14	·29	·19	·05	1·27
16,000	·18	·15	·12	·10	·25	·18	·04	1·02
18,000	·14	·12	·09	—	·20	·16	·02	·51
20,000	·11	·08	·07	—	·16	·12	·02	·51

Hence the consideration of the small quantity of vapour that is disseminated in the upper parts of the atmosphere shows us that inequalities of level on the earth's surface which are insignificant when viewed in relation to the dimensions of the globe become objects of the greatest importance in connection with the atmosphere which surrounds it. Three-fourths of the whole mass of the air is within range of the influence of the highest mountains; one-half of the air, and nearly nine-tenths of the vapour, are concentrated within about 19,000 feet of the sea level, a height which hardly exceeds the mean level of the crest of the Himalaya, while one-fourth of the air and one-half of the vapour are found below a height of 8,500 feet. Thus mountains, even of a moderate magnitude, may produce important changes in very large masses of the atmosphere as regards their movements, their temperature, and their hygrometric state, and especially in those strata that contain the great bulk of the watery vapour, and that have the greatest effect therefore in determining the character of the climate.

The diminished pressure of the atmosphere is not the only cause that tends to diminish weight on the top of a mountain. The farther we are from the earth's centre, the more strongly developed is the centrifugal force, and consequently the force of gravity is proportionately weakened. Owing to the form of the earth being an oblate spheroid, the terrestrial radius gradually increases in length from the pole to the equator (the equatorial radius being 3,962·824 miles, and the polar radius 3,949·585 miles), and the force of gravity therefore diminishes in proportion of the square of the sine of the latitude; or, in other words, as we approach the equator, a shorter pendulum is required to beat seconds. The proportion of this increase of the length of the terrestrial radius is found to be 365 metres for each degree of latitude from the pole to the equator; and this figure being contained about 9 times in 3,404 metres, the quantity by which the earth's radius is lengthened by the addition of the Maladetta; it follows that the earth's radius, taken from the summit of the Maladetta, is equal to the radius of the sea level in a latitude 9° nearer to the equator. In fact, we find that the earth's radius, which in latitude 42° 30′ is in length 6·368,032 metres at the foot of the Pyrenees, is extended to 6,371,436 metres on the summit of Nethou, being equal to that of latitude 33° on the sea level.

MEMORANDA.

The length of the Pendulum to beat Seconds is as follows:—

	Latitude	Pendulum Millimetres	Inches
Spitzbergen	79° 49′ 58″ N.	995·92	39·210
London	51° 29′	994·26	39·139
Paris	48° 50′	993·86	39·124
Pyrenees (foot of)	42° 30′	993·20	39·098

	Height Metres	Feet	Pendulum Millimetres	Inches
Luchon	629	2,063	993·00	39·091
Summit of Nethou	3,404	11,168	992·15	39·055
St. Thomas (lat. 0° 24′ 41″)	—		990·83	39·01

The force of gravity at Greenwich, in feet per second, is 32·1908 = 9·8116 metres.

A body becomes heavier going from the equator to the pole by $\frac{1}{194}$ part.
The *greater* attraction at the pole increases the weight $\frac{1}{590}$ part.
The *diminished* centrifugal force increases the weight $\frac{1}{289}$ part.
Which together makes up the whole quantity $\frac{1}{194}$ part.

Where the gravity is greater, the pendulum is more quickly pulled to the earth; and therefore a pendulum beating seconds requires to be *shortened* as we approach the equator, and *lengthened* as we approach the pole.

The weight of a cubic decimetre = 1 litre of water at 4° cent. = 1,000 grammes.
The weight of a cubic decimetre = 1 litre of vapour at 0° cent. = 0·844384706 grammes.
The weight of a cubic decimetre = 1 litre of dry air at 0° cent. = 1·293223 grammes.
The two last under a barometric pressure of 760 millimetres, their coefficient of dilatation being 0·00367 for each degree centigrade.

	English grains
The weight of a cubic foot of water at 40° Fahr. is	437,272·0
„ of a cubic foot of steam at 40° Fahr. is	257·218
„ of a cubic foot of dry air at 40° Fahr. is	553·77

Light moves 192,000 miles in a second; according to Foucault, 298,000 kilometres = 185,177 miles.

Light travels from the sun to the earth in 8¼ minutes.

Sound travels in dry air, at the freezing point 1,089 feet in a second = 331·92 metres.

Sound travels at 62° Fahr. . . . . 1,123 feet in a second.
„ „ through water . . . 4,708 „ „
„ „ through iron, glass, and some woods 18,530 „ „

The mean distance of the moon from the earth is 237,000 miles = about 60 equatorial radii.

The distance of the sun from the earth is 94½ millions of miles; or 400 times that of the moon.

Earth's diameter, { equatorial, 7925·6 miles = 12754·826 kilometres.
{ polar, 7899·1 miles = 12712·148 kilometres.

Rate of revolution of earth's surface at equator 500 yards per second, or 465·117 metres, more exactly.

Rate of earth's progression in the ecliptic, { in January, 69,600 miles an hour.
18·977 miles per second = 30539·93 metres { in July, 66,400 ,,

Rate of the moon's motion round the earth is 2,300 miles an hour.

Plane of the equator is inclined to the plane of ecliptic at an angle of 23° 27′ 14″.

Plane of the lunar orbit is inclined to the ecliptic at an angle of 5° 8′ 48″.

The sun rotates on its axis in 25 days, 8 hours, 9 minutes.

Earth's equatorial radius, 6,377,267 metres.

Earth's radius in lat. 43° 6,366,946 metres.

1 degree of latitude, or meridian, in latitude 43° = 111080 metres.
1 minute of ,, ,, ,, = 1851·33 ,,
1 second of ,, ,, ,, = 30·856 ,,
1 degree of longitude or parallel in latitude 43° = 81531 ,,
1 minute of ,, ,, ,, = 1358·85 ,,
1 second of ,, ,, ,, = 22·648 ,,

Variation of the compass for Central Pyrenees, i.e. Luz; in the year 1867, 17° 42′ west, decreasing about 7′ annually.

THERMOMETER.

The standard thermometer scale in France is the Centigrade; but that of Réaumur is sometimes adopted. The relations of the three are as follows:—

	Freezing Point.	Boiling Point.
Centigrade	0	100
Réaumur	0	80
Fahrenheit	32	212

A degree centigrade = 1⅘ or 9/5 Fahrenheit.

A degree Réaumur = 2¼ or 9/4 Fahrenheit.

To convert centigrade into Fahrenheit above freezing point, multiply by 9/5 and add 32.

To convert Réaumur into Fahrenheit above freezing point, multiply by 9/4 and add 32.

The boiling point sinks { 1 degree Fahrenheit for every 528 feet of elevation or
on an average { 1 degree centigrade for every 290 metres of elevation.

As we ascend, however, the interval of rise between each degree of fall in the boiling point becomes somewhat longer, as may be seen in Table V.

Section 114.

A great deal of the difficulty of scientific observations, and by far the driest part of the working out of the details, arises from the circumstance of each country preserving a distinct standard of weights and measures. To such an extent does this prevail, that not only in neighbouring kingdoms are the systems of measures altogether different, but the same word, sometimes even in contiguous provinces, has a different equivalent. What can be more confusing, when accuracy is required, than a standard as variable as the stature of the human body, such as the ell and the foot? It is true that with us the length of the standard foot is carefully adjusted by the beat of the pendulum, but it is not so with all nations; and even if it were so, we can scarcely expect that other countries would adopt a standard regulated by the beat of a pendulum at Greenwich Observatory, rather than one situate in their own country. What a difference there is between the foot of different countries is shown in the following comparison of a few only of the principal:—

		English foot
English foot	=	1·000000
French foot	=	1·065765
Russian foot	=	0·992000
Moscow foot	=	0·928000
Vienna foot	=	1·037000
Barcelona foot	=	0·992000
Madrid foot	=	0·915000
Frankfort foot	=	0·933000
Verona foot	=	1·117000

It will be evident that differences such as these become patent in even the most general scientific computations, and lead to much confusion in the intercourse of everyday life. Then again, what can be more vague than such distances as a Stunde or a league?—uncertain whether 17, 20, or 25 to a degree—or such quantities as a long or short hundred?

The French were the first people to endeavour to introduce a uniform method and basis, and reorganise their system of weights and measures. This they did by perhaps what we may consider a somewhat fanciful, as certainly it was a very nice and difficult, operation—the measurement of an arc of the meridian extending over several degrees. The standard arrived at, namely, the metre, or the 10,000,000th part of the quadrant from the pole to the equator, may not be entirely satisfactory, from obvious reasons; the principal being the inequality of the earth's surface, and the variation of the length of a degree with the latitude. Still the French nation has been the first to introduce a uniform system, perfect in its theory, and sufficiently accurate in its application, to meet the requirements of the most eminent mathematicians of Europe; and I think it would not only be fair to them, but conduce much to the advantage of all scientific observers, if the other nations of the world were to adopt the same system. Foremost among the many advantages of the French scale, is its ready applicability to decimal punctuation; and, indeed, I find it so much more convenient than our own that I always employ it in making my own computations. As, however, a change of this kind is not effected without some little trouble, I have in almost every instance given after the metres and kilometres the equivalent in English feet and miles. I have also given tables, by which I believe that anyone will be able to convert for themselves, without difficulty, French metres and kilometres into English feet and miles. By the Metric Act, the 27 and 28 Vict. c. 117, the metrical system has been legalised in England, but unfortunately not being compulsory, it makes but slow progress among the mass of the people.

Section 114.
TABLES OF WEIGHTS AND MEASURES.

The fundamental unit of all measures of length, solidity, and weight, is the metre; the metre being the 10,000,000th part of the arc from the pole to the equator, or the 40,000,000th part of the circumference of the earth.

1 Metre = 39·37079 inches
= 3·2808992 feet } English; or, roughly, 3 feet $3\frac{3}{8}$ inches
= 1·093633 yards   Say 1 yard $3\frac{1}{4}$ inches.

TABLES OF WEIGHTS AND MEASURES.

The multiples are compounded in Greek, as—

| Myriametre = 10,000 metres. | Kilometre = 1,000 metres | Hectometre = 100 metres. | Decametre = 10 metres. |

The sub-multiples, or fractional parts, in Latin, as—

| Decimetre = 0·1 metre. | Centimetre = 0·01 metre. | Millimetre = 0·001 metre. |

1 Kilometre = 0·621 of an English mile = 4 fur. 213 yds. 1 ft. 11 in.
5 Kilometres = 3 miles 0 fur. 188 yds.
8 Kilometres = 4 miles 7 fur. 169 yds.; or, roughly, 5 miles.
1 Lieu de Poste, 25 to a degree, the ancient road measure = 2·422 English miles, or, roughly, 2½ miles = 4,445 metres.
1 Old Spanish League, of 17½ to a degree = nearly 4 English miles.
1 Lieu Marine, 20 to a degree = 3·45 English miles = 5,556 metres.
The old French foot is somewhat larger than the English foot = 1·0657654; or, roughly, $1\frac{1}{16}$ of an English foot ∵ 1,000 French feet = 1,066 English feet.
The ancient French Toise, now never used = 1·949040 metres.
    ,, = 6·3945925 English feet.
    ,, = 6 ft. 4·73 inches.
A Statute English mile = 1,760 yards = 5,280 feet = 69 to a degree = 1609·3 metres.
A Geographical Mile = 2,025·2 ,, = 6,075·6 ,, = 60 a degree = 1852 metres.
A Degree of Lat. in lat. of Greenwich, 51° 28′ 38″ = 364,982 feet = 111,244·6 metres.
A Degree of Lat. in latitude 43° = 364,444 ,, = 111,080·3 metres.
A Degree of Long. in latitude 51° 28′ 38″ = 227,913 feet = 69,466·6 metres.
A Degree of Long. in latitude 43° = 267,495 feet = 81,531·1 metres.
The superficial French measure is the Hectare = 10,000 square metres = 2·471 acres, or, roughly, 2½ acres.
The Centiare = 1 metre squared, or 1 square metre.
The Are = 10 metres squared, or 100 square metres.
The Decare = 31·6228 metres squared, or 1,000 square metres.
The Hectare = 100 ares = 100 metres squared, or 10,000 square metres.
1 English Acre = 4,840 square yards = 4046·7 square metres; and equals a square having each of its sides 69·57 yards or 63·613664 metres.
1 English acre = 4046·7 square metres.
= ·404671 of a hectare, or 40·4671 ares.

The *Journal*, the ordinary measure of surface used by the peasants when speaking of their land (but not legal, or allowed in any deed or public document) varies in different departments.*

---

* Mr. Peyrafitte, the landlord at Argeles, tells me that in a good year 1 journal of land yields 10 hectolitres of maize, or 5 hectolitres of wheat; being at the rate of 7·393 quarters of maize, and 3·696 quarters of wheat per acre. His only rotation of crops are wheat and maize alternate years.
The following extract from the market quotations at Bigorre, Jan. 30th, 1866, will give an idea of the average prices in this country :—

In the neighbourhood of Argelés 1 journal = 18·76 ares or ·1876 of a hectare.
= ·4635596 of an acre English.
1 Arpent des Forêts = about 1¼ acres English.
1·95 Arpents = 1 hectare.
The Square Metre = 1·196 square yards, or 10,000 square centimetres.
The Square Centimetre = ·155 square inch, or rather less than ⅙ of a square inch.
The Cubic Metre, or Stere in the measure of solidity = 35·317 cubic feet. Wood for burning is sold by this measure, the price being about 13 francs per stere.
The Cubic Centimetre = 0·06102 cubic inches.
The weight of a Cubic Centimetre of water, at its maximum density, i.e. at a temperature of 4° Centigrade = 39·2° Fahrenheit, is 1 gramme.
The Gramme = 15·432349 English grains.
The Kilogramme = 15432·349 English grains = 2·204621 lbs. avd. or 2 lbs. 3 oz. 119·85 grains, roughly 2 lbs. 3⅓ oz.
1 litre = 1·760773 of an Imperial pint, = ·88038748 of an Imperial quart or about ⅖ of an English gallon.
1 English bushel = 36·35 litres.

English—Avoirdupois			French	
1 grain		=	·064798956	grammes.
1 oz.	(437·5 grains)	=	28·349540	grammes.
1 lb.	(7,000 grains)	=	453·592645	grammes.
1 cwt.	(112 lbs.)	=	50·802	kilogrammes.
1 ton	(20 cwt.)	=	1016·048	kilogrammes.

	English—Apothecaries' weight			French	
	1 grain		=	6·4798956	centigrammes.
	1 scruple	(20 grains)	=	1·29597912	grammes.
3 scruples =	1 dram	(60 grains)	=	3·88793736	grammes.
8 drams =	1 oz.	(480 grains)	=	31·103496	grammes.
12 oz. =	1 lb.	(5,760 grains)	=	373·241948	grammes.

English—Troy			French	
1 grain		=	·064798956	grammes.
1 dwt.	(24 grains)	=	1·555175	grammes.
1 oz.	(480 grains)	=	31·103496	grammes.
1 lb.	(5,760 grains)	=	373·241948	grammes.

	The Hectolitre	
Wheat	17 fr.	
Mixed Wheat and Rye (Meteil)	14 ,,	
Rye	12 ,,	25 c.
Barley	12 ,,	
Maize	11 ,,	50 ,,
Oats	12 ,,	50 ,,
Hay (100 kil.)	8 ,,	40 ,,
Straw (100 kil.)	5 ,,	
Potatoes	6 ,,	

MEAT.

	Kilog.	
Ox-beef	1 fr.	15 c.
Veal	1 ,,	
Cow-beef		75 ,,
Legs and Necks of Mutton	1 ,,	25 ,,
Wether Mutton		95 ,,
Ewe Mutton		80 ,,

As the heights of mountains are given on French maps in metres, the following little table may assist in reducing them to English feet.

I.		II.		III.		IV.	
Metres	feet	Kilometres	Miles	Millimetres	Inches	Inches	Millimetres
1 =	3·280899	1 =	0·6213	1 =	0·03937	31 =	787·4
2 =	6·5618	2 =	1·2427	2 =	0·07874	30 =	762·0
3 =	9·8427	3 =	1·8641	3 =	0·11811	29 =	736·4
4 =	13·1236	4 =	2·4855	4 =	0·15748	28 =	711·2
5 =	16·4045	5 =	3·1069	5 =	0·1968	27 =	685·8
6 =	19·6854	6 =	3·7283	6 =	0·2362	26 =	660·4
7 =	22·9663	7 =	4·3497	7 =	0·2755	25 =	635·0
8 =	26·2472	8 =	4·9710	8 =	0·3149	24 =	609·6
9 =	29·5281	9 =	5·5924	9 =	0·3543	23 =	584·2
10 =	32·8090	10 =	6·2130	10 =	0·3930	22 =	558·8
						21 =	533·4
						20 =	508·0

In this table, by removing the point one, two, or three figures, the equivalent of the unit is transformed into the equivalent of tens, hundreds, or thousands. Thus in Column I. 3 metres = 9·843 feet.
                                30   ,,   = 98,430   ,,
                               300   ,,   = 984·300 ,,
                             3,000   ,,   = 9,843·000 ,,

E. g. Pic de Nethou is 3,404 metres.
                             3,000   ,,   = 9,843 feet.
                               400   ,,   = 1,312 ,,
                                 4   ,,   =   13 ,,
                                         ─────────────
                                          11,168 feet.

So Column III. may be applied for accurately converting the French barometric scale of millimetres into the English one of inches.

E. g. 760 millimetres = the standard height.
      700   ,,   = 27·550 inches.
       60   ,,   = 2·362   ,,
                       ─────────────
                        29·912 inches.

1 inch = 25·4 millimetres.

Weight of cubic foot of pure water, barometer 30 inches ; thermometer, 62° Fahrenheit = 62·321060 lbs. avoirdupois ; containing a little more than 6 gallons.

	Metres	Eng. ft.	In.
1 English Foot =	0·30479	= —	12
1 French ,, =	0·32484	= 1·066	= 12·789186
1 Barcelona ,, =	0·28266	= 0·992	= 11·904
1 Madrid ,, =	—	= 0·915	= 11·128
1 Spanish Varas =	0·847965	= 2·782	

20 Varas = 1 new Spanish League, of 20 to a degree.

Paris 2° 20′ 22″ E. long. of Greenwich.
    Paris time is 9 m. 21·5 sec. faster than that of Greenwich.

Section 115.

THE PYRENEES FROM WEST TO EAST.

Giving the situation and heights of the principal Peaks and Passes* of the chain.

Latitude N.	Longitude W. of Greenwich	Peak or Pass	Metres	Feet	Where situate
° ′ ″	° ′ ″				
42 57 0	7 0 0	Pic de Penamarela, S.	? 2,880	? 9,450	Asturias.
42 57 0	0 43 40	Pic d'Anie.	2,504	8,216	
42 47 30	0 37 30	Pic d'Aspe	2,503	8,202	
42 55 30	0 29 30	Pic Moudaut or Escarpu, F.	2,808	9,213	
42 53 0	0 29 30	Pic d'Aule, F.	2,410	7,907	
42 41 0	0 24 0	Peña Colorada, S.	2,800	9,186	
42 51 30	0 26 38	Pic du Midi d'Ossau, F	2,885	9,465	
42 45 0	0 28 0	Port d'Izas, S.	2,817	9,162	From Sallent to San Antoine.
42 47 30	0 23 0	Pic de Socques, F.	2,717	8,914	
42 56 30	0 20 0	Pic de Gers, F.	2,613	8,573	
42 56 30	0 17 30	Pic de Gabisos, F.	2,684	8,806	
42 51 0	0 20 0	Arrieu Grand, F.	2,976	9,764	
42 50 0	0 21 0	Som de Scoube	2,825	9,269	
42 48 0	0 19 20	Port de Lavedan	? 2,835	? 9,300	From Val d'Azun to Sallent.
42 48 0	0 18 0	Baletous or Murmuret	3 145	10,318	
42 54 0	0 15 0	Pic Cristail, F.	2,992	9,817	
42 56 0	0 12 30	Pic du Midi d'Arrens,F	2,268	7,441	
42 48 0	0 15 0	Pène d'Aragon	2,865	6,399	
42 48 0	0 14 0	Port de Fonfry	? 2,630	8,624	Val de Labat to Panticosa.
42 48 0	0 11 0	Pic Peterneille	3,020	2,909	
42 55 0	0 8 0	Monné de Cauterets, F	2,724	8,937	
42 47 0	0 9 0	Col d'Aratile	? 2,750	9 023	Lac de Gaube to Val de Serbigliana.
42 46 30	0 8 30	Vignemale.	3,298	10,820	
42 48 0	0 7 0	Col de Vignemale, F.	2,788	9,148	Lac de Gaube to Val d'Ossoue: Gavarnie [C.P. barometer].
42 49 0	0 5 0	Col de Lac d'Estom, F.	? 2,800	9,187	Cauterets to Val d'Ossoue. [C.P. barom.]

* The peaks and passes followed by the letter S are on the Spanish side, and those followed by F on the French side of the frontier. Those not so designated are on the frontier line. The heights are mostly taken from M. Johanne and the map of the État-Majeur.

Latitude N.	Longitude W. of Greenwich	Peak or Pass	Metres	Feet	Where situate
° ′ ″	° ′ ″				
42 50 0	0 2 0	Pic d'Ardidieu, F.	2,988	9,803	
42 49 30	0 0 40	Pic Aubiste, F.	2,791	9,157	
42 47 30	0 0 40	Pic Malle Rouge, F.	2,969	9,715	
42 44 0	0 2 30	Port de Gavarnie or Boucharo	2,280	7,481	
42 43 30	0 2 0	Les Tourettes	3,033	9,921	
42 43 20	0 1 30	Le Taillon	3,146	10,322	
42 41 0	0 0 0	Brèche de Roland	2,804	9,200	
	Long. E.				
42 40 30	0 0 30	Casque de Marboré	3,006	9 863	
42 41 35	0 0 48	Pic de Marboré	3,253	10,673	
42 42 0	0 1 0	Tour de Marboré	3,018	9,903	
42 42 30	0 2 30	Pic Astazou, F.	3,080	10,105	
42 42 10	0 2 30	Pic de Cascade	3,037	9,952	
42 41 10	0 1 35	Cylindre, S.	3,322	10,899	
42 40 37	0 2 5	Mont Perdu, S.	3,351	10 994	
42 46 30	0 1 20	Coumelie, F.	2,600	8,531	
42 44 18	0 1 24	Piméné	2,804	9,200	
42 59 0	0 3 0	Pic de Montaigu, F.	2,341	7,681	
42 43 40	0 1 54	Brèche d'Allauz	2,505	8,215	From Gavarnie to Estaubé.
42 41 50	0 2 24	Port de Pinéde	2,700	8.858	
42 42 15	0 3 57	Cañaou d'Estaubé	2,665	8,744	
42 42 40	0 4 27	Port vieux d'Estaubé	2,762	9,062	
42 42 50	0 4 47	Port de la Canaou	2,800	9,187	
42 42 50	0 7 27	Pic de la Munia	3,150	10.334	
42 44 56	0 8 19	Pic des Aiguillons	2,849	9,347	
42 45 15	0 8 2	Col des Aiguillons	2,752	9 029	From Héas to Aragnouet.
42 45 33	0 7 34	Hourquette de Héas	2,596	8,517	
42 46 8	0 6 30	Pic de Salette	2,960	9,712	
42 47 0	0 8 0	Mont Cambiel, F.	3,173	10.410	
42 47 30	0 8 0	Col de Cambiel, F.	2,595	8,514	From Aragnouet to Gèdre.
42 48 30	0 7 0	Pic Long F.	3,195	10,483	
42 50 0	0 7 0	Néouvielle, F.	3,092	10,145	
42 48 0	0 9 30	Pic Badet, F.	3,163	10,378	
42 56 18	0 8 35	Pic du Midi de Bigorre, F.	2,877	9,439	
42 50 30	0 9 30	Col d'Aure, F.	2,500	8,202	Barèges to Aragnouet in Valley d'Aure.
42 43 0	0 9 0	Port Vieux or de Baroude	2,600	8,531	From Aragnouet to Bielsa.
42 43 0	0 11 0	Port de Bielsa	2,465	8,087	From Aragnouet to Bielsa.
42 42 0	0 14 0	Port de Moudang	2,487	8,159	From Valley d'Aure to Bielsa.
42 42 0	0 16 0	Port de Cavarrère	2,530	8,301	Tramesaigues to Bielsa

Latitude N.	Longitude E. of Greenwich	Peak or Pass	Metres	Feet	Where situate
° ′ ″	° ′ ″				
42 44 30	0 18 0	Pic Aret, F.	2,940	9,646	
42 41 52	0 16 7	Col d'Ordisset	2,414	7,920	Tramesaigues to Bielsa
42 41 50	0 19 30	Port de Plan	2,457	8,061	Vielle Aure to Plan, Val de Gistain.
42 44 30	0 20 0	Pic Fittelongue, F	2,930	9,613	
42 45 0	0 24 30	Pic d'Azet, F.	2,667	8,750	
42 44 20	0 23 42	Pic de Lustou	3,025	9,925	
42 43 10	0 22 22	Port de la Pez	2,466	8,090	Val de Louron to Val de Gistain.
42 43 0	0 25 30	Pic du Midi de Genos, F	2,479	8,134	
42 42 0	0 25 27	Port de Clarabide	2,619	8,592	Val de Louron, Arreau; to Val d'Essera, Venasque.
42 42 0	0 27 0	Pic de Clarabide	2,935	9,627	
42 41 40	0 28 0	Port de Benasqué	2,629	8,625	Val de Louron, Arreau; to Val d'Essera, Venasque.
42 38 30	0 25 30	Pic des Posets, S.	3,367	11,047	
42 42 0	0 28 30	Pic du Port d'Oo	3,150	10,335	
42 42 0	0 29 0	Port d'Oo	3,001	9,846	Lac d'Oo to Val d'Essera, Venasque.
42 42 30	0 30 10	Tus de Montarqué, F	2,933	9,623	
42 42 0	0 31 30	Portillon d'Oo	3,044	9,987	Lac d'Oo to Val d'Essera, Venasque.
42 53 0	0 28 30	Monné de Luchon, F.	2,147	7,044	
42 43 30	0 23 30	Crète de Spijoles, F.	3,049	10,003	
42 45 0	0 28 0	Pic Néré, South, F.	2,840	9,318	
42 41 30	0 32 0	Perdiguères	3,220	10,564	
42 42 0	0 32 10	Col du Quairat	3,002	9,850	Lac d'Oo to Val de Litayroles, Venasque.
42 45 10	0 32 30	Céciré, F.	2,400	7,864	
42 44 40	0 31 0	Hount Sec., F.	2,707	8,882	
42 43 40	0 31 30	Montaroye, F.	2,803	9,197	
42 43 0	0 31 30	Pic Quairat, South, F.	3,059	10 036	
42 42 30	0 32 30	Pic de Crabioules	3,219	10,560	
42 42 30	0 32 40	Pic Intermédiare	3,104	10,184	
42 42 3	0 33 0	Col de Crabioules	3,018	9,900	Val de Litayroles to Val de Lys, Luchon.
42 42 0	0 33 20	Tus de Maupas	3,110	10,203	
42 42 0	0 34 0	Pic de Boum	3,060	10,039	
42 42 0	0 35 30	Port Vieux	2,678	8,785	Val de Lys to Val d'Essera, Venasque.
42 42 20	0 36 0	Pic Sacroux, F.	2,678	8,785	
42 42 0	0 38 0	Pic de Sauvegarde	2,786	9,139	
42 42 0	0 38 30	Port de Venasque	2,417	7,930	Hospice de Luchon to Venasque.
42 43 0	0 39 10	Pic de la Pique, F.	2,393	7,851	
42 42 0	0 39 0	Pic de la Mine	2,767	9,076	
42 39 0	0 38 0	Pic Albe, S.	3,280	10,761	
42 38 30	0 39 24	Pic de Maladetta, S.	3,312	10,866	
42 38 0	0 39 52	Pic de Milieu, S.	3,354	11,044	

Latitude N.	Longitude E. of Greenwich	Peak or Pass	Metres	Feet	Where situate
° ′ ″	° ′ ″				
42 37 16	0 40 42	Pic de Nethou, S.	3,404	11,168	
42 38 30	0 43 20	Pic Fourcanade, S.	2,882	9,454	
42 39 50	0 42 54	Pic de Poumero, S.	2,780	9,121	
42 43 40	0 40 47	Pic Entecade	2,220	7,284	
42 51 30	0 41 0	Bacanère	2,195	7,202	
42 41 30	0 40 30	Port de la Picade, S.	2,424	7,953	Val d'Essera Venasque, to Val d'Artigues de Lin.
42 38 0	0 47 0	Port de Viella, S.	2,456	8,058	Viella to Castaneza.
42 51 0	0 51 0	Port Rouge	2,464	8,084	Valley de Castillon to Val d'Aran.
42 57 30	0 42 0	Pic du Gar, F.	1,786	5,860	
42 57 0	0 46 0	Pic de Cagire, F.	1,912	6,273	
42 51 0	0 48 30	Pic Tentenade	2,300	7,546	
42 50 0	0 52 0	Tour de Crabère	2,630	8,629	
42 34 40	0 52 22	Pic de Montarto	2,940	9,646	
42 50 0	0 54 0	Port de la Rouquette	2,545	8,350	Valley de Castillon to Val d'Aran.
42 49 0	0 56 0	Pic de Mauberne	2,880	9,449	
42 36 0	1 1 0	Col de la Ratière	? 2,600	? 8,531	Viella to Valley de la Noguera.
42 41 0	1 0 30	Col de Beret or Peyreblanque	1,889	6,198	Viella to Valley of La Noguera. Hospice Montgarri.
42 48 0	0 59 0	Porte d'Orle	2,631	8,632	Valley de Castillon to source of La Noguera.
42 48 0	1 3 0	Montvallier	2,840	9,318	
42 46 0	1 5 0	Port d'Aula	2,237	7,231	Seix to Hospice de Montgarri.
42 43 0	1 8 0	Port de Saldau	2,052	6,733	Salau to Val de la Noguera.
42 41 0	1 10 0	Mont des Cuns	2,865	9,400	
42 42 0	1 14 0	Mont Collat	2,884	9,331	
42 43 0	1 16 0	Port d'Aulus or Guillou	? 2,402	7,881	Aulus to Esterri in Valley de Noguera.
42 41 0	1 22 0	Port de Tabascain	2,319	7,609	Vicdessos to Tabascain; Valley de Cardos.
42 40 0	1 23 0	Montcalm	3,080	10,105	
42 39 0	1 24 0	Pic d'Estats	3,141	10,305	(*Chausenque*.)
42 36 0	1 28 0	Port du Rat	2,278	7,474	Vicdessos to Andorre.
42 39 0	1 30 0	Port de Vicdessos	?	?	id. id.
42 39 0	1 32 0	Pic de Siguler	2,901	9,518	
42 39 0	1 33 0	Port de Siguier	2,599	8,234	Vicdessos to Andorre; Val de Rialp.
42 39 0	1 35 0	Mont Rialp	2,747	9,012	
42 38 0	1 38 0	Port de les Cabannes	? 2,680	? 8,793	Les Cabannes to Andorre; Val de Rialp.
42 38 0	1 40 0	Pic Serrère	2,911	9,551	
42 38 0	1 46 0	Port de Fontargente	? 2,460	? 8,071	Ax to the Val d'Andorre; Les Caldas.
42 34 0	1 45 0	Port de Saldeu	2,500	8,202	Ax to the Val d'Andorre.
42 30 0	1 45 0	Port de Morrey	2,556	8,386	Hospitalet to the Val d'Andorre.

Latitude N.	Longitude E. of Greenwich	Peak or Pass	Metres	Feet	Where situate
° ′ ″	° ′ ″				
42 49 0	1 46 0	Pic St.Barthélemy, F.	2,349	7,707	
42 34 0	1 49 0	Col de Puymorin	1,931	6,335	Ax to Puycerda.
42 34 0	1 54 0	Pic Carlitte, F.	2,921	9,584	
42 37 0	1 56 0	Pic Peyric, F.	2,810	9,219	
42 30 0	2 8 0	Col de la Perch, F.	1,621	5,318	Montlouis to Puycerda.
42 27 8	2 8 20	Pic Cambresdase, F.	2,750	9,023	
42 23 15	2 7 7	Pic de Puigmal	2,909	9 544	
42 24 40	2 7 40	Col de Llo	2,550	8,366	From Fontpedrouse to Nuria.
42 26 8	2 8 52	Tour d'Eyne	2,832	9,292	
42 26 0	2 15 0	Roc de Prats	2,845	9,334	
42 28 0	2 16 0	Pic de Jeganne	2,881	9,452	
42 31 12	2 27 26	Canigou, F.	2,785	9,144	
42 24 0	2 26 0	Col d'Ares	1,652	5,420	Prats de Mollo to Campredon
42 27 0	2 52 0	Col de Pertus, or de fort de Bellegarde	420	1,378	Perpignan to Barcelona.

PRINCIPAL LAKES FROM WEST TO EAST.

Principal Lakes from West to East	Height		Superficies
	Metres	Feet	Hectares
Lac d'Artouste	2,092	6,864	50 hectares.
Lac Miguelou	2,267	7,438	35 hectares ?
Lac d'Estaing	1,264	4,147	
Lac de Gaube	1,788	5,866	16 hectares, 40 ares.
Lac de Mont Perdu	2,560	8,399	
Lac Deou or Peyrelade	1,952	6,404	
Lac Bleu	1,958	6,424	49 hectares.
Lac Oncet	2,238	7,343	
Lac d'Aigues-Cluses	2,046	6,713	
Lac Dobert	2,160	7,087	
Lac Domar (200 metres E. of Lac Dobert)	2,202	7,225	
Lac du Cap Long	2,182	7,159	
Lac Doredom	2,000	6,562	60 hectares.
Lac d'Oo	1,497	4,911	39 hectares, 16 ares.
Lac Glacé d'Oo	2,670	8,760	
Lac de Portillon d'Oo	2,650	8,694	
Lac Vert	1,960	6,431	
Lac Glacé de Graouès; or Port Vieux			
Lac Litarolles	2,750	9,022	
Lac Gregonio	2,656	8,714	65 hectares.
Lac Féchan	2,600	8,530	40 hectares?
Etang de Liat	2,000	6,562	30 hectares ?
Lac de l'Isle (near Pic Montarto)	2 200	7,218	100 hectares ?
Lac des Religieuses	2,300	7,546	
Lac Lanoux (Pyr. Orientales)	2,154	7,067	100 hectares ?

FERNS OF THE PYRENEES.

The Cryptogamous plants of the Pyrenees make but a poor figure among the beautiful and varied flora of the mountains, which in the first days of April begin to paint the lower rocks, and linger on the higher summits till the end of August. Still, as many travellers, who would disclaim any general knowledge of botany, are zealous and indefatigable fern collectors, a list of those to be found in the Pyrenees may not be without interest.

I have endeavoured as much as possible to avoid multiplication of species. The error of doing so in any class of plants, and especially in the Cryptogamous ones, must be apparent to any one who has studied them in their habitats on the mountains, where great varieties of climate and soil are continually producing slight modifications, so that the cognate species pass into each other by a series of undefined varieties.

From my own observation I can give forty-two species of ferns existing in the Pyrenees, and perhaps the two doubtful ones, marked with an asterisk, may be included. The short description of each, not intended for the professed botanist, gives, I hope, sufficient diagnostics for the general traveller. I have avoided technical words as much as possible, and give a glossary of the few that were indispensable.

Frond. The leafy portion of the fern. Fronds are said to be **entire** when they are undivided, as in the hart's-tongue; **pinnatifid** when they are deeply divided, the incision reaching inwards towards the rachis, but not reaching it; **pinnate** when they are divided down to the rachis, and in this case each of the distinct leaf-like divisions is called a pinna. When these pinnæ are divided again upon the same plan, the frond becomes bipinnate, and the secondary division pinnules; but if the pinnæ are only deeply lobed, the frond is only bipinnatifid, and the ulterior divisions, segments or lobes. The pinnules may again be divided, when the frond becomes tripinnate or tripinnatifid; and if still more intricate they are called decompound.

Vernation, or prefoliation, is the manner in which the young fronds are arranged before development; being usually *circinate,* when the young fronds are rolled inwards volutely from the apex to the base, and develop by unrolling, or *plicate* as in the Ophioglosseæ, when the young fronds are upright and simply folded, and develop by unfolding.

Stipes. The stem below the frond, usually furnished with membranous scales.

Rachis, or mid-rib. The continuation of the stalk through the leafy portion.

Caudex. The stem or upright stock of a fern.

Rhizome. The stock when creeping, either above or below the surface of the ground.

Spore cases, or Sporangia. Minute roundish-oval one-celled bodies, containing the spores or seed, and for the most part surrounded by an elastic ring, by which they are attached to determinate portions of the veins on the back of the frond, called the receptacle.

Sori. The clusters of seed cases or fructification, which may be round, oblong, or linear. The sori are medial when on the back and centre of the vein between its base and apex, as in the Polypodies and Aspidieæ; lateral, when attached to the side of the vein, as in the Asplenieæ; marginal, when attached to the margin, as in the Pterideæ, or extra-marginal, as in the Hymenophylleæ.

Indusium. The thin skin or membrane which in a large proportion of ferns partially covers, or in some few species underlies, or even contains the seed clusters. It is sometimes also called an involucre, though properly only in the latter case.

An indusium may be attached only by the centre, as in the shield ferns, when it is peltate; by one side, as in the Lastræas, when it is reniform, or more or less linear, as in the Asplenieæ. Sometimes it is hood-shaped, as in the Cystopteris; or urn-shaped, as in the Hymenophylleæ. Sometimes it is underlying, as in the Woodsias, or special and formed by a replicate edge of the frond, as in the Pterideæ.

FILICES.

Tribe I. POLYPODIEÆ. Ferns whose spore cases are collected in circular clusters (*sori*) on the back of the frond, having no indusium.

Polypodium vulgare. Fronds pinnatifid, with lanceolate segments. Sori rather large, yellow or orange coloured; a single row on each side of the rachis or mid-rib. Very common all over Europe; in the Pyrenees growing mostly on old walls and trunks of trees, at an elevation of from 5 to 1,000 metres. Biarritz; Luz; Sia. Perennial, summer.

Polypodium Phegopteris. Beech fern. Fronds pinnate, with the pinnæ opposite and pinnatifid. Sori rather small, near the margin of the lobes. The mid-rib, principal veins, and margin of the fronds more or less hairy; by which this species may be readily distinguished from the smaller specimens of the marsh shield-fern, which it somewhat resembles. Common in England and Wales; found on rocks and in damp woods, especially in the neighbourhood of waterfalls, at an elevation of from 500 to 1,800 metres. Perennial.

Polypodium Dryopteris. Oak fern. Fronds perfectly smooth and of a brilliant green; erect on long purplish stalks, from which they are bent back at an angle; in form bipinnate, or rather *ternate* in three equal divisions. The young fronds make their appearance in March and April; each at first resembling three little balls on wires. These three balls gradually unfold, and display the triple character of the frond. Sori circular and small.

Common in the North of England, in similar situations to the last, but in dryer situations and rather less widely diffused. Cauterets; road to Pont d'Espagne. Perennial, June to September.

Polypodium Calcareum, or Robertianum. Limestone Polypody. A distinct species, but closely resembling the last; from which it differs in the form of the fronds, which are tripinnate and not ternate; i.e. they cannot be separated into three *equal* divisions, but consist of a series of pinnæ gradually decreasing in size; and the young frond never presents the appearance of three little balls, remarkable in the last. The frond also is less bent back from the stalk, and more firm and rigid than that of the oak fern. The fronds of this fern are often variegated, with their general colour of a dull greyish green, caused by a minute scaly or glandular meal, which covers their surface. The P. calcareum only grows on limestone rocks and walls, and, unlike the oak ferns, delights in sunshine rather than shade. Gavarnie; St. Marie, roadside. 500 to 1,800 metres. Perennial, June to September.

Polypodium alpestre. P. Rhæticum. Alpine Polypody. A very fine and rather rare fern, growing in coronets, much resembling those of the Lady Fern, with which it is often intermingled, but its habitat is somewhat higher, from 1,200 to 1,800 metres. It may be recognised by the following characteristics. The sori are small, quite round, and perfectly without an indusium, seated on the back of the anterior branch of the vein, but not at or very near its extremity, while the sori of the Lady Fern, *Athyrium Filix-fœmina*, are more or less (especially the lower ones), kidney-shaped, or in the form of a crescent, attached to the side rather than to the back of the vein, with a distinct indusium attached longitudinally, with its anterior margin raised, free, and split into capillary segments. This fern has a strong tendency to twist up into knots the apices of the fronds. The fronds, which are bipinnate and lanceolate, attain 5 to 8 decimetres. In Britain only found in the Highlands of Scotland. In the Pyrenees, abundant in the wood of La Glère, near Luchon, at a height of about 1,600 metres. To find it, take the path leading from the Hospice de Luchon to the Plateau of the Port de la Glère; a most beautiful walk through the forest, which abounds in many other good plants. Perennial, June to October.

Allosorus crispus. Parsley fern. Fronds two or three times pinnate, on slender stalks, almost without scales. Barren and fertile fronds dissimilar, but growing together in clusters; the outer barren ones 10 to 18 centimetres high, of a lively green, somewhat resembling parsley leaves; the fertile fronds rather taller, with equally numerous oblong segments, and their thin membranous edges incurved over the sori. Sori small and circular, approximate, and ultimately confluent, thus forming a continuous line; no indusium. Fronds annual, appearing in May and perishing in October, the fertile fronds not being well seeded till the middle of August. Common on the schistose and granite rocks, at a height of from 1,200 to 1,400 metres. Héas; Port de Venasque; Lac d'Espingo; Vallée d'Azun. Common on the mountains of the north of England.

Gymnogramma leptophylla. Grammitis leptophylla. Annual maidenhair. An exceedingly delicate and rare fern, differing from most others in being an annual. Fronds in little tufts, 6 to 14 centimetres in height, resembling at first sight very small specimens of the parsley fern. It differs in the fructified fronds being not contracted, and the sori are oblong, not circular, though ultimately confluent, and nearly covering the under surface of the frond. To be found in May and June on the rocks of Cazaril, near Luchon, at 800 metres elevation. In July it is quite dried up. Also in the Pyrenées Orientales, and in the neighbourhood of Nice. In Britain only in the Channel Islands.

The name is characteristic of the fern, the spore vessels having no indusium. Gymnogramma, naked writing; leptophylla, slender leaved.

Tribe II. ASPIDIEÆ. Shield ferns. Ferns having the sori circular, as in the polypodies, but covered when young by a membrane (*indusium*) attached by the centre, or by a point near one side, so that when raised all round by the growth of the spore cases it becomes either peltate or kidney-shaped. The sori spring from the back of the lateral nerves. Those shield ferns that have the indusium attached by the centre are classed under the sub-tribe Polystichum; those where the indusium is attached laterally under the sub-tribe Lastræa; but this character is often inconstant, and difficult to distinguish.

Polystichum lonchitis. Aspidium lonchitis. Holly fern. Fronds pinnate, of a narrow lanceolate form, and rigid; of from 1 to 5 decimetres, growing in coronets from the rootstock. The pinnæ are longest in the middle, tapering to the top and bottom of the frond; their edges jagged all round with sharp prickles, and enlarged at the base on the inner or upper side into a toothed angle or lobe. The pinnæ are always entire, and never cut through to the mid-rib so as to form pinnules, as in the Aspidium aculeatum. Fructification mostly on the upper half of the frond. The sori are large and circular, crowded in two rows on either side of the mid-rib.

The holly fern is common on almost all the mountains of the Pyrenees, at an elevation of from 1,600 to 2,200 metres. At a lower altitude it is in vain to look for it, and its place then seems to be taken by the Blechnum boreale. I have never found these two ferns growing side by side at the same altitude. This fern seems to grow indifferently on limestone and schist, and more sparingly on granite rocks. It grows, however, to the largest size on limestone. I have fronds from the cirque of Gavarnie, measuring 7 decimetres; Col de Tortes; Port de Venasque. Perennial, June to November.

In Britain the holly fern is very rare, found only on the highest mountains, and it is not easy to cultivate.

Polystichum aculeatum. Aspidium aculeatum. Prickly shield fern. A fern in shape somewhat resembling the last, but it is bipinnate, the pinnæ being deeply incised, and the fronds are rather more broadly lanceolate and less rigid. On shady banks common in England, and wooded parts of the Pyrenees. 600 metres. Pic du Gar. June to September.

Polystichum lobatum. Aspidium lobatum. A fern much resembling the last in character, but with broader fronds, and the sori rather smaller and less regular. Range, 800 to 1,600 metres. Granges of Héas; Vallée de Burbe, near Luchon.

Polystichum angulare. Aspidium angulare. Common shield fern. This is the more ordinary and normal form, the fronds being still broader, and approaching angular. It is found at a height of from 100 to 1,300 metres, and is common in England. All these last four ferns seem to be mere varieties of the same species, modified according to climate, from the A. lonchitis, the most northern, which has the most pointed and prickly form, to the A. angulare, the most southern, whose fronds are the most leafy and broadest. At any rate the last three scarcely merit the distinction of species. They are called shield ferns from their circular peltate indusium.

Lastræa filix mas. Aspidium filix mas. Polystichum filix mas. Male shield fern. Length of frond usually 5 to 10 decimetres; bipinnate, with the segments of the pinnæ regularly oblong and very obtuse. The fronds form a coronet, the main stalk being very shaggy, with thick brown scales. Sori large, round, and very distinct, with a peltate, or rather kidney-shaped indusium. One of the commonest of European ferns; in the Pyrenees from 500 to 1,300 metres. Gavarnie. July to September.

Lastræa cristata. Aspidium cristatum. Polystichum cristatum. Var. **Lastræa uliginosa.** Crested buckler fern. Allied to the male shield fern, but the frond is less massive, and the pinnæ less regular, broader, and wedge-shaped, with their pinnules much more serrated, and decreasing in size rapidly to the apex. Sori circular, with reniform, deciduous indusium. Length of frond 3 to 9 decimetres. Wood of La Salenque, on the south flank of the Maladetta, at 1,400 metres. Perennial, July and August.

Lastræa spinulosa. Aspidium spinulosum. Polystichum spinulosum. Broad shield fern. Fronds 3 to 8 decimetres, erect, and broader than those of the last, from which also it differs in having the segments much more divided, being tripinnatifid, stipes having broad, ovate, pale brown scales. Fronds in height 3 to 9 decimetres, somewhat resembling the Lady Fern and Alpine Polypody, but distinguished from them by the shape of the fronds, which are more triangular, the lowest pair of pinnæ being not much shorter than the centre ones, and more accurately by the sori, which are circular, with a reniform smooth-edged indusium, as in the male shield fern, although much smaller, and when mature the indusium often disappears. Above Artias; Upper Valley d'Aran. July to August.

Lastræa Dilatata. Aspidium dilatatum. L. fœniscii. Broad prickly-toothed fern. Very similar to the last, from which perhaps it should not be distinguished even as a sub-species. The colour of the frond is a dark green, with the fructification occupying the whole of the under side. The pinnæ taper gradually from the base to the apex, being generally spreading, and set nearly at right angles. The stipes is a full third as long as the frond, being densely covered with pointed oval smooth scales, which extend some way up the rachis. The indusium examined by a lens is reniform, fringed with stalked glands. This Lastræa is a very common mountain fern throughout the Pyrenees, in shady situations, from 1,000 to 1,500 metres.

Lastræa rigida. Aspidium rigidum. Rigid shield fern. Fronds bipinnate, of 2 to 5 decimetres, with the pinnæ set nearly at right angles, and each pinnule deeply divided, so as to appear nearly pinnatifid. The lobes are toothed, the lobes being broad, and scarcely mucronate; a diagnostic by which this fern is distinguished from its congeners. The stipes is unusually thick at the base, and very densely clothed with large pale red scales, which continue, though less abundant, up the entire length of the rachis. Numerous minute sessile glands are scattered over the frond, especially on the under side, giving it a somewhat glaucous appearance. By this, as well as its more rigid habit and inferior size, this plant is at once distinguished from the *L. filix mas.*

This fern seems confined to limestone rocks, in mountainous districts. Rare in England, on the mountains of Yorkshire. In the Pyrenees and Jura more common. Pic du Gar; Vallée d'Ossoüe, near the Saute de l'Espagnol. July to September.

Lastræa thelypteris. Aspidium thelypteris. Polypodium thelypteris. Marsh fern, or female shield fern (θῆλυς πτερίς).

Fronds pinnate, of 2·7 decimetres, with long smooth stalks, unprovided with scales. The pinnæ are set at wide intervals, slightly drooping, and pinnatifid, with perfectly smooth edges, of a thin membranaceous texture, of a pale green colour, and quite glabrous; by which this fern is known from the beech fern, which it rather resembles; as also by the segments of the pinnæ gradually tapering, those next the stem being the longest. On the fertile fronds the margins of the segments are convolute.

Sori almost confluent, ranged in two parallel lines, with a minute circular and evanescent indusium. A widely-spread though local fern in Britain. In the Pyrenees not common, but found growing along with the crested fern in damp mountainous places and peaty soils.

Lastræa oreopteris. Aspidium oreopteris. Polystichum oreopteris. Scented fern.

The mode of growth, in circular tufts, and shape of the frond, much resemble that of the filix mas; from which it may be known by the lighter colour, especially of the stalk; by the lobes of the pinnæ being quite entire and not serrated, with the small sori in a line near the margin, as in the L. thelypteris; and by the shape of the indusium, which is a small ragged minute involucre, sometimes to be seen near the centre of each cluster of capsules, but often evanescent. The shape of the frond is long-lanceolate, gradually diminishing from about two-thirds of its length to the very base, the lower pinnæ being remarkably short, and nearly triangular in form. This attenuation of the frond towards the base is a peculiarity by which this fern may be recognised from the L. thelypteris, from which it is further severed by having the stipes or stalk below the pinnæ, short, and nearly hidden by pale brown scales, sometimes continued on the lower portion of the rachis. It is further distinguished by the minute resinous glandular dots on the under side of the fronds, which give a powerful though not very agreeable odour when rubbed, and in consequence of which property this species was originally named by Linnæus *Polypodium fragrans.*

Lastræa oreopteris grows throughout Europe, but is confined to that continent. In the Pyrenees it is not uncommon, being found chiefly in damp mountainous places and woods, at an altitude of 1,300 to 1,600 metres.

Very common in Scotland, South Wales, and the lake districts of England.

Pic d'Antenac, in the wood; Vallée de Lys, ascent to Lac Vert; road to Pont d'Espagne; Cauterets. July to September.

Cystopteris. Bladder fern.

Delicate ferns with twice or thrice pinnate fronds, and a sinuous midvein. Sori circular or oblong, on the under surface, enclosed when young in a very thin hood-shaped membrane or indusium, attached half-way between the mid-vein and extremity, and opening outwards into a cup with much jagged edges, ultimately disappearing. The name is derived from the Greek, and signifies bladder fern, in allusion to the inflated indusia.

Cystopteris fragilis. Brittle bladder fern. Fronds tufted and erect, oblong-lanceolate in their general outline, bipinnate, of a bright grass green, 1 to 3 or even 4 decimetres in height. Lower pairs of pinnæ generally opposite, ovate-lanceolate, and the pinnules pinnatifid, with their segments deeply indented. Stipes slender, brown, and very brittle; slightly scaly near the base. Sori numerous, and nearly circular, scattered irregularly on the back of the frond. This fern is very inconstant in form and size. One variety occurs with the fronds once or twice forked, named by Mr. Lowe *C. furcata.* This pretty and fragile, but very hardy, fern has a very wide range in the temperate regions of Europe, extending to high latitudes. In the Pyrenees it is perhaps the most common of all the ferns. There is scarcely a rock or wall on which it may not be found growing at an elevation of from 300 to 2,000 metres. Perennial, May to October.

A variety of the C. fragilis has been found in two places in Scotland, which has been elevated to a species under the name of **C. Dickieana.** I believe I have found the same form under some moist rocks in the ascent to the Lac Bleu, near Bigorre. I give the characteristics. The frond is ovate-lanceolate, pinnate; the pinnæ are rather crowded, and overlapping each other; they are ovate, deflexed, and pinnatifid; the pinnules or lobes are crenate. The sori are situate near the margin. The stipes is short, and scaly near the base, with the caudex, or crown from which they spring, strongly defined.

Cystopteris Alpina. C. regia. Cyathea incisa. Alpine bladder fern. A species very closely allied to the last, but the fronds are narrower, being lanceolate, and almost tripinnate; the pinnæ are shorter than those of the C. fragilis, being ovately triangular, obtuse, and *not lanceolate*, unequal, and ranged in pairs nearly opposite. The pinnules are ovate and deeply pinnatifid. The general colour of the frond is of a darker green than the last species, and the fructification more dense, covering the back of the frond. Rare in the Pyrenees. Cirque de Gavarnie; Lac Bleu, near Bigorre, at a height of about 1,800 metres. Perennial, August.

Cystopteris Montana. A rare and beautiful species, readily distinguished from the two last by the triangular form of its bright green fronds, which in shape resemble those of the oak fern. Fronds triangular, and tripinnate, 1 to 3 decimetres, including the long slender stalk. Pinnæ spreading, and almost at right angles to the rachis; the pinnules being oblong-obtuse, and the segments incised, or dentate, by which it may be at once recognised from the oak fern. Sori small and circular, and when the frond is immature, generally accompanied by a slender white torn indusium. Stipes as long as the frond, springing from the long creeping black rhizome at irregular distances.

In moist Alpine situations in Northern and Arctic Europe; in Britain only on two of the Scotch mountains.

In the Pyrenees it grows at an elevation of from 1,550 to 1,700 metres, on mica schist and limestone, especially on the ledges of dripping rocks. Lac d'Oo, on W. side, in a hole under a rock, 200 metres above the lake; Pic du Gar, above St. Béat; Cirque de Gavarnie, in a cave above right bank of the stream, just below the cabane; and very abundant and fine on the opposite, i.e. left, bank of the stream, in the Houle de Marboré, under the rocks. The stock is perennial, but the fronds deciduous, perishing in autumn. They are in fruit from June to August.

Woodsia. A small interesting family of dwarf ferns, chiefly confined to cold climates, or growing in crevices of rocks at high elevations. Its distinguishing character consists in the peculiar structure of the involucre, which is inserted under the spore clusters, the attachment of which it surrounds, while its margin is split into a number of articulated capillary segments, resembling white chaffy hairs, which intermingle with the capsules, and partially conceal them. The stipes or stalk is jointed at 3 to 5 centimetres above the caudex or rootstock, having a tendency to break off at this point; above this joint the stipes, as well as the rachis, are hairy.

Woodsia hyperborea. Northern Woodsia. Fronds 5 to 15 centimetres, narrow-lanceolate, pinnate. Pinnæ triangular and pinnatifid, with rounded obtuse segments, the lobes being 5 or 7 in number; lower pinnæ distant and alternate, above crowded together, the longest ones being towards the middle of the frond. Stipes of a pale reddish-brown colour. Sori circular, placed near the margin.

This rare Alpine fern is found in Northern and Arctic Europe, Asia, and America, on a few of the loftiest mountains of Great Britain, and on the great mountain chains of central and Southern Europe. In the Pyrenees its range of elevation is from 2,200 to 2,600 metres, growing in clefts on schistose rocks. I have found this fern only in three places: among the rocks at the foot of the Cascade d'Oo, where it is very scarce, the seed having been probably brought down from the higher mountains; on the Col des Aiguillons, above Héas; and lastly, in abundance on the rocks of graphitic schist, N. of the Granges of Castaneza, forming the col between these and Malibierne.

Perennial, July and August.

Woodsia Silvensis. Pointed-leafed Alpine Woodsia. This species of Woodsia, if distinct, is very cognate to the last. It differs in having the fronds broader than those of the W. hyperborea, and the pinnæ more oblong and less blunt, being deeply divided into 7 to 9 obtuse or crenated lobes. It also has the under surface of the fronds densely scaly. This form much resembles some forms of Cystopteris, but is to be distinguished by an examination of the involucre with a lens; by the articulated stipes and more or less hairy rachis; and by the pinnate division of the frond, which, although the pinnæ may be profoundly pinnatifid, is never, I believe, bipinnate, as with the Cystopteris. This form of Woodsia, in Britain, is rather less rare than the last. In the Pyrenees the two grow together on the rocks of Malibierne.

Tribe III. ASPLENIEÆ. Spleenworts. Fronds once, twice, or thrice pinnate or forked; usually rather stiff, though slender, and often small; differing from the Aspidieæ in having the spore clusters linear or oblong rather than circular, attached to the side rather than to the back of the vein, and the indusium which covers them attached longitudinally rather than transversely. The ferns of this tribe are generally at once recognised by their rigid habit, and peculiar dark green smooth appearance.

Athyrium Filix-fœmina. Asplenium Filix-fœmina. Lady fern. A large and most elegant fern, with the massive rootstock and circular tuft of fronds of the male shield fern; but more divided, the stalk less scaly, and the sori different, being lateral and not medial to the vein. Fronds 5 to 10 decimetres, of a vivid green, broadly lanceolate, bipinnate, with the pinnæ alternate, and their segments sharply toothed. Sori shortly oblong, spreading over the whole of the under side of the frond; with the indusium persistent, and in form somewhat crescentic, being attached by the concave interior side, with its anterior margin raised, free, and split into capillary segments.

Common in moist and shaded situations at 500 to 1,400 metres; above this height it is replaced by the Polypodium alpestre, which in aspect it closely resembles; Gavarnie, among the rocks by the stream, 400 metres below the inn; Vallée de Burbe, near Luchon. Common in England. Perennial, July to September.

Asplenium fontanum. A. Halleri. Rock spleenwort. Fronds 8 to 20 centimetres, of a narrow lanceolate form, bipinnate; the broadest part above the middle, tapering to the apex and base. Pinnæ obovate and deeply notched, with from 3 to 6 angular mucronate teeth. Rachis and frond smooth, of a shining green. Sori small; 2 to 4 on each pinnule, confluent and covering the whole under side of the frond. A very pretty fern, most rare in England, but growing profusely in some parts of the Pyrenees, though only on the calcareous rocks. It seems to require shelter, and a certain amount of heat. Its elevation is 700 to 1,500 metres. Pas de l'Echelle, on the road from Luz to Gédre, to the left of the road in ascending; on the rocks by the pathway leading from Boucharo to Torla, between the cascade of St. Helene and the Echelle de Torla, very abundant; rocks south of Plan, on the road to Saraviglia (in Spain), very fine; Val d'Essera, on the road from Hospice de Venasque to Venasque; 15 minutes above Hospice de Luchon on route to Port de Picade; Monné, near Bagnéres de Bigorre. Perennial, July to October.

Asplenium Adiantum nigrum. Black spleenwort. Fronds

bipinnate, tripinnate in the lower part, of a dark shining green and firm consistence, 1 to 3 decimetres long, including the stipes, which is dark purple, or black at the base. The form of the frond is triangularly elongate, and drooping; the pinnæ alternate, gradually decreasing, and less divided from the lowest pair to the point; the lowest pair being always the longest. Sori narrow-oblong or linear, approximate to the mid-rib; eventually becoming confluent, and covering nearly the whole under-surface of the frond. Very common in every country in Europe, on banks by the road-side and in woods, especially on sandy soil, at elevations of from 100 to 900 metres. Biarritz; St. Sauveur; Luchon. Perennial, June to November. This is a very hardy fern. The fronds come to maturity about the end of September, but continue vigorous through the winter, till replaced in the spring. The forms known as *acutum* and *obtusum* seem scarcely to deserve the name of varieties; the only difference being in the shape of the frond.

Asplenium lanceolatum. Fronds usually bipinnate, sometimes only once pinnate, in length 1 to 3 decimetres, lanceolate, of a light clear green.

This fern seems intermediate between the Asplenium adiantum nigrum; and the Asplenium fontanum. From the first it may be distinguished by being once or twice pinnate, never tripinnate; the pinnæ are set at right angles to the radius; and the lowest pair are always shorter than the second, so as to give the frond a lanceolate rather than a triangular appearance. It is also more erect in its growth and never drooping. When simply once pinnate, this fern much resembles the Asplenium fontanum; but the lobes are larger, and the outline of the frond less gradually tapering to the base, the second or third pair of pinnæ being equal to any in length, and the rachis, or leaf-stalk, is brownish towards the base, not green throughout. Sori oblong, and ultimately confluent, and covering the frond.

This fern occurs in Britain, but usually near the sea-side: common in the western part of Jersey. In the Pyrenees, I have only found it at Biarritz, and on the slaty rocks of Cazaril, near Luchon. May to Sept.

Asplenium trichomanes. Common spleenwort. False maidenhair. Fronds narrow, pinnate linear, 5 to 20 centimetres, with numerous pinnæ, irregularly ovate, and crenate on the edges. Radicles black. Stipes or foot-stalk wiry, of a blackish purple; and the rachis or leaf-stalk, though green in young fronds, with maturity changes to the same colour. Sori in a single row on each side the mid-rib of the pinnæ, arranged obliquely, linear in form, and eventually becoming confluent in two series, which, however, very rarely unite over the mid-rib. The clusters are 10 or 12 in number. This fern is very common on walls and hedge-sides in the North of England, as well as in the Pyrenees, at elevations from 100 to 1,600 metres; above this last height it is replaced by the A. viride. Gavarnie; Luchon. June to October.

Asplenium viride. Green spleenwort. Fronds narrow, linear pinnate, with numerous pinnæ roundish-ovate, or lozenge-shaped. Rachis and stipes entirely green, or brown only at the base.

This fern, though not nearly so common as the last, and rare in England, is not unfrequent in the Pyrenees on moist rocks, at elevations of from 1,500 to 2,000 metres. It resembles the last in its form and its manner of growth, but it is at once known by its green rachis. Its colour is of a livelier green and its habit less rigid. The

o

pinnules also are somewhat shorter, and broader on the upper side, and their edges are cut into round lobes, and not serrated. The sori are linear, usually 6 in number, ranged obliquely in a row on each side of the mid-rib, eventually becoming confluent, and forming a solid bright ferruginous mass on the centre of the pinna and concealing the mid-vein. Cirque de Gavarnie; Val d'Ossoüe; Col de Tortes; Col de Gourzy; near Eaux Chaudes; Col d'Espingo; Pène de l'Heris, and Monné. July to October.

Asplenium Ruta muraria. Wall-rue spleenwort. The normal form of the frond is triangular and bipinnate, the pinnæ being alternate, and the pinnules of varied form, but mostly somewhat diamond-shaped; with their interior margin generally serrated or crenate. Length of frond 5 to 15 centimetres, including the stipes, which is generally longer than the frond. Sori elongated, 2 to 5 on each pinnule, eventually confluent, and covering the whole under surface of the frond with a dark brown mass of seed. Indusium fringed at the margin, and evanescent. Found throughout the whole of Europe. Very common on old walls and rocks at an elevation of from 500 to 1,400 metres. Bridge at or near Luchon; Pragnères; Luz. Perennial, June to November.

Asplenium septentrionale. Forked spleenwort. Fronds linear and forked, divided into two or three segments. The separation of the stipes and frond indistinct. Length of frond 5 to 15 centimetres, usually growing horizontal from a perpendicular surface, and rooted in a cleft of the rock. Sori in a continuous line attached to each fork, but ending by covering the whole under surface. This fern is indigenous to Great Britain, but there very rare, and confined to the highest mountains. In the Pyrenees it is abundant at an altitude of from 660 to 1,600 metres, on the rocks, schistose, and granite, but preferring the latter. It is easily recognised by its forked grass-like form, the under part of the frond being thickly seeded in July, August, and September. This fern is strictly evergreen; the fronds remaining green all through the winter: they are of a rich dark green. Gavarnie; Luchon; very abundant on the rocks above St. Mamet.

Asplenium germanicum. Asplenium Breynii. A. alternifolium. Alternate spleenwort. Fronds linear and pinnate, of 4 to 12 centimetres. Pinnæ alternate, the lower ones being three-lobed. Colour of the stipes dark at the base and green above. Pinnæ a pale green. Sori narrow-linear, ultimately confluent, and covering the whole under side of the lobe. The edge of the indusium is smooth and even, that of the A. septentrionale being sinuous, and that of the A. Ruta muraria being jagged.

This plant seems intermediate between the alternate spleenwort and the wall-rue spleenwort, and in a well-chosen series of varieties from each, it would be not easy to define the limits.

A. Ruta muraria, Germanicum, and Septentrionale form a group very distinct from the other European species of Asplenium, inasmuch as their pinnules want the mid-vein, which is always present in the rest. They have on this account sometimes been placed in a distinct genus, *Amesium*.

The A. germanicum is the rarest of the British ferns, and nowhere found in abundance. It is said to be confined to Europe. It grows in the same localities and together with the forked spleenwort, which it exactly resembles in its choice of situation and manner of growth; though its general habitat seems to be rather less elevated, from 650 to 1,100 metres, and it generally prefers a southern aspect. It thrives best in the interstices of stone walls. I have found it in several places on the granite rocks and low granite walls in the neighbourhood of Luchon; also on a wall by the side of the road over the Portillon below the Spanish douane. It also grows on some rocks above Argeles, in the Vallée de l'Esponne, and near Cauterets. Perennial,August to October. If planted in a pot, this fern should be wedged horizontally between pieces of stone with a little fine peat earth, and water introduced very sparingly.

Scolopendrium vulgare. Hart's-tongue spleenwort. Fronds tufted, undivided except in monstrous forms, glabrous and broadly linear, cordate at the base. Length 2 to 4 decimetres; colour bright green; smooth and shining on the upper surface. Stipes and rachis scaly. Sori linear and parallel in two rows set obliquely to the mid-rib; but not extending either to the mid-rib or the margin of the frond. Indusium attached along either side of the sori, opening in a longitudinal fissure down the centre.

A graceful evergreen fern, widely spread, and very common, though not found at any great elevations. In the Pyrenees it abounds along the streams and hedge-sides at from 100 to 800 metres. Abundant at Bigorre, Biarritz, and Argeles. This fern is subject to great varieties of form; the fronds being sometimes crisped or forked. Plants taking this form may be seen growing near the Grotte de Judios, S.E. of Bagnères de Bigorre. Perennial, June to November.

Ceterach officinarum. Grammitis ceterach. Asplenium ceterach. Scolopendrium ceterach. Scaly ceterach. Fronds lanceolate, pinnatifid, and coriaceous, 5 to 15 centimetres, with broadly oblong or rounded lobes, green and glabrous on the upper side, but the under side thickly covered with brown scarious scales, which completely conceal the sori until they become very old. Stipes short and covered with scales. Fronds numerous, growing in tufts on rocks and old walls. Very common in central and Southern Europe; but limited to the western and northern counties of England; rare in Scotland; in the Pyrenees abundant, from 300 to 900 metres, on old walls. Perennial, May to October.

Tribe IV. BLECHNEÆ. Fronds pinnatifid or pinnate, sometimes simple in tropical species. Sori linear, ranged on each side of the mid-rib of each segment and parallel to it. Indusium attached along the outer edge of the sorus, opening towards the mid-vein.

Blechnum boreale. Blechnum spicant. Common hard fern. Fronds tufted, of two kinds. The fertile fronds in the centre, 1 to 8 decimetres, erect, linear and pinnate, gradually tapering towards the apex and base, having narrow-linear pinnæ, with reflexed margins, and their under side entirely occupied by the two linear sori. The lower half of the stem is purple, smooth, and shining, furnished on each side with minute rudimentary pinnæ. The barren fronds are shorter and more spreading, with oblong linear flat lobes, a shorter stalk, and the segments broader and not reflexed. The fertile fronds contract much after seeding, and ultimately die away, while the barren fronds remain green through the winter.

This fern is generally distributed over Europe, found in all parts of England, and in the Pyrenees it is very common at elevations of from 800 to 1,800 metres. Cauterets, road to Pont d'Espagne; Lac d'Oo. Perennial, June to October.

Tribe V. PTERIDEÆ. Mid-vein distinct, lateral veins anastomosing at the margin, forming a marginal vein. Sori linear, attached to the margin of

o 2

the frond, with an indusium attached to the inner side of the marginal vein, and opening inwards with its edges split into fibres. The name Pteris is derived from πτέρον, a wing, in reference to the form of the fronds of this genus, which are usually stiff, and very branching.

Pteris Aquilina. Common brake. Fronds triangular and very compound, twice or thrice pinnate. The pinnæ are pinnate, and the pinnules pinnatifid, with the segments obtuse, and for the most part entire. The sori are ranged in continuous lines along the margins of the upper segments of the pinnæ. Stipes attached underground to the rhizome, and deeply buried. The underground portion of the stalk is velvety, stout, and of a dark brown colour; if cut transversely in rather an oblique direction, the section presents a regular figure, said to represent an oak tree, and called King Charles in the oak. By others it is supposed to resemble a spread eagle; hence the specific name, '*Aquilina*,' given by Linnæus. The frond is killed by the first frosts of autumn, however slight they may be; it instantly turns to a deep brown colour, but remains erect and perfectly undecayed during the whole winter. Very common from 200 to 1,400 metres. Gavarnie; Argeles. Perennial, June to September.

Cheilanthes odora. The specific name of this fern does not describe it, as I have never observed in it any smell. The generic name, *Cheilanthes*, taken from the Greek 'lip-flowering,' is more happy, as the sori form a continous line along the outer edge of the divisions of the frond.

Frond 4 to 12 centimetres, seldom exceeding 8; glabrous, ovate-lanceolate, and bipinnate, of a light vivid green colour on the upper side. The pinnæ are most delicately cut; obtuse oblong, with rounded and very entire lobes, ranged opposite along a dark red rachis, which is set with hirsute scales. The rhizome is somewhat tufted, and the stalks are very brittle. This is an exceedingly pretty little fern, not known in England, and very rare in the central Pyrenees, though, I believe, more common towards the shores of the Mediterranean, especially in the neighbourhood of Mentone. Rocks of Cazaril, near Luchon. It is found growing both on the schist and limestone, but is most abundant on the latter, growing in the chinks, at an elevation of about 850 metres. Perennial, May to October.

Tribe VI. ADIANTUM. Ultimate divisions of the frond leaf-like, without a mid-vein. The sori on the under side at the extremity of the lobes, which form an indusium by being replicate. The fronds of the genus Adiantum have a curious property of repelling moisture, hence the name, ἀδίαντον, in the Greek meaning not moistened.

Adiantum Capillus Veneris. Cheveux de Venus. True maidenhair. Fronds 1 to 2 decimetres, mostly of an elongate triangular frond, but lax and irregular; twice, sometimes thrice, pinnate; the branches or pinnæ being alternate, and on these are the pinnules, also alternate, fan-shaped, and each on a distinct petiole. The rachis or principal stem is throughout naked, shining, and nearly black. The fronds are mostly fertile, the exterior margin of each pinnule being divided into lobes, and the terminal portion of these is bleached, membranaceous, and reflexed, with the spore cases attached on its under surface. From this plant capillaire is made.

Common in Southern Europe, but in Britain only in the south-western counties of England, South Wales, and Ireland. This very graceful fern will grow almost on any soil. schist, limestone, or sand, but not on granite. It requires continual moisture, and considerable heat, though with shade; so that it is not a common fern on the mountains: its favourite habitat is the entrance of moist caves and wells. Its elevation in the Pyrenees is from the sea-level to about 700 metres. Biarritz; Pierrefitte, among rocks, on right of the old road leading to Cauterets; also on the right of the new road, near the stone marking the 7th kilometre from Argeles; under the wall of the ruined château of Vidalos, 3 kil. from Argeles; on some rocks to the right of the road leading to the Port de Peyresourde, 2·2 kil. from Luchon. To reach these rocks, mount a wall and cross a steep little bit of field. The hart's-tongue fern grows abundant in the same place.

Tribe VII. HYMENOPHYLLEÆ. A group of dwarf ferns, with semi-transparent membranaceous fronds, for the most part more resembling mosses than ferns, two of which, H. Tunbridgense and H. Wilsoni, are natives of Great Britain, the former only occurring in the Pyrenees. The name is derived from the Greek *hymen*, a membrane; and phyllon, a leaf.

Hymenophyllum Tunbridgense.* Tunbridge film fern. Fronds 4 to 10 centimetres, pinnate and lanceolate, of a smooth pellucid, membranaceous texture, olive green in colour, with a very slender stem. The segments are divided into 3 to 8 linear-oblong lobes, which appear minutely toothed when seen through a lens. Sori extra-marginal, enclosed in a two-valved involucre. This involucre consists of two flattish orbicular valves, finely toothed on their upper edge, placed on the extremity of a vein, at the base of the segments or lobes, on their inner edge, and projecting beyond them. This fern inhabits moist, rocky and shady situations, covering the damp rocks and trunks of trees. It is said by M. Fourcade to grow at the Cascade de Juzet, near Luchon, and Philippe gives it on the Col de Tortes. The cognate species, H. Wilsoni, or *unilaterale*, though the more common form in England, is not, I believe, found in the Pyrenees. It is distinguished by the form of the valves, which are rather ovate and convex, and their edges entire, not serrate; the involucres also are generally turned in an opposite direction to the pinnæ.

Tribe VIII. OSMUNDEÆ. Fronds once or twice pinnate, the leafy part barren, the fructification consisting of clustered spore cases, arranged in simple or compound panicles. Sori without any indusium, opening by a vertical fissure.

Osmunda regalis. Royal fern. Fronds 8 to 20 decimetres, or even more, growing in tufts, usually bipinnate with the fructification at the top of the frond, usually also bipinnate, each spiked-like branch representing a segment of the frond. The caudex or rootstock is large and perennial, the fronds of remarkably quick growth, making their appearance in May, and dying away with the first winter frost. Found throughout Europe; in Great Britain chiefly confined to the Western counties. It grows luxuriantly in wet woods, and in marshy places, especially by the side of shallow lakes, but it will not grow at any great elevation. Lac de Lourdes, very abundant on the SE. side; wood of Rebenac, near Pau; village of Melles, near St. Beat. Perennial, June to September.

Tribe IX. OPHIOGLOSSEÆ. A small family of herbaceous ferns, bearing a single barren frond, and a terminal fruiting spike, distant, but on the same stem. Spore cases without indusium.

Ophioglossum vulgatum. Common adder's-tongue. Stem solitary, from 1 to 3 decimetres in height, consisting of a sterile oval or ovate-lanceolate entire leaf, and a terminal spike, bearing the spore cases, which are sessile, and closely ranged in two rows. The barren frond is shorter than the fertile one, and forms a sheathing foot-stalk. The stipes is erect, smooth, cylindrical, and succulent, and the roots white and brittle, very different to the radicles of other ferns. The adder's-tongue is found in England. It grows in moist meadows, preferring a strong loam or clay. In the Pyrenees it does not seem to grow at an elevation exceeding 700 metres. Meadows of the Elysée-Cottin, near Bigorre, and also on the road to the Pène de l'Heris, near the limekiln above Asté. Stipes perennial, but the fronds produced annually *exterior* to the base of the last year's frond. June to July.

Ophioglossum lusitanicum. Dwarf adder's-tongue. This fern is scarcely distinguishable from the last except in size, but the diminutive form is very constant in all situations, and it has a further marked distinction in its early fructification. The fronds make their appearance at the end of December, by the end of January they are in full fruit, and die off in March. The fronds are in height 3 to 6 centimetres, and the shape of the barren leaf is more narrow than that of the common species, being linear-lanceolate. In Britain it occurs only in the Channel Islands. It requires a light sandy soil. Abundant in the meadows of Bilhéres, near Pau. Varieties of this species sometimes occur having two barren leaves instead of one. December to March.

Botrychium lunaria. Moonwort. Fronds herbaceous and solitary, the sterile and fertile branches being distinct, 5 to 20 centimetres.

This species has the stem of the adder's-tongue, but the terminal spike is branched, forming a panicle, and the globular spore cases, although sessile, are quite distinct, forming alternate clusters; the barren leaf also is wholly different, being pinnate, and divided into 4 to 7 pairs of obliquely fan-shaped or half-mooned segments, of a thick consistence. The margin of these is entire, or slightly crenate, occasionally partially fertile. The roots and rhizome of moonwort much resemble those of adder's-tongue, but the rudimentary plant for the next year is enclosed within the stem, and not exterior to it, as in the latter. The generic name, from *botrys*, a cluster, is descriptive of the arrangement of the sori; and the specific name, *Lunaria*, of the form of the leaves. This fern is widely diffused over Europe, affecting dry, open, and elevated pastures, its strong succulent roots always creeping horizontally amongst the radicles of the surrounding grass. In the Pyrenees its range of elevation is from 1,800 to 2,200 metres. Port de Gavarnie; Valley of Serbigliana, on the south side of the Vignemale; Castaneza. Perennial, May to August.

ALPHABETICAL INDEX.

	PAGE
Accous	19
Adour, R.	55
Ahu-ky	21
Aiguillons (pic and col des)	67
Albe (pic)	123
Alfred (col)	125
Allanz (brèche)	51
Amélie-les-Bains	151
Ancizan (col d')	62
Ancou (col d')	17
Andorre (val d')	139
Anie (pic)	20
Antenac (pic)	116
Aouhe (col d')	55
Aragnouet	64
Aran (valley of)	134
Aranais (col des)	124
Aratile (col d')	30
Arbizon	61
Ardidieu	37
Argeles	27
Ariège	133
Arras (valley of)	44
Arreau	61
Arrens	22
Arribit (gorge of)	25
Arrongé (col d')	90
Arse (cascade of)	137
Artias	130
Artigues de Lin	105
Artouste (lac)	26
Aspe (valley of)	19
Aspin (col o')	61
Astazou (brèche d')	40
Aubert (col d')	70
Aubiste (pic)	37
Aude, R.	147
Aula (port d')	135
Aulus	136
Aumar (lac d')	70
Aure (val d')	61
Ax	138
Azun (port d')	23
Azun (valley of)	22
Bacanère	112
Badet (pic)	69
Badet (col)	68
Bagnères de Bigorre	56

	PAGE
Bagnères de Luchon	71
Baletous (pic de)	23
Barèges	52
Barèges (port de)	70
Bausen	109
Bayonne	5
Bazert (croix de)	61
Becibère (lac)	132
Bergons (pic de)	36
Betharran	27
Biarritz	5
Bielsa (port de)	66
Bielsa (town)	67
Bious-Artiques	15
Bleu (lac, de Bigorre)	56
Bleu (lac, de Luchon)	95
Bondellos (col de)	17
Bordeaux	4
Bosost	108
Boucharo	44
Boum (pic de)	96
Bouncou (val de)	96
Bourg d'Oueil	118
Bourg-Madame	148
Brèche de Roland	39
Broussette	17
Cabannasse	145
Cadéac	64
Caillaouas (lac)	89
Caldas (port de)	131
Caldas de Bohi	130
Cambiel (col)	68
Cambiel (pic)	69
Cambo	21
Cambredase (pic)	148
Canaou (port de)	51
Canaou (d'Estaubé)	51
Canfranc	19
Canigou	149
Cap de Guerri	132
Cap Long	69
Capsir	147
Carlitte (pic)	146
Castanéza (valley of)	107
Castets	63
Cauterets	28
Cazaril	116
Cazaux	79
Céciré	93

	PAGE
Cerdagne	143
Clarabide	88
Conflens	135
Cotieilla (pic)	107
Coumelie	49
Couplan (gorge of)	68
Couradilles (pic de)	114
Crabère (pic)	133
Crabioules (pic de)	86
Croix Blanche	60
Cylindre	43
Dax	9
Demoiselles (cascade des)	98
Doredom (lac)	70
Eaux Bonnes	14
Eaux Chaudes	15
Enfer (cascade d')	94
Enfer (port d')	89
Entecade (pic de l')	115
Escoubous (lac d')	53
Espingo (lac)	83
Esquierry	92
Estaing (lac d')	27
Estats (pic d')	137
Estaubé (cirque of)	51
Estom (lac)	49
Estom-Soubiran (lacs)	49
Eynes (val d')	148
Fanlo	46
Foix	138
Fontarabie	8
Formiguères	147
Fos	110
Fourcanade (pic de)	124
Gabas	19
Gabisos (pic de)	22
Garen	90
Gargas (grotto of)	111
Garonne (course of)	102
Garonne (source of)	135
Gaube (lac de)	29
Gavarnie	37
Gèdre	68

INDEX.

	PAGE
Ger (pic de)	16
Gistain	65
Glacé (lac of Port d'Oo)	83
Glaire (lac)	71
Glère (port de la)	99
Gourgouttes (lac de)	99
Gours-Blancs	89
Gourzy	16
Gregonio (lac)	126
Gripp	55
Héas	49
Hermittans (pic)	90
Hospitalet	139
Iseye (col d')	19
Jaca	19
Judios (grotte de)	59
Jurançon	13
Juzet	112
Lanoux (lac)	145
Larboust (valley of)	79
Laruns	14
Las Bordes	105
Lavedan (valley of)	27
Lès	109
Lescun	20
Liat	133
Libones	65
Lisey (col de)	33
Litayrolles (col)	85
Llaurenti	147
Llo (col de)	148
Long (pic)	69
Loudervielle	62
Lourdes	60
Louvie-Juzon	13
Lustou (pic)	65
Lutour (valley of)	48
Luz	34
Lys (val de)	93
Maladetta	123
Malibierne	127
Marboré	42
Marcadaou	47
Mauleon-Barousse	120
Maupas (tus de)	96
Maylen	117
Mayrenge	117
Medassoles	93
Melles	134
Mendé (col de)	134
Mérens	139
Miguelou (lac de)	26
Moines (col des)	19
Monné (de Bigorre)	57
Monné (de Cauterets)	33
Monné (de Luchon)	117
Monségu (pic de)	90
Montaigu	59
Montarqué (tus de)	84

	PAGE
Montarto (pic de)	131
Montauban	112
Montcalm	137
Montjoyo (col de)	103
Montlouis	149
Montrejeau	61
Montvallier	136
Moudang (port de)	66
Munia (pic de)	52
Néouvielle	70
Néré (pic)	91
Nethou (pic de)	120
Nethou (village)	107
Niscle (col de)	52
Nuria	148
Œil de Joucou	105
Olette	149
Oncet (lac d')	53
Oo (lac d')	79
Oo (port d')	82
Ordincède	58
Ordino	144
Ordisset (col d')	65
Orthez	13
Ossau (valley of)	13
Ossoüe (valley of)	47
Oueil (valley of)	117
Paillas (port de)	135
Paillette	19
Paillole	62
Pales de Burat	113
Panticosa	18
Paoules (cabane of)	84
Parisiennes (cascade de)	98
Pau	13
Pedrous (pic)	145
Peña Blanca	104
Pène de l'Heris	58
Pène Taillade	56
Perche (col de la)	149
Perdiguère	85
Perdu (Mont)	40
Perpignan	149
Peyrelue (col de)	17
Peyresourde (port de)	61
Pez (port de la)	66
Pic du Midi d'Arrens	27
Pic du Midi de Bigorre	53
Pic du Midi d'Ossau	15
Picade (port de la)	102
Pierrefitte	28
Pierrefitte (col de)	63
Pierre St. Martin (port)	24
Piméné	43
Pindorle (puits de)	58
Pinède (port de)	41
Pique (valley of the)	112
Pla de Beret	135
Pla Guilhem	151
Plan	65
Plan (port de)	65
Plan d'Aube	31
Poey la Houn	23

	PAGE
Pont d'Espagne	30
Pont de Mahomet	122
Pont du Roi	110
Portet	143
Portillon (de Bosost)	109
Portillon (d'Oo)	85
Portillon (lac de)	84
Port vieux (d'Estaubé)	51
Port vieux (de Luchon)	97
Posets (pic des)	128
Poujastou (pic de)	114
Prades	149
Prats de Mollo	151
Preste (la)	151
Puigmal	148
Puycerda	143
Puymorins (col de)	144
Pyrenees (structure of the)	9
Quairat (pic)	88
Querigut	147
Rencluse	121
Rhune (la)	8
Rialp	145
Rio Bueno (lac de)	126
Rioumajou	65
Sacroux (pic)	97
Saint-Aventin	79
Saint-Barthélemy (pic)	138
Saint-Béat	110
Saint-Bertrand-de-Comminges	111
Sainte-Engrace	21
Saint-Girons	133
Saint-Jean-de-Luz	8
Saint-Jean-pied-de-Port	21
Saint-Lary	63
Saint-Lizier-d'Ustou	137
Saint-Marie (baths of)	120
Saint-Néré (chalets of)	119
Saint-Paul	117
Saint-Sauveur	35
Saint-Savin	27
Saint-Sebastien	8
Saldeu (port de)	139
Salenque	127
Sallent	17
San Julia de Loria	140
Sarrieste (col de)	119
Saucède (col de)	22
Sauvegarde (pic de)	102
Seculejo (lac or d'Oo)	79
Seintein	133
Senet	107
Sia (cascade and bridge)	37
Siguier (port de)	144
Somport (col de)	19
Superbagnères	92
Suyen (gourgue de)	24
Tabascain (port de)	137
Tarbes	2
Taillon	41

INDEX.

	PAGE		PAGE		PAGE
Tentenera (col de)	46	Turmes (cabane of)	84	Venasque (port de)	100
Torla	44			Venasque (town of)	105
Tortes (col de)	22			Vernet	149
Toulouse	3	Urdos	19	Vert (lac)	95
Tourmalet (col de)	55	Urgel	140	Vicdessos	136
Tramesaigues (near Gripp)	55	Ussat	138	Vicdaillet	107
Tramesaigues (valley d'Aure)	64	Ustou (port de)	137	Viella (town and port)	108
				Vielle Aure	64
Tramesane (lac de)	131			Vignemâle	31
Trou de Toro	101	Venasque (baths of)	104	Vignemâle (col de)	47
Troumouse (cirque de)	50	Venasque (hospice of)	104	Villefranche	149

BOOKS AND MAPS
FOR
TRAVELLERS.

THE NORTH-WEST PENINSULA OF ICELAND;
Being the Journal of a Tour in Iceland in the Summer of 1862. By C. W. Shepherd, M.A. F.Z.L. With a Map and Two Illustrations in Chromolithography. Fcp. 8vo. 7s. 6d.

BEATEN TRACKS;
Or, Pen and Pencil Sketches in Italy. By the Authoress of 'How we Spent the Summer.' With 42 Lithographic Plates, containing about 300 Sketches. 8vo. 16s.

HOW WE SPENT THE SUMMER;
Or, a 'Voyage en Zigzag' in Switzerland and Tyrol with some Members of the Alpine Club. From the Sketch-Book of one of the Party. Third Edition, re-drawn. In oblong 4to. with about 300 Illustrations, price 15s. cloth.

A GUIDE TO SPAIN.
By H. O'Shea. Post 8vo. with Map, 15s.

GUIDE TO THE PYRENEES,
For the use of Mountaineers. By Charles Packe. With Maps, &c. New Edition, enlarged, now ready.

THE COMMERCIAL HANDBOOK OF FRANCE.
By Frederick Martin, Author of 'The Statesman's Year-Book.' With 3 Maps. Crown 8vo. 7s. 6d.

GUIDE TO THE EASTERN ALPS,
By John Ball, F.L.S. M.R.I.A. late President of the Alpine Club. Post 8vo. with Maps and other Illustrations. [*In the Press.*

GUIDE TO THE WESTERN ALPS,
Comprising Dauphiné, Savoy, and Piedmont; with the Mont Blanc and Monte Rosa Districts. By the same Author. With an Article on the Geology of the Alps by M. E. Desor. Post 8vo. with Maps, &c. 7s. 6d.

London: LONGMANS, GREEN, and CO. Paternoster Row.

GUIDE TO THE OBERLAND AND ALL SWITZERLAND,

excepting the Neighbourhood of Monte Rosa and the Great St. Bernard; with Lombardy and the adjoining portion of Tyrol. By the same Author. Post 8vo. with Maps, &c. 7s. 6d.

FLORENCE, THE NEW CAPITAL OF ITALY.

By CHARLES RICHARD WELD. With 23 Woodcut Illustrations. Post 8vo. 12s. 6d.

PEAKS, PASSES, AND GLACIERS;

a Series of Excursions by Members of the Alpine Club. Fully Illustrated with Maps and Engravings.

FIRST SERIES. Edited by JOHN BALL, M.R.I.A. F.L.S. Square crown 8vo. 21s. or, 16mo. (*Travelling Edition*) 5s. 6d.

SECOND SERIES. Edited by EDWARD SHIRLEY KENNEDY, M.A. F.R.G.S. 2 vols. Square crown 8vo. 42s.

NINETEEN MAPS OF THE ALPINE DISTRICTS, from the FIRST and SECOND SERIES of 'Peaks, Passes, and Glaciers.' Square crown 8vo. in envelope-portfolio, 7s. 6d.

MAP OF THE CHAIN OF MONT BLANC,

From an actual Survey in 1863–64. By A. ADAMS-REILLY, F.R.G.S. M.A.C. Published under the authority of the Alpine Club. In Chromolithography on extra stout drawing paper 28 in. by 17 in. price 10s. or mounted on canvas in a folding case, 12s. 6d.

ALPINE CLUB MAP OF SWITZERLAND

And the Adjacent Countries on a Scale of $\frac{1}{250000}$ (four miles to an inch), from Schaffhausen on the North to the Southern Slopes of the Val D'Aosta on the South, and from the Orteler group on the East to Geneva on the West, constructed under the immediate superintendence of the Alpine Club, and Edited by Mr. R. C. NICHOLS, F.R.G.S. Engraved by ALEXANDER KEITH JOHNSTON, F.R.G.S. [*In preparation.*

*** The FIRST SHEET, being the North-West portion of Switzerland, and comprising Bâle, Lucerne, Interlachen, Grindelwald, Bern, Freiburg, and Neuchâtel, is expected to be ready in June.

London: LONGMANS, GREEN, and CO. Paternoster Row.

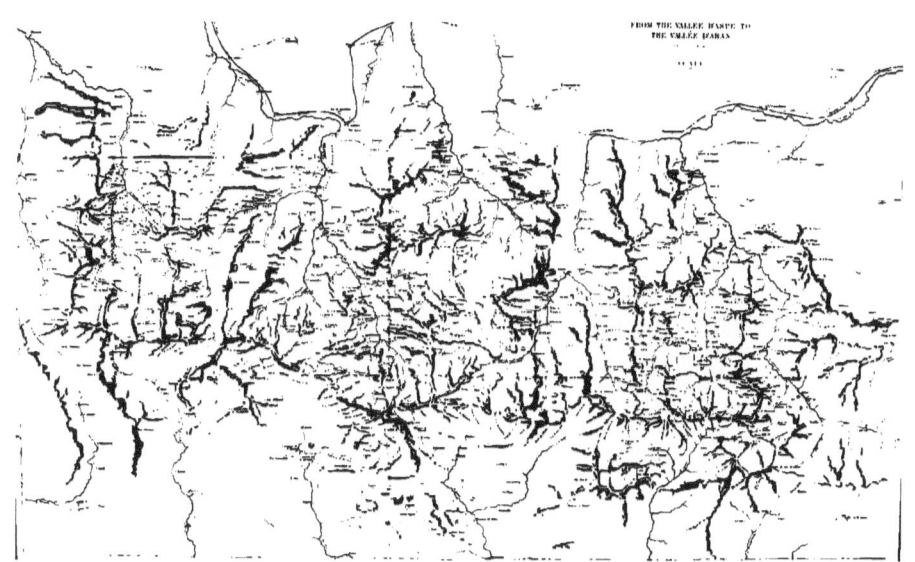

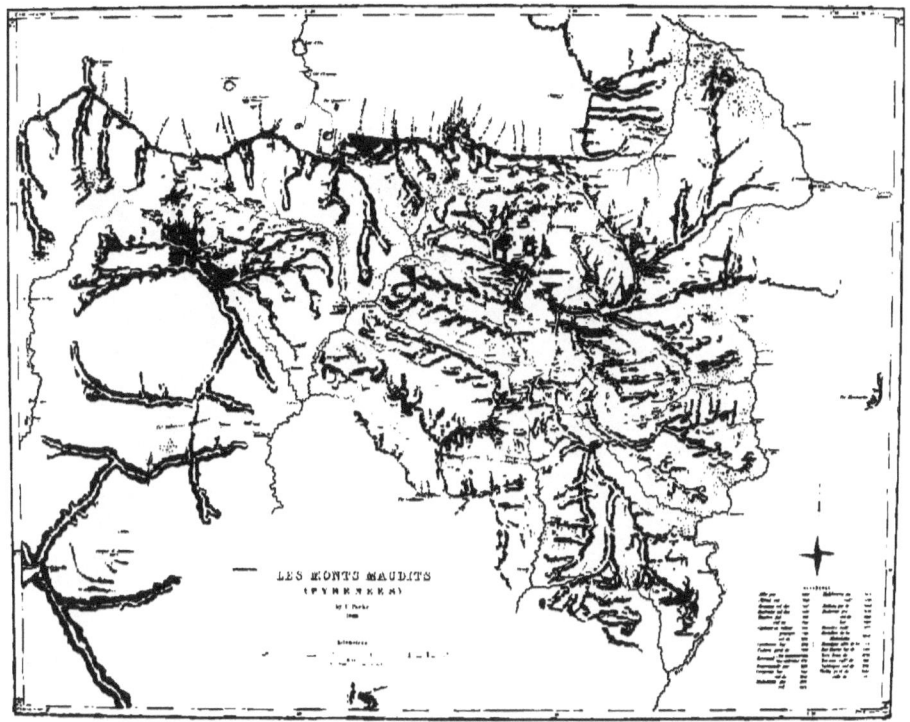

www.ingramcontent.com/pod-product-compliance
Lightning Source LLC
Chambersburg PA
CBHW031733230426
43669CB00007B/336